GREAT VEGETABLES FROM THE GREAT CHEFS

GREAT VEGETABLES FROM THE GREAT CHEFS

Baba S. Khalsa

Introduction by M.F.K. Fisher

Chronicle Books • San Francisco

Printed in The United States of America.

Library of Congress Cataloging in Publication Data

Khalsa, Baba S.
Great vegetables from the great chefs / Baba S. Khalsa ;
introduction by M.F.K. Fisher.
p. cm.
ISBN 0-87701-607-0
1. Cookery (Vegetables) 2. Cooks—United States—Interviews.
I. Title.
TX801.K53 1990
641.6′5—dc20 89-25336
CIP

Distributed in Canada by
Raincoast Books
112 East Third Street
Vancouver, B.C. V5T 1C8

10 9 8 7 6 5 4 3 2 1

Chronicle Books
275 Fifth Street
San Francisco, California
94103

Dedication

TO MY mother, Rosalind Lasser, who instilled the love of good food in my heart. To my wife, Krishan, who nurtures and supports me every day of my life. And to Yogi Bhajan, who taught me the meaning of life.

Table of Contents

Preface **xii**

Acknowledgments **xvi**

Introduction **xvii**

Part One

JEAN BANCHET **3**

JULIA CHILD **6**

CRAIG CLAIBORNE **9**

MARION CUNNINGHAM **12**

ROBERT DEL GRANDE **15**

MARCEL DESAULNIERS **18**

JOHN DOWNEY **21**

DEAN FEARING **24**

SUSAN FENIGER AND MARY SUE MILLIKEN **28**

M.F.K. FISHER **33**

MICHAEL FOLEY **36**

LARRY FORGIONE **40**

JOYCE GOLDSTEIN **42**

VINCENT GUERITHAULT **46**

JEAN JOHO **49**

MADELEINE KAMMAN **53**

EMERIL LAGASSE **56**

ROLAND LICCIONI **61**

BRUCE MARDER **64**

TONY MAY **68**

MICHAEL MCCARTY **71**

FERDINAND METZ **76**

MARK MILITELLO **81**

MARK MILLER **84**

GORDON NACCARATO **90**

PATRICK O'CONNELL **93**

BRADLEY OGDEN **99**

JEAN-LOUIS PALLADIN **102**

CINDY PAWLCYN **105**

GEORGES PERRIER **111**

RICHARD PERRY **113**

ALFRED PORTALE **117**

WOLFGANG PUCK **121**

STEPHAN PYLES **127**

JEAN-JACQUES RACHOU **132**

THIERRY RAUTUREAU **135**

LESLEE REIS **138**

SEPPI RENGGLI **142**

MICHEL RICHARD **146**
JACKY ROBERT **149**
MICHAEL ROBERTS **152**
ANNE ROSENZWEIG **155**
JIMMY SCHMIDT **158**
CORY SCHREIBER **162**
JOHN SEDLAR **165**
GUNTER SEEGER **169**
PIERO SELVAGGIO **173**
JACKIE SHEN **177**
ANDRE SOLTNER **180**
JOACHIM SPLICHAL **183**
JEREMIAH TOWER **186**
BARBARA TROPP **190**
MAURO VINCENTI **194**
JEAN-GEORGES VONGERICHTEN **197**
ALICE WATERS **201**
JONATHAN WAXMAN **205**
JASPER WHITE **208**
BARRY WINE **212**
ROY YAMAGUCHI **216**

Part Two

APPETIZERS AND FIRST COURSES **223**

Provençal Artichoke Tarts / Leslee Reis
Warm Asparagus with Pine Nuts and Orange Rind / Jackie Shen
Asparagus with Hazelnut Vinaigrette / Joyce Goldstein
Cèpe Carpaccio with Garlic Chives / Gunter Seeger
Sautéed Wild Mushrooms with Sage Sauce / Thierry Rautureau
Onion and Black Olive Tarts / Michel Richard
Stuffed Zucchini Flowers with Tomato Marinade / Thierry Rautureau
Dialogue of Vegetables / Gunter Seeger
Filo Pastry Stuffed with Ratatouille and Tofu / Thierry Rautureau
Tarragon and Chervil Cream Sauce / Thierry Rautureau
Crispy Tarts with Vegetables / Leslee Reis
New Mexican Sushi / John Sedlar
Vegetable Timbales with Peanut Sauce / Seppi Renggli
Spring Vegetable Gyoza *with Roasted Red Bell Pepper Butter Sauce* / Roy Yamaguchi
Vegetable Fritters with Chick-Pea Batter / Susan Feniger and Mary Sue Milliken
Grilled Mozzarella in Romaine with Sun-dried Vinaigrette / Mark Militello
Pan-fried Goat Cheese with Roasted Peppers / Thierry Rautureau
Caponata / Mauro Vincenti

SOUPS **247**

Vegetable Stock / Vincent Guerithault
Artichoke and Asparagus Soup / Roland Liccioni
Cafe Provençal's Avocado-Cucumber Soup with Fresh Oregano / Leslee Reis

Chef Vincent's Cream of Avocado Soup / Vincent Guerithault
Cold Beet and Buttermilk Soup / Cindy Pawlcyn
Black Bean Soup with Hot Green Chutney / John Downey
Yellow Bell Pepper Soup / Mauro Vincenti
Carrot and Red Pepper Soup / Alice Waters
Carrot and Zucchini Vichyssoise / Mauro Vincenti
Mexican Cauliflower Soup / Joyce Goldstein
Puree of Celery Soup / Seppi Renggli
Cream of Celery Root Soup / Georges Perrier
Grilled Corn Soup with Ancho Chili and Cilantro Cream / Stephan Pyles
Shoe Peg Corn and Peanut Soup / Marcel Desaulniers
Cream of Eggplant Soup / Emeril Lagasse
Mexican Corn Chowder / Susan Feniger and Mary Sue Miliken
Curried Green Bean Potage / Madeleinie Kamman
Mushroom Won Ton / Seppi Renggli
Butternut Squash Soup / Leslee Reis
Roasted Tomato and Chili Soup / Mark Miller
Pappa al Pomodoro / Mauro Vincenti
Herbed Tomato and Onion Soup / Marcel Desaulniers
Tomato Stew / Michael Roberts

SALADS **275**

Avocado, Pasta, and Tofu Salad with Chinese Sesame Seeds / Jackie Shen
Braised Belgian Endive and Baby Bok Choy / Michael Roberts
Grilled Eggplant and Chili Salad / John Downey
Strawberries, Belgian Endive, Macadamia Nuts, and Balsamic Vinegar Dressing / Marcel Desaulniers
Wild-Mushroom Salad / Michael McCarty
Pasta Salad with Baked Eggplant and Tomatoes / Jeremiah Tower
Curly Endive with Warm Grilled Vegetables and Mustard Dressing / Marcel Desaulniers
Arugula and Fried Okra Salad with Roast Corn Vinaigrette / Stephan Pyles
Yellow and Red Plum Tomato Salad with Avocado Oil and Basil / Michael Foley
Wild Rice Salad with Cilantro Vinaigrette / Gordon Naccarato

SIDE DISHES **291**

Artichokes, Cauliflowerets, and Mushrooms à la Barigoule / Andre Soltner
Moroccan Carrots / Joyce Goldstein
Celery Root and Apple Puree / Richard Perry
Fava Bean with Olive Oil, Garlic, and Rosemary / Alice Waters
Involtini de Melanzane / Piero Selvaggio
Eggplant and Mushrooms with Blue Cheese Glaze / Cory Schreiber
Fennel with Tomato / Cindy Pawlcyn
Braised Leeks with Saffron and Sherry Vinegar / Cory Schreiber
Potatoes and Onions Roasted with Vinegar and Thyme / Alice Waters
Potato and Yam Gratin with Artichoke and Peppers / Jimmy Schmidt
Pommes Paillasson / Georges Perrier
Shiitake Provençale and Grilled Shiitake Mushrooms / Vincent Guerithault
Salsify and Fava Bean Ragout / Michel Richard

Sauteed Butternut Squash with Rosemary / Richard Perry
Zucchini alla Menta / Piero Selvaggio
Zucchini "Vermicelli" with Two Sauces / Roy Yamaguchi
Zucchini with Almonds / Craig Claiborne
Ratatouille / Jean-Jacques Rachou

MAIN COURSES **311**

Poivron Jardiniere / Jean Banchet
Texas Cream Corn with Apple-Pecan Fritters / Dean Fearing
Grilled Mongolian-style Eggplant Napoleon with a Caper, Garlic, and Basil Butter Sauce / Roy Yamaguchi
Corn "Risotto" with Okra and Shiitake Mushrooms / Michael Roberts
Haricots Verts à la Repaire / Ferdinand Metz
Salsify Pancakes with Grilled Portobello Mushrooms / Mark Militello
Potato and Cheese Enchiladas with Red Chili Sauce / Robert Del Grande
Wild-Mushroom Chili / Richard Perry
Pipe Organ of Baby Zucchini Stuffed with Vegetable Purees / Jacky Robert
Summer Vegetable Tart / Leslee Reis
Vegetarian Club Sandwich / Susan Feniger and Mary Sue Milliken
Ratatouille Tamales / Vincent Guerithault
Warm Vegetable Stew / Jeremiah Tower
Sauteed Shiitake Mushrooms with Zucchini, Tomatoes, Spinach, Broccoli and Thyme / Marcel Desaulniers
Potpourri of Vegetables Pot au Feu-Style with Chervil Butter / Jean Joho
Napoleon of Vegetables / Joachim Splichal
Gratin des Capucins / Georges Perrier

PASTA AND RISOTTO DISHES **341**

China Moon Cafe Pot-browned Noodle Pillow Topped with Curried Vegetables / Barbara Tropp
Fresh Fettuccine with Spring Asparagus and Goat Cheese / Emeril Lagasse
Fettuccine with Sun-dried Tomatoes, Basil, Broccoli, and Goat Cheese / Jackie Shen
Wild-Mushroom Pasta / Alfred Portale
Poblano and Black Bean Linguine with Ancho Threads, Oaxaca Cheese, and Mexican Vegetables / Dean Fearing
Shittake and Sour Cherry Ravioli / Barry Wine
Vegetable Ravioli in a Tomato Coulis with Fresh Herbs and Romano Cheese / Bruce Marder
Cannelloni of Eggplant Caviar Glazed with Goat Cheese and Zucchini Juice / Jean-Georges Vongerichten
Grilled Vegetable Lasagne / Gordon Naccarato
Yam Ravioli with White Truffles / Gunter Seeger
Spring Vegetable Risotto with Orzo / Anne Rosenzweig
Risotto con Asparagi / Piero Selvaggio
Corn and Peppers Risotto / Piero Selvaggio
Green Risotto / Richard Perry

PIZZAS AND BREADS **367**

Pizza / Alice Waters
Focaccia with Caramelized Onions, Gorgonzola Cheese, and Walnuts / Joyce Goldstein
Chili Corn Bread / Mark Miller

RELISHES AND INTERMEZZOS **377**
Salsa Verde / Mark Miller
Avocado-Corn Salsa / Vincent Guerithault
Marinated Onions and Cucumbers / Richard Perry
Jicama-Melon Relish / Stephan Pyles
Roasted Fresh Pimiento and Saffron Butter / Mark Miller
Wild-Mushroom Chutney / Marcel Desaulniers
Fresh Vegetable Granites / Jean-Louis Palladin

Credits **386**

Index **387**

Preface

I'LL NEVER forget the day I tasted real food for the first time. It was in Los Angeles in 1969. On a dare I had gone to a vegetarian restaurant called the Source. Most of my life I had eaten the typical American diet of hamburgers, French fries, and cokes; what few vegetables I ate were frozen or canned. So I ordered one of the few things on the menu that sounded a little like what I was used to: "high-protein cereal." Composed of hand-ground wheat berries, oats, chia seeds, and sunflower seeds, it was served with raw milk and topped with organically grown almonds, raisins, bananas, and apples. I looked for the sugar bowl, but I had to settle for raw organic honey.

The first spoonful was a consciousness-raising experience. The pure, elemental power of that cereal, undiluted by the refining processes that strip food of its taste and nutrition, gave me a surge of energy through my entire body. I was like Popeye with spinach. I had never tasted whole grains full of nutty goodness; milk with rich, dairy flavor the way it came from a cow; fruit grown for taste and not looks. The food was full of a living vitality whose reality I could not deny. I felt as though I had just tasted food for the first time in my life. Right then, I decided to change the way I ate. What had started as a lark became a way of life. Like a baby coming into the world, I could never go back to where I had been.

It took a couple of years of experimentation to choose the right ingredients for a nourishing diet. Through the teachings of Yogi Bhajan, an Indian spiritual master, I settled on a diet primarily of steamed vegetables, mung beans and rice, yogurt, and fruit juices.

But after eating this way for a while, I found myself growing bored with my nutritious but limited diet. About that time I began hearing of some young chefs who were using their classical training to create marvelous breads, salads, soups, side dishes, and even main courses out of the same foods that I ate: vegetables, grains, nuts, fruits, and dairy products. I sampled Michael's, West Beach Cafe, La

Toque, Trumps, and Valentino and found I didn't have to be bored to be healthy.

I learned that great chefs could cook great vegetarian meals. So whenever I heard about such a person, I would call him or her in advance and ask for a special vegetarian meal, usually a *menu dégustation*, or a tasting, of some of the chef's favorite meatless dishes. Wherever I travel I use this same modus operandi. And over the years I've found that wherever they cook, great chefs love to please and are happy to meet the challenge of doing something different and doing it well. Most of these chefs are not vegetarians, but they all recognize how dining habits in America have changed, and they are making healthful food that is as delicious as the heavier fare we associated with fine dining in years past. The chefs in this book have been instrumental in helping Americans change their eating habits. And this revolution in American dining is what this book is all about.

Americans are eating a lighter, more healthful diet emphasizing foods higher in fiber and lower in fat. Such a diet includes more vegetables, fruits, grains, and legumes, and less meat and fewer eggs. As a result, we can expect to live longer, healthier lives. More of us will live to fulfill our dreams while avoiding the nightmares of heart disease and cancer.

A corps of young chefs, many of them Americans without the traditional apprenticeship required of European chefs, have elevated this country's cuisine to world-class status. By introducing us to dishes and ingredients previously unknown or long forgotten in America, these chefs have awakened millions to the pleasures of good food.

Unfettered by tradition, America's great chefs are free to create and improvise. And, in the main, they have chosen to move away from richly sauced, heavier foods, cooked for long periods of time and elaborately presented, in favor of lighter dishes, quickly cooked and well seasoned so that the natural flavors and colors burst forth.

Outstanding American food is alive and well in restaurants all across the country. In the past, New York and San Francisco were the sole repositories of haute cuisine establishments, but great restaurants can now be found in nearly every city in America. And a new phenomenon, the great restaurant in an out-of-the-way place, is taking hold. Just as Michelin gives three stars to French restaurants in places its vaunted tires might fear to lose their tread, similar accolades could go to great dining establishments in Aspen; Washington, D.C.; and Santa Barbara.

For this book, I've selected chefs who are making what I consider to be a significant contribution to the growth of American cooking and who have a way with lighter cuisine.

None of the chefs in this book cook in vegetarian restaurants. But they all can work wonders with vegetables, one of the most difficult things to cook right. Chef after chef told me how rewarding it is to get vegetables to come out perfectly, and of the danger of cooking them a moment too long and losing their priceless color and crunch. Not only have these chefs mastered the art of cooking vegetables, they have learned to create vegetable dishes that are innovative and exciting.

Restaurant chefs, of course, don't have a lock on great vegetable recipes. Helping make great food accessible to America's home cooks are the "cooks without restaurants" I have included. Without their imaginative taste and their lucid writing styles, which have demystified recipes for the home cook, fine light dining might have been a privilege reserved for the well-to-do.

Many of the restaurants featured in this book are expensive, and understandably so. After all, these chefs use only the finest ingredients; organically grown produce is costly. And in order to offer you a peak dining experience, many owners have spent a lot of money to create an environment pleasing to the eye, in areas where rents are high and where a professional level of service is expensive.

But if you don't want to spend the money to visit the restaurants mentioned in this book, you have only to choose from the recipes on these pages to recreate some of the best cooking in America.

The recipes in this book are divided into chapters according to course. Any of the dishes would be a superb accompaniment to your favorite entree, or you may wish to compose a meatless meal for a vegetarian friend or as a change of pace for you and your family. Due to the pioneering efforts of these chefs, you should be able to find most of the ingredients in these recipes in grocery stores, small ethnic markets, farm stands, and open-air markets around the country.

Today a growing number of small, local farmers and marketplaces are selling produce that rivals that of Europe. Growers like Tom Chino and Bob Cannard in California, Mike Michaels in Illinois, and hundreds of others from Texas to Florida to Long Island provide chefs with produce that was unheard of only a few years ago. In Union Square in New York, in Santa Monica, California, and in cities and towns all over America, farmer's markets are selling produce that was picked that morning; homemade cheese, preserves, and baked goods; and fresh-pressed cider—all exploding with flavor.

One of the great joys of writing this book was getting to know the featured chefs, who are some of the most caring people around. They all work too long and too hard for the compensation they receive. But the vast majority of them, successful though they are, aren't in it simply for the money. They want to please their customers by creating what pleases them.

You'll see what I mean if you choose to dine at some of these restaurants. Call ahead and ask to speak to the chef. Don't be shy. People do it every day. Tell the chef what you like to eat and see if there's something on the menu that appeals to you. If not, ask if he or she wouldn't mind preparing something special for you and your party.

Whether you just sit back on a rainy afternoon with a cup of tea to read about these fascinating people, or whether you tie on the apron and go into the kitchen with their recipes, I hope this book brings satisfaction and enjoyment to your body, mind, and spirit.

Acknowledgements

I GRATEFULLY acknowledge the help of many friends without whom this book would not have been possible. Susan Grode, who surely must be the finest and most trustworthy entertainment lawyer in the business worked tirelessly to find just the right publisher. That turned out to be Chronicle Books where Nion McEvoy gently and with much humor helped me find just the right shape for my manuscript which he lovingly guided to publication. He was ably assisted by Annie Barrows, Bill LeBlond and the entire staff at Chronicle, all of whom were a pleasure to work with. Chronicle was wise to enlist the services of Carolyn Miller who helped prune my manuscript with consciousness and care. Har Har Kaur Khalsa, Sat Bachan Kaur Khalsa, and Maharaj Kaur Khalsa performed flawlessly the tedious but absolutely essential tasks of transcribing and typing.

Sat Shakati Singh Khalsa, whose boundless enthusiasm for this book and limitless appetite for good food spurred him to test the recipes, was a constant inspiration. I deeply value his friendship and support. I'm still amazed by how much time he put into this project. Ably assisting him in the kitchen were his wife Sat Shakati Kaur Khalsa and Sewa Kaur Khalsa.

Many thanks to Sadhu Singh Khalsa both for his appreciation of fine food and his companionship during the research and writing of the book.

Finally, I will always be grateful to a great chef and good friend Akasha Kaur Khalsa who was always available for advice and consolation. I could not have completed this book without her.

Introduction

M.F.K. Fisher

IT IS never an easy job to write a good introduction to a book that is good, and it is especially hard to keep any kind of sense or even nonsense straight in a book that mentions the person who is trying to write a good sensible introduction.

I happen to be mentioned in this book by Baba Khalsa, and no matter how much I can protest that what he says about M. F. K. F. is fulsome or foolish or both, I like what he's written. Fortunately, though, I like everything else he's written in *Great Vegetables from the Great Chefs*. I know some of the other people he writes about in it, and I agree completely with him. With them, he does not seem even slightly silly or overloaded with adjectives of praise, so I can only hope that what he said about me will seem practical and right to them.

As for the recipes, they are clearly written, if I permit myself to judge the many of them that I have read. I know that I am something of a fanatic on the subject of how, if not when, to write what used to be called receipts, to be followed by any cook in the world, with edible results. In fact, I once wrote a small treatise on the subject, called appropriately enough "The Anatomy of a Recipe," and I fully believe that it should be printed in solid gold letters somewhere in this world so that it will be read, learned, and inwardly digested by anyone who graduates from a school of cookery, as well as every person who ever sets pan on stove or fork into mouth. I admit that I have not cooked every recipe in *Great Vegetables from the Great Chefs*, but the ones I have looked at do pass the severest of all musters.

In other words, they are eminently makeable. What is most important, though, about everything in this book, whether it be about what to eat or why or even how, is the true enthusiasm with which it is presented. My friend Baba's pure, gay spirit is evident everywhere, and I am sure that it is what makes his book such a pleasure to read, and (I admit) even to write about. There is a kind of candor about it, which is never naive, but which springs from real experience. And since I know a little about Baba, I feel that there is a kind of religious excitement about it which few of us attain, although most of us can appreciate and envy anyone who has gone as far as this man has in accepting his lot as a hungry human being.

Part One

JEAN BANCHET

Chicago, Illinois

"I'm crazy."

ASK ANY of Jean Banchet's fellow chefs and they'll smile and agree with what he told me about himself: "I'm a crazy guy." It probably did seem crazy to locate Le Français in out-of-the-way Wheeling, Illinois. And it may have seemed even crazier for him to sell the fabulously successful restaurant, which he recently did, to Roland Liccioni, the former chef at Carlos'. Though Banchet has decided to take a much-needed rest, I won't forget the dining experiences I had when he was chef at Le Français.

The first time I went to the restaurant, I wondered if I was in the right neighborhood when I passed the practice batting cage and the high-speed Go-Kart track; then I entered a forest preserve. Why would Jean Banchet locate his restaurant way out here, in the middle of nowhere? Because the world-renowned chef wanted a visit to Le Français to be premeditated. To dine at Banchet's, you had to make a special effort and plan to devote your entire evening to the event.

The standards Banchet set were ingrained in him at an early age: at thirteen he was a cooking prodigy in the Triosgros Restaurant. In three decades of fourteen-hour days and seven-day weeks, Jean Banchet learned that there is no easy path to greatness.

Banchet worked long hours and even traveled to train others as he was trained. In fact, it was nearly impossible to interview him for this book. In one week he took *separate* trips to Paris and to Geneva, returning each time to carry his normal load at Le Français. Through my own sloth I slept through the only sliver of time he had for an interview, which left me feeling like a space shuttle missing the reentry window, doomed to float endlessly in the asteroid belt. So we ultimately settled for a telephone interview during which he, naturally, continued to work.

But I didn't get to know Jean Banchet through an interview. I learned about him through his exquisite food. When I entered the convivial atmosphere of the wood-beamed room, the first thing I noticed was what a good time everyone, including the staff, was having. There were so many staff, in fact, that they seemed to outnumber the customers. If the occasion had been a football game they'd have been penalized for having too many men on the field. And it was the staff, going through their presentations with the unlikely combination of military precision and motherly love, that made a long evening seem short and lively.

The special quality of the restaurant began long before the menu was ever presented: waiters brought tray after tray and cart after cart of artfully prepared offerings.

From a selection of some twenty different pâtés and terrines I chose one composed solely of vegetables. The terrine rippled with color. It was a seventeen-jewel movement composed of carrot, eggplant, *haricots verts*, asparagus, mushroom, broccoli, hearts of palm and artichoke, zucchini, red, green and yellow peppers, celery, tomatoes, turnips, and leeks—all banded with a thin strip of carrot and presented in a truffle vinaigrette striated with red bell pepper sauce.

Banchet describes his cuisine as "classically modern," and my next choice, a Roquefort pâté, combined time-honored technique and unusual ingredients. A slab of pale green cheese of bone-chilling richness was imbedded with nuggets of white Roquefort and chunks of walnut. The walnut flavor was picked up by a light hazelnut vinaigrette that adorned the accompanying greens. Without the lettuces the pâté would simply have been too rich to eat.

Banchet's cooking has been called too rich. But by choosing carefully, it was possible to strike a balance at Le Français. One way to effect this was to order the fresh vegetable platter as a main course. Says Banchet: "Vegetables help you digest rich sauces and bring color to some otherwise plain-looking meat or fish. But I like to eat them alone, as well. For me they are good food. They're just plain good for you."

It may be hard for some to opt for bland-sounding veggies over the romance of a succulent rack of lamb, noisettes of veal, and other rich selections. But when the silver dome lifts on a selection such as I am about to describe, no one would regret such a choice. And consider this fact: The vegetables prepared by Banchet were no less satisfying than the other entrees, and due to their lightness, they left room for the finest desserts in the Midwest. Too often we eat a heavy meal at a great restaurant and don't have room for dessert. That's a shame, but a worse one is to eat the pastry anyway out of a sense of duty and end up feeling stuffed.

The kaleidoscope of vegetables Banchet prepared for me that night featured such delights as tiny squashes hollowed out and filled with mushroom duxelles and squash puree. There were cherry tomatoes speared with asparagus, and yellow squash filled with minced vegetables in an intense, marjoram-laced tomato *coulis*. A peppery succotash of julienne celery, leeks, and carrots was memorable, as was a wild rice–zucchini flan. Baby carrots poached in honey and splayed into a tiny fan was a sweet treat. And sautéed onion strands with zucchini and eggplant wafers made a savory tomato-less ratatouille. There were only a few bites of each selection, and as I grazed my way happily along, my waiter exhorted: "Don't give up!"

The staff continued to take a great interest in my happiness long after they presented their cavalcade of specials at the outset. Friendly without being familiar, caring but not obsequious, the staff seemed always to be there when I needed them.

But you would expect that from a restaurant whose chef-owner trained at La Pyramide as well as at the Hôtel de Paris in Monte Carlo, and who worked alongside Paul Bocuse. Jean and his wife, Doris, presided over what many thought was the finest French restaurant in America. Time will tell if new chef-owner, Roland Liccioni, can carry on Banchet's proud tradition. He certainly has everything going for him—not the least of which is his own reputation as a fine French chef. And his new venue provides the perfect showcase for his considerable talents.

Jean Banchet will no longer work behind the swinging windowed doors to the kitchen. The toque is passed. Banchet will be missed, but when you think of the intensity with which he worked for so many years, you realize that he's deserving of a much-needed rest.

Banchet will do some consulting now, and may eventually open another, smaller restaurant. But whatever he does, he'll do it in that single-minded way. For Banchet only knows how to do things one way—all the way. And he has an intensity that may seem crazed to some.

"I *am* a crazy guy. All I do is work and sleep. I'm passionate about cooking. I'm always trying to improve—to get better, fresher ingredients and make better and better food. That's what I'm all about."

MAIN COURSES

Poivron Jardinaire

JULIA CHILD

Cambridge and Santa Barbara

Teaching The Way to Cook *in the Twenty-first Century*

JULIA CHILD: a household name. Though she is a prolific cookbook author, she came to most of us through the medium of television. And I daresay many who had not the least interest in cooking became interested in the art after being captivated by the woman with the lilting voice.

Like Milton Berle, Lucille Ball, Alistair Cooke, and Johnny Carson, it seems she has been in our homes forever. In fact, if you asked most people, they'd be hard put to tell you when there *wasn't* a Julia Child brightening the tube that all too rarely brightens our homes. Was it 1949, '55? Let's see—when did she start her cooking shows?

The fact is, Julia Child made her first television appearance in 1961 to promote *Mastering the Art of French Cooking*, which she co-authored with Simone Beck and Louisette Bertholle. During the promotion of the book, Child was invited to appear on Boston's WGBH program, "I've Been Reading." She caught the attention of the channel's producers, who were fascinated by the way she could explain, in simple language, how to whip up impressive French dishes.

In the summer of 1962 Child was asked by the station to do three pilot shows, which were well received. "The French Chef" went on the air on February 11, 1963, and her cooking shows were a fixture on television for years afterward.

We tend to take her for granted, what with re-runs and books (seven to date, and she swears, "I'm never going to write another one after my next is published!"). But I wonder what the state of home cooking and even restaurant cooking (Patrick O'Connell first learned to cook from her books) would be in this country if there'd been no Julia Child. The food on thousands of American tables and the quality of life in general is much better because of her.

Julia was in China during World War II, serving in the OSS, a forerunner of the Central Intelligence Agency. She became interested in cooking there out of a need for survival.

"The only place to eat was the army mess. The army cooks were terrible. Dried potatoes, canned tomatoes, and water buffalo. Awful! I was with a lot of sophisticated people who'd eaten all over. I'd never been to France and knew nothing about French food. I was interested in cooking but really didn't know how to do it. Then I met my husband, Paul, there, and after we returned to the States, we got married. That spurred me to learn to cook.

"I did a lot of cooking from *Gourmet* magazine and from the *Joy of Cooking*. But it wasn't until 1948, when the United States Information Agency, for whom my husband now worked, transferred him to France, that my whole attitude changed. I really began to appreciate food and classical cooking. I began to learn something about it. For example, until that time, I'd never baked a cake. So I enrolled in the Cordon Bleu, met my French colleagues, and joined a French ladies' gastronomical club."

Julia became enamored of French cooking after the first bite. She quickly immersed herself in the world of cuisine. After studying at the Cordon Bleu and with private cooking teachers, and after her experiences in Le Cercle des Gourmettes, she decided to commit her life to learning about and teaching cooking. With Beck and Bertholle she opened a modest cooking school in a rooftop kitchen on rue de l'Université. They called it L'École des Trois Gourmandes. And from their experiences at the cooking school they wrote the first volume of *Mastering the Art of French Cooking*, which was published in 1961.

Six books and hundreds of television shows later, Child's style has changed: she has broadened her scope beyond French cooking. As she stated in her fourth book, *From Julia Child's Kitchen*, published in 1975, "Now, in addition to French concerns, I feel free to delve into New England chowders, Belgian cookie doughs, personal fruitcakes, curries, pastas . . . putting my cooking vocabulary to work in all directions. I hope, in turn, if you are not already of the same persuasion, to encourage the same attitude in you, my fellow cook."

"My fellow cook." The phrase says a lot about Julia Child and her approach to teaching. In all of her books and videos she conveys the feeling that the two of you are in the kitchen together, figuring out how to cook this thing or that. It is her total ease of manner and accessibility that endear her to us. She makes us feel that she's on our side.

Julia Child often demystified the art of French cooking just by translating it—"*Coq au vin* is a chicken stew. . . . *Soubise* is plain old rice cooked with onions. . . . "—without abandoning the traditional principles that elevate French food from the mundane to the sublime. "If you have a good classical background," she explains, "you can adapt yourself to anything at all. The main thing is to know how to cook.

"When you buy vegetables fresh and cook them lovingly," she states in Volume II of *Mastering the Art of French Cooking*, "you may find yourself more renowned for your remarkable zucchini stuffed with almonds than for your spectacular *crêpes Suzette*."

Vegetables are one of the most significant components of Julia's cuisine. "Vegetables mean good food to me," she told me. Then, she asked rhetorically, "I think they're one of the most important things in life, don't you? We eat a lot of them. Last night as a matter of fact, potatoes, corn off the cob, sautéed and very good, asparagus, and for dessert, some fresh ripe melon. That's all we had. It was a perfect meal. We're not vegetarians, but very often we eat vegetables for our whole dinner. The main thing is that they must be good and fresh and that you prepare them in such a way that although they're nutritious, you're careful not to go too far and take all the pleasure out of food."

Recently Julia made a six-tape video cassette series. *The Way to Cook*, released in 1985, is Julia Child on demand. Pop in the cassette on vegetables, for instance, and you get instruction on not only how to cook but how to buy and store vegetables like asparagus, broccoli, green beans, eggplant, and green peas. You also learn how to braise celery, leeks, and endives, how to make risotto, and how to cook potatoes *dauphinoise*.

Her most recent book, *The Way to Cook*, is a far-ranging, fully illustrated culinary guide based on, among other sources, the cassettes, her "Good Morning America" and "Dinner at Julia's" television appearances, and her articles for *Parade* magazine.

The book promises to be the *Joy of Cooking* of the twenty-first century, presenting the distilled knowledge of forty years of studying, teaching, and writing. With it, Julia Child will have left a legacy we can treasure for years to come.

CRAIG CLAIBORNE

East Hampton, New York

The World Is His Kitchen

THE JUNE sun is hot in East Hampton. It's the kind of weather Craig Claiborne has been waiting for. He ambles out to the garden and pulls open a pod bursting with plump peas, still warm from the sun. He loosens them with his finger and then tilts his head back and lets them trickle in. Ahh—summer has arrived, bearing nature's bounty.

"There's nothing better than going to a garden and eating the first crop of June peas right out of the shell. In fact I like all vegetables when they're raw or undercooked. Of course where I came from, in the South, during my childhood they had vegetables overcooked. In fact they'd cook them for hours. I still like that. It's just part of my emotional heritage to appreciate foods from my childhood."

It was a Mississippi childhood redolent of chopped onions, green pepper, celery, and garlic cooking on his mother's stove—a smell that Claiborne remembers as "distinctly Southern." Part of that Southern cooking heritage was soul food. That's where the overcooked vegetables came in. Mustard greens, turnip greens, or collards were cooked in salted water for hours. Before the vegetables were even served, Claiborne would have a bowl of the rich green cooking liquid known as "pot likker," with a buttery hunk of corn bread fresh from his mother's oven.

Over the course of the thirty-one years he served as food editor for the *New York Times*, Craig Claiborne grew to love vegetables and salads of every kind from the cuisines he encountered during his travels around New York, America, and the world.

"I have a passion for beets. I love vegetable salads with sliced beets, sliced artichokes, sliced mushrooms, and greens in a light vinaigrette. And what could be better than a fresh vegetable soup, like a plain, simple tomato soup with green peas? My memories of places always bring to mind a wonderful vegetable dish I had there. I remember when I was in Italy last month I went to a restaurant where they served me a dish of small slices of zucchini sautéed till al dente in olive oil and browned with garlic, then covered with almonds, pepper, and romano cheese. It was the best thing I had in Italy."

Craig Claiborne's life has been filled with peak eating experiences that are recorded in a large number of articles and books. He tested and published under his byline nearly nine thousand recipes during his tenure with the *Times.*

Among his one hundred favorite recipes chosen from those thousands are savory vegetable preparations as humble as a grits and Cheddar cheese casserole and as grand as *timbales au Roquefort.* But Claiborne is a Southerner, and his tastes gravitate toward the foods that form part of his emotional life. He rhapsodizes over black-eyed peas cooked until tender and served with a mild vinaigrette like those made by Patrick O'Connell. And Claiborne adores fritters, even when they're made with garbanzo flour and go by the name of *pakoras*, or are made with rice flour and called *tempura.*

I interviewed Claiborne at his city apartment near Carnegie Hall. "I don't much like New York," he told me, "but I have to be here on business." Although New York was the scene of his many years with the *Times*, the daily writing grind was stressful, and he was not always on the best of terms with the paper. He is happier at his country home, which reminds him of his youth. During our conversation he told me of his circuitous journey from rural Mississippi to his job as food editor at the *Times.*

His mother ran a boardinghouse in Sunflower, Mississippi, where she cooked those wonderful Southern treats Claiborne remembers. He enrolled in the journalism school at the University of Missouri, and after graduation, served in the navy during World War II, when he experienced both French and Moroccan cuisines. But it was not until his dissatisfaction with a post-war public relations job in Chicago that Craig Claiborne, at age thirty-two and broke, combined the two great interests in his life: food and writing.

"I hated every minute of the public relations job. I felt like a male prostitute. So to escape Chicago I volunteered for the Korean conflict. When I was in the Pacific on a remote island base I had time to think, and I asked myself: 'What are you going to do when you grow up?' I replied: 'I like to cook and I like to write.' So I decided to go to Switzerland and attend the Professional School of the Swiss Hotel Keepers Association. I took a six-month course in table service and classic

French cooking. Then I came back to New York, with a well-developed palate, but desperate and poor."

After a short stint at *Gourmet* magazine, Claiborne got back into public relations. But this time it was with a woman he much admired while they were at *Gourmet*. Their major account was Fluffo, a yellow knock-off of Crisco, and one of the first people he called on to promote the product was Jane Nickerson, the food editor at the *Times*.

He jokingly told her he coveted her job, and when she revealed that she was resigning for family reasons whether they found a replacement for her or not, Claiborne boldly applied for the position. A male food editor was unheard of at the time, yet he got the job in 1957 and launched a career that has brought fulfillment to his life and a first-class education about food to ours. Craig Claiborne became a food editor when we as a nation knew little and cared even less about food. We were living in the wonderful world of chemicals and progress, and we had learned how to mass-produce tasteless, beautiful-looking food. We had left behind the farm stand and our regional foodstuffs and regional dishes for TV dinners and fast-food chain restaurants.

But Claiborne remembered what good food was. And when a great restaurant appeared, it was he, the first of America's great restaurant reviewers, who told us about it. He witnessed the rebirth of American regional cuisine in small, independent restaurants that had rediscovered good, farm-fresh produce. And he told us about it in a way that made us want to run to the kitchen or to a recently reviewed restaurant and eat, eat, eat!

When he was told he had the job at the *Times*, Claiborne was frightened.

"They called me and said I had the job. I went walking on Fire Island, and I started to cry. 'My God, what will I write about?' I thought. Then I saw a bluefish, and I thought, 'I'll write about how to cook bluefish.' Then I thought, 'What'll I write about tomorrow?' After the third day, there was always something to write about."

A phenomenal twenty cookbooks later, Craig Claiborne still has not run out of things to write about. *The New York Times Cookbook* is a classic, and Claiborne, often in conjunction with confrère Pierre Franey, has written on such diverse topics as Chinese cooking, cooking with herbs and spices, and gourmet diet dishes.

Now nearing seventy, and free from the confines of writing a daily column, he is busier than ever. "I've got *too* much going on now. I'm about to go to China with the American ambassador to Burma. When I get back I'm going to do some endorsements for a knife manufacturing firm. I've just finished a diabetic cookbook for a pharmaceutical company. And I've got to get back to Italy. I travel a lot, both for business and pleasure."

It's reassuring to know that the man who helped bring about a new way of cooking and eating in this country is still learning about and creating wonderful food.

SIDE DISHES
Zucchini with Almonds

MARION CUNNINGHAM

Walnut Creek, California

Mom's Home Cooking

HOME COOKING: what we all love, what sustains us, what we long to come back to after a vacation, a business trip or a restaurant binge that may have been glamorous, possibly even nourishing, but hardly ever nurturing. Probably no one has added more to our knowledge of home cooking than Marion Cunningham, who, before she was James Beard's assistant, a cookbook writer, a columnist, a teacher, and a restaurant consultant, was a home cook. The reason, I think, she has been so successful in all those venues is that she relates everything she does (even in her consulting work for Berkeley's short-lived breakfast restaurant Bridge Creek) to celebrating the pleasures of the home table.

"I think there's nothing more welcome than the simplicity of home cooking. The rather rustic, hand-crafted preparations are wonderful. They're the things that sustain us. I think somehow that kind of food is steadily the most nourishing of all."

I met with the gracious and lovely silver-haired cook in her ranch-style house surrounded by stables in the San Francisco Bay Area community of Walnut Creek. The house was sweet and spicy with the warm fragrance of the freshly baked date-nut muffins and brown sugar muffins she had set out on the kitchen table. It was just like coming home.

Cunningham is a cook, not a chef. She has never attended the Cordon Bleu or the Culinary Institute of America, and she hasn't cooked on the line at a restaurant. For many years she was a wife, mother, homemaker, and community contributor. "The one thing I truly loved to do was cooking and baking. I used to do it for big groups, PTA sales, that sort of thing. Then there was a rash of cooking classes in the sixties, for non-pros, taught by women who loved to cook. I took every class available in the Bay Area. I decided I would start teaching because I loved the classes so much."

During this period Cunningham journeyed north to Seaside, Oregon, to take classes from James Beard in the summer of 1962. She and Beard became friends. They corresponded during the year, and he asked her to come back the following summer to help him. She continued to assist Beard for the next eleven years, traveling all over the world with him for four to five months out of every year.

"James Beard was one of the most compatible companions in the world. His memories and passion for food were vast. He had wonderful taste. His memory could give him foods that went back to his very beginnings. He remembered what he ate, and then he'd conjure up these dishes. His memory was incredible. He made so much sense to us, with his catholic appreciation of food. He wasn't biased. He could love a hot dog as much as foie gras. Good was good. It didn't matter."

It had never entered Marion's head to write until one day Beard, the dean of American cooking, suggested to her that she revise the classic collection of American recipes, *The Fannie Farmer Cookbook*. It took five years to complete the revision, which was published in 1979. She followed it with *The Fannie Farmer Baking Book*, which was all her own material.

Baking has always been Cunningham's love. To further advance the knowledge of the art, she has recently formed a group of professional bakers called Baker's Dozen, to share ideas and discuss the mysteries of why baked goods rise and fall. The group also discusses baking equipment and ingredients, and functions as sort of a job bank for bakers. There are chapters in the San Francisco Bay Area and

Los Angeles, and, before long, there will probably be a baker's dozen of Baker's Dozens. Nothing would please Marion more.

"I love baked goods. The kind you simply make at home with ingredients that you personally handle and know is the best. The bread course is the most important part of the meal, next to dessert. For myself, I could eat bread and salad, skip the rest, and go to dessert!"

Of course, Marion loves vegetables, outside of salads as well as in. She loves them quick cooked until just tender with a little salt, pepper, butter and herbs, or slow cooked in the oven, as in the case of root vegetables, which she serves with roasts and chicken dishes.

Marion Cunningham knows that sooner or later, we all come home. And when we do, we want to know what to cook. Though the temptation might be to recreate in the home what we had in some great restaurant (one of the inspirations for this book) she would have us do otherwise:

"I believe the important thing about food is sharing. That is the priority. It's the people you're with. Next, the food should be simple in preparation, appropriate to where it's being presented. If you're doing it at home, have home cooking. Don't try to copy a restaurant. Let that treat be enjoyed when appropriate. There are so many dishes that people don't know how to make any longer."

To that end, Cunningham is doing another revision of the venerable *Fannie Farmer Cookbook*. She'll be going for easier-to-prepare recipes tailored to our busy lives.

"Times have changed. It's amazing. It'll be over eleven years in 1990 since the previous [book] came out. We've veered away from a lot of the French hotel-type sauces and richness. I think we're more committed today to doing the regional-type cooking, homey dishes. We all know the word *busy* is in everyone's life. The effort is to do dishes that aren't complicated. Health plays a large part in today's thinking about food. It means a reduction in fat wherever possible. It certainly is going to be written with an eye to the way we're all currently trying to eat. It's vastly different."

The interview and the muffins ran out at about the same time. I thanked my hostess, and as I moved down the hall past a wall of James Beard and Julia Child photos and related memorabilia, Marion commented: "I'm going to keep on teaching cooking classes and writing, because one of the great joys about cooking for almost all of us who are more than just casually committed to it is the return you get from people you've shared it with. To me that's really a joy. There's something very nice about it. When somebody gets in touch with you about a recipe they've tried. People are very thoughtful that way."

Sounds just like Mom.

ROBERT DEL GRANDE

Cafe Annie, Cafe Express, Houston, Texas

Low-to-the-Ground Cuisine

IF YOU had asked someone to describe Texas cooking a few years ago he or she might have mentioned chicken-fried steak or enchiladas. But that's not the kind of cooking Robert Del Grande is talking about when he says: "If you were blindfolded and carried in here, and hadn't been told where you were, I'd want you to know you were in Texas just by tasting my food."

Robert Del Grande is redefining what we think of as Texas cooking. "We try not to have any preconceived ideas about food. That can make your cooking too staid. If you avoid the 'this is how it should be done' sort of thing and just think in terms of the flavors on the plate, you end up cooking much more successful dishes. I'll give you an example. Most chefs think every entree has to have some meat on it. That's categorical thinking. But if you just think in terms of flavors and suddenly notice, 'Hey, there's no roast beef or something on this plate,' you realize it's okay because you have plenty of other things working on the plate already."

Such unconventional thinking comes from a man who became a chef in an unconventional way. Although he has a Ph.D. in biochemistry, Del Grande began cooking at Cafe Annie after graduation for some extra money and for the fun of it. He became so fascinated by the transformation of foods during the cooking process that he decided to trade in his lab apron for a kitchen apron. Now he is the chef and general partner at Cafe Annie, but he hasn't lost his scientific approach.

"We consider ourselves natural biologists in our approach. We look at the product, study how it grows, then develop a cooking style that is congruous with that. In other words, we don't make vegetables, for example, do things they really don't want to do. We probably wouldn't carve any sculptures out of potatoes, if you know what I mean. Instead, we work pretty closely with the inherent flavors in the produce itself. It's a very low-to-the-ground approach to cooking."

It's a kind of cooking that creates dishes such as barbecued sweet potato slices with forest mushrooms and walnuts. The chef cuts thick, round slabs from the heart of the sweet tuber, boils them until tender, and then bastes them with a tangy sauce before slapping them on the grill. The result is a sweet, smoky, tender delight that plays well against the crunchy walnuts and earthy mushrooms.

"Because we don't have a categorical approach to cooking, we'll barbecue something like sweet potatoes in spite of the fact that most people say you would only barbecue meat or chicken. So we come up with an incredible dish that no one ever considered. When you think in terms of flavor and not category, you succeed."

Succeed he has. Cafe Annie remains wildly popular despite the depressed economy in Baghdad on the Bayou. With its dark wood booths, red leather banquettes, and black and white checkered floor, the bistro is pleasant and welcoming.

Part of the cafe's appeal is also due to the warmth of the hostess, Robert's wife and general manager, Mimi, whose genuine welcome does not emanate from the well-established Texas tradition of practiced politesse regardless of circumstance. She's just naturally happy you've come to the restaurant and wants you to know it.

But it's the food that makes the place endure. One recent dinner began with a sweet potato torte in a yellow-pepper cream sauce. A sweet, perfumey essence radiated from the torte, which was studded with jewel-like nuggets of colorful peppers. A *pico de gallo* sauce, chunky with onion and tomato, added just the right digestive heat.

Next came a small, flat, golden semolina cake scented with nutmeg and topped with melted mozzarella, Parmesan, and goat cheese and a confetti of red, green, and yellow peppers. Firm, red-ripe Texas tomato slices and sweet, sassy picked red peppers on the side provided flavor, texture, and color balance.

In recent years pasta salads have come and gone. But few are as special as Del Grande's room-temperature noodles with a pumpkin seed–basil pesto. A sprinkling of Reggiano Parmigiano was the finishing touch.

Though Del Grande is self-taught, he admits to certain influences in his cooking, especially "those French guys like Bocuse, Troisgros, and Gerard in their *nouvelle cuisine* days." I think one of "those French guys" may have influenced our main course, which reminded me of a dish I had at Guy Savoy in Paris in the heyday of *la nouvelle*. Savoy's dish was sauteéd wild mushrooms set adrift in cream sauce, surrounded by a tangy parsley puree. Del Grande's consisted of a long green chili stuffed with soft, savory potatoes, placed in the center of a plate of ivory cheese sauce and surrounded by a pungent mahogany puree of pasilla chili.

Also very French was Del Grande's decision to serve the salad course at the finish, though the salad itself was very Texas: big juicy coral-colored wedges of pink grapefruit, tossed with crunchy jícama and peppery radishes in a yogurt-mint dressing.

Del Grande thinks of vegetables as a challenge to his creative powers.

"We've always done a lot with vegetables. From the cook's standpoint it's the greatest area of diversity. So our dishes are actually conceived with the vegetables in mind. The vegetables aren't something we add to the dish at the last minute. They have an integral part to play in the flavors of the dish. For me, the vegetables make the meat or fish even more interesting because of that play of flavors.

"Our best dishes are the ones that work if you take the fish or meat away and just leave the sauce and the vegetables. The dish itself is complete without the meat. In fact I've had many customers who tell me they'd order a dish again, just for the vegetables."

Del Grande notes that his emphasis on vegetables is having a pronounced effect on his suppliers as well as his customers. Instead of bringing in everything from California, Robert has encouraged growers in Houston and south central Texas to provide him with a steady supply of high-quality lettuces, vegetables, and fresh herbs.

When Del Grande's customers taste his vegetables at Cafe Annie or his two casual salad-and-sandwich Cafe Expresses, they demand the same quality from their food stores. So, Del Grande finds, "We look at what we're buying and what's in the supermarket, and the gap is definitely narrowing."

With chefs like Del Grande in the vanguard, the gap between the world's great cuisines and our American cuisine is narrowing as well.

MAIN COURSES

Potato and Cheese Enchiladas with Red Chili Sauce

MARCEL DESAULNIERS

The Trellis, Williamsburg, Virginia

The Inner Voice

FOR SOME of us there comes a time in our life when economic necessity forces us to give up the thing we love—the thing we have worked for all our lives and the thing we want most. Artists do it. Actors do it. We have to survive. Yet deep inside, a voice keeps calling.

One day while Marcel Desaulniers was selling frozen codfish and French fries, he heard a voice calling deep inside him. His travels in the food brokerage business had taken him to California, where he saw what Alice Waters was doing with fresh produce. He had seen what Larry Forgione was achieving in New York. These were the most exciting developments he had witnessed during his twenty years in the food and restaurant business. And he knew he could cook like that too.

Although Desaulniers had given up the restaurant business he loved, he couldn't ignore his inner voice. In 1980, he and two partners opened the Trellis. In a city that is a destination, the restaurant too has become a destination. A whopping four hundred to five hundred tourists pack the place daily. At night, the locals come out to enjoy some of the most inspired cooking on the eastern seaboard. The lunch menu is simple fare, lots of salads and sandwiches as well as items from the grill. It's fresh, well prepared, tasty, and fast. Tourists don't linger over lunch when the bus is waiting to take them to the next historic site. But at night, the Williamsburg gentry promenade down cobblestoned Duke of Gloucester Street to a different kind of restaurant, where Desaulniers cooks the dishes for which he is famous. What were casual rooms by day are transformed into intimate dining chambers decorated in varying degrees of fancy, the most romantic being the forest-green Trellis Room.

The customers in Williamsburg, even the full-time residents, aren't demanding.

"Strangely enough, our customers don't ask for much. They want comfortable surroundings, decent food, good service. We could have put a prime rib restaurant here and been successful. Customers haven't really dictated what we're doing. It's more our philosophy and style which have developed over the years."

The style is an eclectic, seasonal one drawing on foods and culinary traditions from Virginia's colonial past as well as from around the world. So you'll see on the same menu an eighteenth-century-inspired offering of black-eyed-pea cakes next to a contemporary salad of strawberries, Belgian endive, and macadamia nuts in a balsamic vinegar dressing.

It is in such innovative salads that Desaulniers really shows his stuff. He likes to combine flavors that may have never gone together, such as his romaine, grapefruit, and walnut salad with Stilton in a port wine vinaigrette.

His appetizers show the same sort of inspired creativity: the chewy, sweet and spicy treat of fresh herb tagliatelle with oranges, red onions, and black pepper butter is satisfying enough to take the place of an entree.

My ideal meal at the Trellis would be one of the chef's appetizer pastas as a main course, preceded by a platter of housemade breads alongside one of his imaginative soups. In spring, when Virginia's first asparagus come to market, Desaulniers pairs them with locally grown shiitake mushrooms in a creamy soup. Dip in a hunk of herbed Irish soda bread, and you don't even need a spoon. Or fritter away a Southern summer afternoon with Desaulniers's fluffy corn and tomato fritters and a tureen of chilled Stilton and pear soup. In the relative cool of the evening, two vegetables whose flavors peak when the mercury does join forces in a formidable roasted corn and smoked tomato soup.

After a hard day of slogging through museums, battlegrounds, and historic sites, a salad, soup, and bread may only whet your appetite for an entree like Marcel's grilled asparagus with tomatoes, braised black-eyed peas, and basil rice.

The Trellis features an all-vegetarian entree both at lunch and dinner, calling it "the garden selection." When I last dined there the selection for that day was a mélange of naturally smoky Anasazi beans from New Mexico, with pine nuts, golden chanterelles, oyster mushrooms, grilled spinach, and succulent oven-roasted cherry tomatoes. With a dish like this, it's obvious the chef realizes that vegetarians do not live by tofu alone. The garden selection has proved popular with omnivores seeking a change of pace as well. The Trellis sells ten to twelve of the platters a night.

It wasn't easy getting fresh, local produce in out-of-the-way Williamsburg. But by developing a relationship with small farmers in the area, Desaulniers was able to secure a steady flow of nature's bounty, from shiitakes to squashes.

Despite his name, the chef with the contagious, ear-to-ear grin is a Rhode Island native, who was trained in the United States at the Culinary Institute of America when it was still located in New Haven. He has maintained his old school ties by serving on the alumni advisory board of the Institute and being a featured speaker at graduation ceremonies. Marcel sees the CIA as one of the few places a young American with extra-virgin olive oil in his or her veins can learn his exacting craft.

"The problem is, we need a lot more quality chefs to help raise the standard of cooking here in America. It was hard getting started here in this outpost in the early seventies, and it's still hard today. The problem is too many graduates are succumbing to lucrative offers from large corporations and multi-unit operators. If all the chefs with great potential go that route, it'll inhibit their potential and stifle their individual development."

We are fortunate that Desaulniers listened to that inner voice and went back into the kitchen. And it's good to know that he has made individuality pay off, both in the success of his restaurant and in the development of his own brand of highly personal cuisine.

SOUPS

Shoe Peg Corn and Peanut Soup

Herbed Tomato and Onion Soup

SALADS

Curly Endive with Warm Grilled Vegetables and Mustard Dressing

Strawberries, Belgian Endive, Macadamia Nuts, and Balsamic Vinegar Dressing

MAIN COURSES

Sauteed Shiitake Mushrooms with Zucchini, Tomatoes, Spinach, Broccoli, and Thyme

RELISHES AND INTERMEZZOS

Wild-Mushroom Chutney

JOHN DOWNEY

Downey's, Santa Barbara, California

Making the Connection Between Earth and Plate

THE COLDEST place in the world? That's easy. An English bathroom. In fact the whole bloody island is pretty cold and dank. That may explain why the British kept leaving for places like Khartoum and Zulu Africa, where they remained even after being roundly thrashed by the locals.

Now that the days of colonialism are over, where does the perpetually frigid Englishman go to find warm weather? One popular destination is the former thirteen colonies, where one Englishman I know has become a pretty accomplished board-sailor and surfer. My Surrey surfer friend also works for a living, if you can call doing anything in Santa Barbara "work." He operates a little forty-eight-seat storefront restaurant called Downey's. Many people who eat there feel that John Downey is the best chef between Los Angeles and San Francisco.

One thing John Downey didn't leave behind in his homeland was his English love for gardens. He elicits from the rich California earth all manner of vegetables and herbs, from peppers to borage. He's too busy cooking to spend as much time gardening as he'd like, so he also relies on small, local organic farmers to bring him produce warm from the earth, which he quickly, and with as little meddling as possible, turns into wonderful preparations.

"There's a connection between the earth and the plate, if you will. I like to make that connection. The bountiful earth, what comes out of it—I like to get that on the plate with as little interference as possible. It's a gut feeling.

"One farmer brought me some lovely Romanesque broccoli the other day. The head forms a sort of beautiful spiral. All I wanted to do was take that lovely broccoli, steam it, and put a sliver of cold butter on it. I served it with grilled lamb that night, and to me the broccoli was the best part of that dish. People ordered it for the lamb, but the broccoli was really what I was enjoying doing."

But Downey doesn't just do individual vegetables. In fact, one of the most satisfying and wonderfully complex dishes I ever had was his grilled vegetable salad. Red bell peppers, onions, and eggplant were grilled to perfection, chopped, and dropped into a chili vinaigrette lively with cider vinegar, jalapeño, herbs, Worcestershire, and mustard, where they were joined by raw Anaheim chilies and left to marinate, ceviche-like, until the flavors softened and rounded to perfection.

Simplicity is the rule at Downey's. "I want to let food be represented in as natural a state as possible. So what I try not to do is have a lot of overly complex dishes. What used to happen a lot is that food would get masked by too many interfering flavors." Downey, like most chefs, has been through a lot of phases.

"Today I have a sort of earthy philosophy. I get a lot of joy out of looking at something simple and representing it that way. I want to present flavors that just naturally belong together."

Downey wants to use raw ingredients that yield the most flavor, color, and vitality. And he knows where he can get them. "Local organic farmers have better produce and better varieties. The big farmers are looking to grow things that mature fast, can be picked underripe, and ship well. Unfortunately, flavor has been on a back burner for years. It seems to me flavor is the last thing big farmers are concerned about. But the local farmers are concerned. They'll grow different varieties and ask us which ones we liked. Then they'll make plans accordingly for the following year. It's nice."

Several of the chef's best dishes are based on beans, such as his black bean soup with hot green chutney. He is fond of mashing soft, fermented, dried and salted soybeans for his spicy, but not hot, shallot dressing, which he generously smears over fanned California avocados, served butter ripe. Another legumbrous wonder is the warm lentil salad scented with defanged garlic, rosemary, thyme, and bay leaf, and dressed with balsamic vinegar and olive oil.

Downey waxes poetic on the subject of vegetables: "When I was growing up, my dad used to take care of this garden for a British colonel. There were all these different varieties of apples—things you can't get now. That's a sad passing. I find it sad we're losing old varieties of apples. I mourn this passing. Some people are trying to save these old strains. And I'm trying to do a lot for food what those people are doing for apples. I think I'm hanging in there, sticking with what are possibly not popular ideas or concepts. But I want to do my best to educate my customers. Unfortunately, we've gotten away from the soil in this country. I think there's a whole generation of people who don't appreciate the finer flavors God has given us. They're buying supermarket junk because that's all they've been taught, and they're missing the whole point of what we have. I encourage people to grow food and then eat what they've grown. To get into it and sink their teeth into it.

"My wife's uncle brought me a huge box of kale. It was beautiful. I browned some onions, put in some chanterelles, and at the last second added the blanched kale. All those flavors just belong together. It was wonderful. People said they didn't even realize you could eat kale. They thought it was something you garnish salad bars with. Well, I could sit down to a big plate of that and skip whatever else is on the plate. Because the vegetables are as important as the piece of meat or fish on the plate, and we give them as much attention as what some people might consider the main part of the entree. I'm not a vegetarian, but there are lots of times when I turn my back on the meat and fish. You need a break sometimes."

Downey dreams, like so many Englishmen before him, of taking a real break someday and getting farther away from civilization. His recently lightened and whitened restaurant features a daily changing menu with only a few choices. But Downey wants to simplify his life further.

"Right now I'm making a living. But later on in life I won't be as concerned about that. Maybe down the road we'll move somewhere smaller, more remote, near the coast. I like the open air, being in touch with nature. I could just cook what I enjoy cooking. I'd get a big kick out of having a small restaurant, and offering just one prix fixe dinner a night. No choices. I'd just say: 'This is dinner. Would you like to come?' "

Yes, I would. So, John, when you move, please leave a forwarding address.

SOUPS

Black Bean Soup with Hot Green Chutney

SALADS

Grilled Eggplant and Chili Salad

DEAN FEARING

The Mansion on Turtle Creek, Dallas, Texas

The Good Karma Chef

IN 1869, the Russian scientist Mendeleyev made one of the most important discoveries in the history of chemistry. He found that elements (basic things that you can't break down any further, like gold, iron, carbon, and oxygen) were related in families with certain other elements. What was astonishing was that when he created a table of these elements it turned out that they were related to each other vertically as well as horizontally—much like a crossword puzzle. The first time I saw the Periodic Law and Table of Elements, at age sixteen, I was moved by the realization that there was a superior intelligence creating and organizing everything in the universe.

In 1989, I was served a simple-sounding pumpkin and corn soup at the Mansion on Turtle Creek. The superior intelligence of Chef Dean Fearing had to be divinely inspired to create a soup so complex in its construction of flavor and texture, with elements interrelating on so many levels, that it immediately brought to mind the wonder with which I first beheld the Periodic Table.

On a textural level, the soup offered carefully calibrated levels of crunch from cranberry to zucchini to corn to pumpkin seed, with all those ingredients relating to each other and to other elements in the bowl, like cumin and pepper, in terms of flavor. The result was a creamy chowder whose flavor seemed to change from sweet to spicy to earthy, depending on the everchanging interplay of ingredients.

For Dean Fearing, texture is as necessary to a dish as is flavor. "I think texture is very important. The French would do a meat, a sauce, and maybe a vegetable puree. Everything was pretty much the same texture, across the board. That could get kind of boring. I think the fact our food is fun to eat texture-wise is one reason we do so well. The crunch adds texture and pull to the dish. That's why we do a lot of things with vegetables like potatoes, corn, asparagus, jícama, and sweet potatoes. If you just had meat and sauce, there'd be no dimension, no depth.

The vegetables add that great crunch only they can provide that accents whatever's on the plate."

The Mansion on Turtle Creek is the jewel in the crown of the Hunt family's Dallas-based Rosewood Hotel chain. From the outside it looks like a Moorish-style temple, with towers, arches, and balconies all done in varying pale shades of marble and granite. Inside, the sweeping staircases, glass display cases, and magnificently appointed dining room with ornate ceilings are all done on a grand, Texas-size scale.

During his tenure at the Mansion, Fearing has established a national reputation and could, without difficulty, open his own place. Yet he chooses not to.

"I learned a lot from the French. Not only did I learn the basics and the style, I also learned from the French that longevity makes a master restaurant. A lot of these French chefs put in nearly twenty years at one restaurant before they ever think of moving on. It's a development, an education, learning about the people who work here and the people who eat here. I'm just starting to see the effects after being here four years. It doesn't come overnight. I'm very happy where I'm at. I love the Mansion. It's a wonderful place, and I love the management I'm affiliated with."

Fearing has the best of both worlds here. He has none of the financial worries attendant to running a restaurant. The hotel takes care of that, as they do the office work and other non-creative drudgery. Meanwhile he has the freedom to go out on the road, doing charitable work and promoting his national cooking reputation, which in turn promotes the Mansion because everyone knows where he's based. His cookbook is *his* cookbook, but it's called *The Mansion on Turtle Creek Cookbook*, so everyone benefits when he promotes it. In his view, it's a reciprocally beneficial arrangement, and the good things that come to him as a result are, in his words, his "good karma."

I had the good karma of dining at the Mansion recently, and though Fearing was on the road at the time, so well trained is his staff that every dish was flawless.

I can usually tell the character of the kitchen from the contents of the bread basket. If they care enough to serve house-made breads, the same kind of consciousness will be at work in the other dishes. So when I was brought a basket of nutty whole-wheat bread and a crunchy caraway rye that would have done a New York deli proud (even if the bread did contain uncharacteristic, but delicious, raisins) I knew I was in for a good night.

Then the aforementioned pumpkin and corn soup came, followed by roast

Texas corn with bourbon cream, served with smoked peppers and apple-pecan fritters. The slight scent of bourbon was just enough to enhance the smoky, chewy corn, which played nicely off the peppers and crunchy fritters.

The linguine and Mexican vegetables with cilantro–pumpkin seed pesto was a refreshing pasta dish crunchy with jícama, barbecuey black beans, onions, and yellow and green peppers, and just the right amount of spiciness.

Fearing's from Kentucky, and his roots are showing when he brings forth such Southern fried specialties as fritters and hush puppies. The pecan-smoked onion hush puppies, as well as a sweet pecan-persimmon version, would keep a hound dog quiet even in the presence of a red fox.

The creamed French green beans with shiitake mushrooms, topped with crunchy "tobacco" onion rings, were an elegant yet homey side dish with a warming jalapeño bite.

There were other treats: maple-glazed sweet potato balls, crusty sweet onion pudding, and the like. I could have ordered even more, but I was full. And I was full of admiration for a chef who cooks in an area where, when I lived there in the fifties and sixties, French fries were about the only vegetables available.

"It all comes down to good food," notes Fearing. "I want to serve people good food and have them leave here happy. All of them. Let's say Joe Texas comes in here and wants steak and potatoes. I want to make sure he's happy enough and that he has great food and a great time here. If someone wants Oriental food, or vegetarian food, that's what we'll do to make them happy."

Fearing knows what foods make *him* happy. "I eat no meat at home. I believe the human body doesn't need to consume a lot of meat. Vegetables, grains, and fruits are so important to us from a health point of view. They're a part of our daily body makeup. These foods are lighter and cleaner so the food goes right through you. That's how food should be. In and out. Otherwise the body just gets consumed with the heavy intakes of meat and other foods. I'm not a vegetarian, but I agree with the philosophy a lot. Maybe one of these days when I get out this business I can be one."

Though often placed in the Southwestern school, his classical training at the Culinary Institute of America and at Cincinnati's old-line French restaurant, La Maisonette, won't let Fearing stray too far away from the basics. While he does call on the bounty of chilies and other local ingredients so characteristic of the Southwest, his menus feature dishes from all over the map and in each case are faithful renderings of ethnic cuisines.

"The press categorized me in a certain way, like calling me a Southwest chef, and we do use jícama, chilies, mango, papaya, and the like. They add a lot of flavor and color to make things stand out on the plate. But we do a lot of other things. That's because I travel around, eat at different restaurants, and make things people have never had here before, often with a lot of ethnic connotations. I love Oriental food, curries, and chutneys like mint or tamarind. They're wonderful. I want to teach my customers what some of these cuisines are like, especially if people have had, say, a bad curry somewhere. You gain a lot of people's trust if you serve them something new and they like it. So I want our food to taste good and if possible to have a little spice, a little life to it, so the flavors curve around the back of the tongue and hit you after you swallow it."

You certainly couldn't call the menus I saw "Southwestern." There were indeed some Mexican dishes, but they were often juxtaposed to offerings like fried "pot stickers" with a cucumber-ginger vinaigrette, and fettuccine with vine-ripened tomatoes, artichokes, basil, and Romano cheese. Yes, there were enchiladas and tomatillo rice, but you could have just as easily ordered a vermicelli-peanut salad with a Thai sweet chili sauce. For my part, I simply ordered every item under the vegetables, potatoes, and pasta section. Granted it was more than any three people could eat, but despite the generous size of the portions I only rang up a $20 tab. And I couldn't finish everything. I wish more restaurants had a menu section like this. But then, there aren't that many restaurants interested in making great vegetable dishes. Yet.

Fearing gives vegetables a major role in his cuisine. "There's a certain romance I have with vegetables. I think it's real important that you deal with them as though they were as important as any other item in the kitchen. I want the vegetables to come to the table crisp and at their peak of flavor. That will enhance whatever people are eating and help them leave here happy. And if they leave here happy, they'll remember the experience and how good the vegetables tasted, and more importantly, how good they made them feel."

MAIN COURSES

Texas Cream Corn with Apple-Pecan Fritters

PASTA AND RISOTTO DISHES

Poblano and Black Bean Linguine with Ancho Threads, Oaxaca Cheese, and Mexican Vegetables

SUSAN FENIGER MARY SUE MILLIKEN

City, Border Grill, Los Angeles
Border Grill, Santa Monica, California

Searching

WHEN MARY Sue Milliken, fresh out of Washburne Trade School, applied for a position at Chicago's posh Le Perroquet, Jovan Treboyevic looked at her credentials and immediately offered her a job as hat-check girl. She politely declined and persevered until she made it into the kitchen a year later with the job of preparing vegetables. A year later, she and Culinary Institute grad Susan Feniger, who had joined her in the kitchen by then, were running the kitchen in the executive chef's absence. Milliken left to direct a kitchen in Deerfield, Illinois, while Feniger went to Ma Maison to work with Wolfgang Puck.

They were out of touch at this time and, unbeknown to each other, both left for France at about the same time to study *la technique*. Mary Sue went north to Restaurant d'Olympe in Paris while Susan headed for the Riviera to cook at L'Oasis with Louis Outhier. They ran into each other quite by accident in Paris and made a mutual vow to open a restaurant together one day.

Their chance came about six months after their return to the States. Susan Feniger was back at Ma Maison, but had started working part-time at a little coffee shop on Melrose called City Cafe. The twelve-by-thirteen kitchen sported little more than a hotplate, but it was heaven for Feniger and for the thirty-five lucky patrons who daily packed the narrow storefront. For the first time, she could try out her own ideas, even if they took hours to come to fruition. Mary Sue soon joined her, and the partnership was born.

Now, seven years later, City Cafe has become Border Grill, serving home-style Mexican cooking. And the food the two women pioneered at City Cafe lives on at City, a cavernous room three times the size of its progenitor—an always-packed, always-open restaurant that serves the most imaginative food in Los Angeles and has become one of the city's most successful. The food, always satisfying, continues to improve, growing lighter and more flavorful all the time. And the partnership?

During an interview Feniger and Milliken smiled at each other when I asked the question. "It's better than ever and constantly growing. We have difficult times now and then, like anyone else, but we talk it out, and it keeps things from becoming situations. We teach each other lots of things. We depend on each other a lot, and appreciate what the other one brings to the partnership."

Although I was interviewing two very different people, it seemed at times that, like the differing flavors in their curries, they had blended into one. Each anticipated what the other was going to say, and they occasionally finished sentences for each other. Never once was there a trace of ego or jealousy, nor did they compete for attention. And it was clear that the success of their endeavors was due to a selflessness and an ease with one another's views that is refreshing in a world where many seek to dominate their business partners.

The ultimate beneficiary of this attitude is the restaurant patron who comes to Border Grill or City. There are enough things on either menu to satisfy anyone's hunger or thirst. And the restaurants are almost always open. You can come for lunch, the afternoon menu, or dinner. And if you're just in the mood for something to drink, be it a marvelous tea like chilled spiced Yogi Tea, biting lemon-ginger, or something stronger, you can stop by during their "Don't Worry, Be Happy Hour."

The widely traveled chefs have brought back the authentic tastes of distant ports of call, making the menu at City the most international in town. And although there is plenty of improvisation and substitution, the renditions are authentically seasoned and bear the unmistakable qualities of their origins.

City's Indian dishes are a perfect example. The appetizers include potato *bhujias* with mint chutney and yogurt, as well as Poona pancakes with yogurt, tomato, and cilantro. The *bhujias* are deep-fried, tingly wonders—sort of like a cross between French fries and spicy hush puppies. The pancake is made by grinding basmati rice and urad beans, and allowing them to ferment in the restaurant's own yogurt. The resulting batter is poured on the griddle and cooked to a golden turn.

Continuing in the Indian vein, Feniger and Milliken came up with the City vegetarian plate, which is composed of buttery dal, basmati rice, a curry, and some other surprises. Having eaten in Indian homes for the past twenty years I can testify that the City's vegetarian plate tastes as though it were prepared by natives. And the authentic Indian *naan* breads, cooked for a few nanoseconds in the restaurant's own arm-singeing, 1000° tandoor oven, are indistinguishable from those I had on the steamy subcontinent.

Apparently most of City's habitués approve of the vegetarian plate. In a city where people are known to go for the jugular, it's the restaurant's biggest seller. "We sold 589 vegetarian dinners in November—way more than any of our other eighteen entrees," they told me.

But City's menu isn't weighted toward Indian food. The second biggest seller is perfect vegetarian fried rice, chock-full of colorful vegetables and served with grilled marinated eggplant. For appetizers, you can go Italian with gnocchi Parmesan, or Greco-Roman with smoky roasted sweet peppers with feta and basil. And for an entree, you'll probably never taste a better pasta dish outide of Italy than vegetable vermicelli, a heaping bowl of angel hair pasta with julienne oyster mushrooms, carrot, zucchini, yellow squash, tomato, and turnip. Add a mash of garlic and freshly cracked black pepper, some chives, and a little lemon juice to bring out the sweetness in the veggies, top with Parmesan and you've got a dish fit for Pavarotti. Every entree is served with the house's vegetables of the day, which are well seasoned, never overdone, and often feature the restaurant's justly famous sweet parsnip chips.

One popular lunch selection is the three-salad plate. A recent trio consisted of a ratatouille with wonderful Provençal olives, rice vinegar–marinated grilled eggplant, and pasta salad of rigatoni with three pestos. There was a semi-traditional basil pesto with walnuts pinch-hitting for pine nuts, a sun-dried-tomato pesto, and a black-olive pesto.

For all-American fare, I don't think you can top City's lunch sandwiches. Most notable is the City vegetarian club. Whole-wheat sourdough bread is grilled and slathered with mustard-horseradish mayonnaise, then flipped over and stacked on top of, in descending order: roasted red pepper slices, arugula, a middle slice of the grilled bread smeared with olive tapenade, avocado, cracked pepper in lemon vinaigrette, tomato, and a bottom slice of bread spread with baba ganough. Served with caraway cabbage slaw and shoestring potatoes, it would make Dagwood Bumstead's mouth water. Yet some think the avocado and tomato sandwich with goat cheese and cucumber, served open faced on a long French roll, is even better.

With such an eclectic menu, there could have been design problems. Should the place look like the Taj Mahal, or the Piazza San Marco, or Yankee Stadium? They solved that dilemma simply, by putting in tanger-red wood chairs, painting opposing walls olive-lime and white, and opening the doors. There's no art. You either look out a huge picture window into the parking lot (how's that for "city?"), or at some interesting architectural angles and cut-outs in the walls, or at the food, which is, of course, the whole point.

Although the inspiration for City's food comes from many countries, it rarely comes from countries that don't exist such as Frapan or Italifornia. You can find that kind of food lots of places in Los Angeles. But Feniger and Milliken want to be true to the traditions of food, even if they take a little liberty with some of the ingredients.

Naturally, one of the biggest ethnic influences in Los Angeles is that of Mexico. While most Mexican-inspired restaurants serve familiar combination plates or bizarre new Southwestern creations of their own fertile imaginations, Feniger and Milliken take a more difficult course. At Border Grill, they faithfully reproduce the foods of today's Mexico. The chefs took frequent trips to Mexico to learn the mysterious ways of chilies, rich cheeses, thirst-quenching fruit drinks, and unusual salsas. And as they have done after staying in other countries, Feniger and Milliken have uncannily reproduced our southern neighbor's food in our native land.

It starts with the drinks: iced hibiscus tea, a tangy tamarind drink, rice-based creamy *horchata*, and cinammony hot chocolate. If you have any doubt you are way below the border it is quickly dispelled in the trio of authentic salsas, especially the mahogany-colored chipotle sauce. You dip thick crunchy chips into chipotle sauce as well as fiery green tomatillo and *pico de gallo* sauces. But try not to fill up on these openers, for it would be a shame to miss the finest green corn tamales in the Western Hemisphere. Made fresh daily of corn *masa* punctuated by corn kernels, the two-inch tamales are wrapped in corn husks, steamed, and served with rich Mexican sour cream and the confetti-like cilantro, tomato, and onion *pico de gallo*. Other offerings are soft tacos filled with either savory potatoes and peppers or cactus and beans. A decidedly different chile relleno is stuffed with three Mexican cheeses including jack, *enchilado*, and *añejo*. And no meal at Border Grill seems complete without *huaraches*, black bean-filled thick, fresh corn tortillas. The list goes on and on: rich *rajas*, which sound Indian, but are actually red bell peppers, poblano chilies, and onions cooked in cheese and cream; and *chilaquiles*, day-old tortillas stewed in a rich, spicy sauce.

The kitchen serves a fine avocado cocktail, a grown-up parfait of alternating layers of creamy green avocado and red cocktail sauce, served in a tall glass with a long spoon. If you just can't decide, there's always the Border Grill vegetarian plate with some or all of the above, plus simply grilled, simply delicious vegetables and cooling jícama-watercress salad. Like the vegetarian plate at City, it's just about the best seller at Border.

Such ingenious vegetable dishes require two things: a real love and understanding of what vegetables can mean to cooking, and great produce. City and Border Grill have them both.

Feniger: "We trained in classically French kitchens where there wasn't that much emphasis on vegetables. But we're both really into it. I don't have meat that often, and when I have it, I have three or four vegetables with it. And when I eat at City, I always order the vegetarian plate."

Milliken: "I think they make the meal really well balanced and rounded: the different textures, the crunch, the flavors they bring to a meal. I think it comes from our natural cravings. Susan and I are really responsive to what we crave for ourselves. A lot of cooks aren't responsive to their own cravings, their own inner desires of what they want to eat. Your body tells you, if you listen to it."

Over the years, Susan and Mary Sue have developed close relationships with their suppliers. And now that the volume of City has given them buying power, they can afford to buy more expensive organically grown produce.

"A lot of our produce gets picked in the morning and delivered a few hours later. One of our favorite suppliers is the Horticulture Therapy Program for Veterans. They're trying to rehabilitate themselves. It's so great. The people are wonderful. They nurse something from seed to fruit, then they deliver it and see the faces of all our cooks light up."

To keep their staff light and happy, whether it's by cooking them Christmas breakfast and playing them a Mexican piano duet or by giving a helping hand to someone who thinks he or she has ruined five gallons of something, Susan and Mary Sue are always there giving support.

That's because success and money aren't the ultimate for them. "We're searching. We're searching to find the balance in life. And in this process, you can't stop. I think a lot of cooks are cooks because it's a good escapist profession. It's such hard, demanding work that you can get completely drowned in it and never have to face who you are. We're always searching for something better, in business, in our personal lives, and our social lives. Some better balance. I don't think you find it, necessarily," said Feniger wistfully. "It's the process of doing it until the day you die," she smiled, "and then maybe starting all over again."

APPETIZERS AND FIRST COURSES
Vegetable Fritters with Chick-Pea Batter

SOUPS
Mexican Corn Chowder

MAIN COURSES
Vegetarian Club Sandwich

M. F. K. FISHER

Sonoma County, California

My Fair Fisher

"IT'S SILLY to ask" (as I had just asked her), 'Now, what do you think about vegetables.' M.F.K. said. "I don't *think* anything. I just know they *are*. They *are* the way I *am*. Vegetables are essential to life. Very beautiful things. I can't imagine living without them. But I'd rather never have a vegetable than have them the way most people cook them. They cook the hell out of them until they're mush. Vegetables add grace when cooked with dignity. When they're not cooked with dignity, they're just an adjunct to a meal. And vegetables are too beautiful to simply be treated as a part of a meal."

For much of Mary Frances Kennedy's life, vegetables weren't treated with the dignity she felt they deserved. Then, soon after she first married in 1929, on the way from Paris to Dijon, a turning point occurred in her life. And it involved, of all things, a vegetable dish.

It happened in the September-sunny courtyard of the old post-hotel in Avallon. During a forgettable lunch, she was served an unforgettable dish: potatoes served as a separate course. It was a kind of soufflé, hot and puffy, topped with delicately browned Parmesan cheese and fresh herbs. Mary Frances couldn't believe it. Here was a vegetable dish, all by itself, not on the same plate with meat or served as a side dish. And it was served with dignity and grace.

This was the first truly French restaurant in which Fisher had ever dined. The moment stayed with her. And from then on, in a land where food and life were intertwined like the branches of an old grapevine, she began to observe and write about the role of food and dining in our lives. About how humans act, and have acted through history, with a surplus of food and with food denied them. Food, like the tamboura in Indian ragas, ever-resonating, and sometimes dominant, vibrates through all our lives, through all time, as M.F.K. Fisher observes it, providing not only the nourishment to sustain our body but the inspiration to uplift our soul.

The many dishes she savored in her years in France, Switzerland, and the United States, and during sea changes in between, became the focus for a writing career that began in 1937 with the publication of *Serve It Forth* and continues today. M.F.K. Fisher is one of the few authors I can think of whose entire catalogue of works is still in print. For when she writes about food, she is not just writing about food. She is writing about life.

After eight decades of writing about the beauty and, at times, the ugliness of humanity, Mary Frances Kennedy Fisher's own beauty still shines forth. As I sat, in her book-filled rustic ranch house overlooking the green Sonoma foothills, we talked of her life and her writing.

When I later confessed I was in love with her, she admonished, "You're in love with words." But who wouldn't love phrases such as: "Her prose is straightforward, occasionally anecdotal, with short bits of Dickens stuck in like raisins in a bun." Or, in describing chocolate: "When it is round and bubbling and giving out a dark blue smell, it is done." Or her description of a loaf of fresh baked bread: "It was naked, like a firm-hipped woman, without benefit of metal girdlings. It came out . . . round and brownly even and filled with an honorable savor."

One of M.F.K. Fisher's most quoted passages is from her introduction to *The Gastronomical Me*, which was published in 1943. In it she tackles head on those who criticized her for writing about something as apparently frivolous as food when there was a war going on. Her rejoinder:

> "People ask me: Why do you write about food, and eating and drinking? Why don't you write about the struggle for power and security, and about love, the way the others do? . . .

> "It seems to me that our three basic needs, for food, security and love, are so mixed and mingled and entwined that we cannot straightly think of one without the others. So it happens that when I write of hunger, I am really writing about love and the hunger for it, . . . and then the warmth and richness and fine reality of hunger satisfied . . . and it is all one.
>
> "I tell about myself, and how I ate bread on a lasting hillside, or drank red wine in a room now blown to bits, and it happens without my willing it that I am telling too about the people with me then, and their other deeper needs for love and happiness. . . .
>
> "There is a communion of more than our bodies when bread is broken and wine drunk. And that is my answer, when people ask me: Why do you write about hunger, and not wars or love?"

The best way to learn about this wonderful woman is to read her. If you are passionate about food and its relationship with life, read everything she has ever written, and your life will be enriched.

We all have food memories that will stay with us as long as we live. For me, it's the smell of my mother's apple dumplings, rushing at me as I first walked in the door from school. For M.F.K. Fisher one such memory is the warm peach pie with country cream she and her sister and father enjoyed after a long, hot ride over the California hills. Here is another of her memories, from *How to Cook a Wolf*:

> "It was a nice piece of toast, with butter on it. You sat in the sun under the pantry window, and the little boy gave you a bite, and for both of you the smell of nasturtiums warming in the April air would be mixed forever with the savor between your teeth of melted butter and toasted bread, and the knowledge that although there might not be any more, you had shared that piece with full consciousness on both sides, instead of a shy awkward pretense of not being hungry."

One of Fisher's main convictions about food is that we should eat as naturally as possible. She rails against worthless white bread and chemically altered foods, and, long before it was fashionable, kept a compost pile with which she fed her own vegetable garden.

"I believe more firmly than ever in fresh raw milk, freshly ground whole grains of cereal, and vegetables grown in organically cultured soil. If I must eat meats, I want them carved from beasts nurtured on the plants from that same kind of soil."

During her writing career M.F.K. Fisher has mused over the sad habit of meat eating. At the time of our interviews she wasn't eating any, because she knows how quickly eating green and vital things will purify her body of all the medicine given her during a recent and harrowing stay in the hospital. She knows vegetables can restore and renew her. As she wrote in *Serve It Forth*: "There are many ways to love a vegetable. The most sensible way is to love it well treated. Then you can eat it with the comfortable knowledge that you will be a better man for it, in your spirit and your body too, and will never have to worry about your own love being vegetable."

As the sun was setting over the Sonoma hills she said, "You have to have a feeling for vegetables. They're so *beautiful* to touch and to look at. I feel strong just looking at them or thinking about them. I feel better. Purer."

We have many hungers, which we can appease lovingly or mindlessly. M.F.K. Fisher helps us understand what it means to be well fed: "All men are hungry. They always have been. They must eat, and when they deny themselves the pleasures of carrying out that need, they are cutting off parts of their possible fullness, their natural realization of life, whether they are poor or rich" (*How To Cook a Wolf*).

MICHAEL FOLEY

Printer's Row, Foley's, Chicago, Illinois

Not Just Another Pretty Face

BENEATH MICHAEL Foley's rugged good looks is an introspective man deeply concerned about food, both on an individual and a national level.

"I think people have forgotten the fact that they spend 20 percent of their life eating. There should be more pleasure involved. It shouldn't just be opening your mouth and throwing something in. I think if they thought more about what they put in their mouth they would put a higher demand on those people purveying and distributing food. You'd get better quality overall, and a fresher market with more current produce and lots more variety."

Michael's mission, as he sees it, is to draw on the culinary traditions of the ethnically diverse city of Chicago, to create from them a refined cuisine rich in the foods of the American Midwest.

Michael's restaurants have a look of unpretentious comfort and elegance. During a recent meal at Printer's Row, I sat in a rich red patent leather banquette in a wood-paneled room lit by silk-shaded lamps. Purple paisley cafe curtains framed the rain-streaked windows; scenes of the hunt were hung on sorrel-colored walls.

Vegetable soup sounded bland but soothing, so I took a chance. It was a subtle masterpiece. In a crystal-clear golden broth floated vegetables the size and shape of small marbles: zucchini, yellow squash, carrots. All had been sautéed to enrich their flavor before their immersion in the consommé. With a hunk of crusty bread spread with rich Midwestern butter, it was a perfect first course.

For a main course I chose a sesame pasta terrine with portobello mushrooms and pesto in a tomato cream sauce. The nutty flavor of the pasta hit first, followed by the earthy mushrooms, the tomato, and finally the pesto. There was nothing contrived about it, as is typical of Foley's cuisine.

"I'm not into the 'flutter of doves' approach to cooking. I like food to be simple and well prepared. There are classical ideas and combinations, and that's why dishes have lasted in their present form throughout history."

Foley was brought up in a restaurant family: his grandfather and father ran Ray Foley's Steakhouse in Chicago. It was basically a meat-and-potatoes operation, but you can't run a successful restaurant for thirty years without a solid grounding in the basics. Michael got his unofficial training at the family restaurant, and after completing graduate studies at Cornell's Hotel and Restaurant School he returned to Chicago to gain practical experience in running small, personalized restaurants. He augmented his training by frequent visits to France, where he polished up his French technique with Alain Chapel and Paul Bocuse.

In 1981 Foley struck out on his own and opened Printer's Row as a forum for his own expression. This was no small undertaking in a city of formula restaurants that open locations as fast and are as impersonal as the city's ubiquitous currency exchanges. But Foley felt there was something more important he had to say with food than "Thank you: please pay your server."

"A restaurant should become a good source for cooking, not just entertainment. It should help educate and help people think about food and where it comes from. But most restaurants aren't interested in that. I mean, you don't need to educate anybody if you're constantly opening new places. Oftentimes a restaurant becomes successful if it's one that achieves a lot of volume. Naturally you need volume and dollars to stay afloat, but at the same time, restaurants these days lack depth. They're very surface oriented. I would hope to have some influence on that."

Foley is having considerable influence on the entire Midwestern dining scene, having been one of the first American chefs in Chicago to receive national recognition. And now that Foley's has been launched, Michael can concentrate on making the most of both his restaurants.

"I want to take my menus and bring food into a more realistic light. Sometimes fine dining isn't necessarily realistic. You take people out of their natural environment and put them in a coat-and-tie environment. I'd like to do something that would bring a lot of flexibility to our menu. I want to be able to shower people with a lot of seasonal, traditional fare from a lot of different cultures."

That was obvious from a light lunch I had at Foley's, a restaurant that caters to Michigan Avenue professionals. Though it occupies a spanking new building, Michael has created the clubby atmosphere of a between-the-wars restaurant in gracious, chandelier-lit tones of beige and brown, with walls adorned with landscape and still-life paintings. The comfortable, greenery-filled restaurant plays forties music, but the food is strictly eighties. My salad of poached pear with curly endive, Gorgonzola, and walnuts was rich, yet light, leaving plenty of room for a dish of creamed fennel and onions.

Foley sees the future of fine dining as linked to the small farmer. When restaurants can get great raw materials from right in their own backyard, chefs are inspired to create dishes that expand our culinary horizons. "People are pretty amazed when they taste simply a combination of steamed vegetables. They don't really know the flavors, because up till now, you couldn't get good fresh produce in places like

Chicago for most of the year. Now that you can, we can do all sorts of wonderful things, and we have. We've actually made asparagus ice cream, carrot sorbet, tomato sorbet."

To make the greatest impact on public awareness about food, Foley would like to have a small farm of his own, but that's difficult given the day-to-day financial demands of running two restaurants. But the chef sees the need for a real change in farming and how society and the farmer view that occupation. "We don't have great distribution right now for the farmer. And because we aren't getting to market all the great things he's capable of producing, even the farmer doesn't have respect for what he does. A lot of the time that has to do with how he himself has maintained his own education and flexibility. I mean there are ways of maintaining a good farm if you understand what to grow, but today, through ignorance, even though there's a lot of technology, there's a lot of waste. There's overgrowing of some products and gross undergrowing of others. We're going to have to iron this out in the next ten years, because as we move to a vertically integrated society, farm land is going to get cut thinner and thinner, and what's grown there will have to be really good.

"I hope in the future you'll see the local growers producing fresh vegetables all the time, right through the entire growing season, instead of the way it's been for the last ten to fifteen years where it's only coming out of three basic growing regions. In the prime season, the farmers in your region will produce because it's practical, sensible, and as fresh as you can get."

Local organic farms like Ladybug and purveyors like Tom Cornille provide Foley and other Chicagoland chefs with produce every bit as good as Tom Chino's in California. The short, intense growing season in this cold climate intensifies flavors. And a better-educated dining public has raised the common denominator about what they want out of a restaurant and what constitutes good, wholesome food. With that kind of support from the grower and customer, Foley's inventive yet homey style of cooking will continue to influence the way we eat.

SALADS

Yellow and Red Plum Tomato Salad with Avocado Oil and Basil

LARRY FORGIONE

An American Place, New York, New York

All-American

AT AN American Place, you can taste our country's past, present, and future. In a time when American cooking is becoming every bit as good as any other cuisine, Forgione links us with our history and points the way toward our culinary future. Not long ago, fine food was thought to be French. When we dined, we headed for palaces of butter and cream and ate serious food with a French accent. But there were still a few people who quietly, but consistently, championed our own home cooking, and one of them was James Beard. Like Walt Whitman, George Gershwin, and Mark Twain, Beard captured America for us so that we can always cherish it. So it was natural that when Forgione decided on the direction for his cooking he turned to Beard as a mentor.

Forgione had trained with Michel Bourdin at the Connaught in London and with Michel Guerard at Regine's in New York. That taught him great technique, but nothing about cooking with American foodstuffs. While training in Europe he was very aware of how good the raw materials there were, and he wondered why American chefs didn't have as much to work with.

By the time Forgione landed his first executive chef's job at the River Cafe, he knew exactly what he wanted to do. If there weren't great products readily available to cook with, he'd go out and encourage somebody to find them or grow them. And if there wasn't much guidance as to how to best bring out their flavor, he'd get James Beard to show him. At first, Beard was reluctant. But after over a year and a half of persuasion, he agreed to help Forgione.

Meanwhile, Larry had developed a whole network of fishers, farmers, and foragers to provide him with the foodstuffs he needed to re-interpret American cooking: fish and shellfish from New England, organically raised Charolais beef from Michigan, mushrooms from the Northwest, baby vegetables from Long Island.

Cottage industries, like American Spoon Foods in Michigan, sprang up all over the country to support enterprises like Forgione's.

By 1983, he had opened An American Place on Lexington Avenue, and Beard had become a "friend and mentor," according to the menu. And Forgione's cooking had become fully realized. Now, after five years, during which time his cuisine has become, in the chef's words, "a little more country-ish, a little more homespun," Forgione has taken one giant step and is preparing to take another.

First, An American Place has relocated to the former site of the Ritz Restaurant on Thirty-second at Park. He has gone from forty-four seats at his old Lexington Avenue location to three times that number.

Even more ambitious will be Forgione's new country restaurant on a forty-two-hundred-acre estate near the Hudson. What will it be like? "We're not going to do a displaced New York City restaurant in the country. We're going to do a country restaurant, which is what I've always wanted to do."

In the countryside Forgione plans to raise sheep, grow vegetables, and cook in a big brick fireplace as well as on a restaurant stove.

Forgione says proudly, "When it reopens, it will be the oldest restaurant in America. The place we're renovating was built as a restaurant and operated from 1703 to 1790. It's been in my partner's family (we're talking AT&T here) since 1850. I'm looking forward to it, to the whole setting, to everything!"

Garden produce has always been important to Forgione's cuisine. "Vegetables certainly play as important a role in eating as fish, meat, or poultry. I know at home, on a Saturday evening, we've gone to the farmstand, and all we'll have is fresh vegetables sautéed in some olive oil. And at the restaurant we've certainly taken many steps forward with the whole focus of vegetables. We always create dishes that incorporate vegetables instead of just shuffling them off to the side. When I design a plate, I do so from garnish to finish. If we put some little buttered baby corns on, or some broccoli flowerets or a compote of vegetables, those things are as important as anything on the plate."

That was obvious from a recent dinner at An American Place, which began with a basket of warm black-pepper brioche and nutty whole-wheat bread. Then came ravioli stuffed with mushrooms, pumpkin, and snow peas—succulent gems topped with juicy strips of butternut squash and crunchy roasted pumpkin seeds. Instead of a cream sauce, Forgione served the vegetable pillows in a simple but tangy apple cider vinegar sauce.

The sweetest Brussels sprouts this side of Brussels shared a plate with golden baby beets, snow peas, and asparagus, and crispy leek fries that were better than any onion rings. Autumn squash with sautéed apples, potato blintzes, sautéed spinach, Portland hot slaw, and forest mushroom croquettes completed the feast.

An American Place will always be a joy for city folk, but Forgione has done the right thing in putting his heart in the heartland. The country is where we all like to be—to breathe, to stretch, to renew. And Forgione's return to the country-grown foods of the past will surely affect the cuisine of the future.

JOYCE GOLDSTEIN

Square One, San Francisco, California

Food Evangelist and Jewish Mother at the Same Time

TALK TO anyone who has worked with Joyce Goldstein, and they'll smile as they reminisce about being taught, and mothered, by the radiant owner of San Francisco's Square One. The food coming out of the warm, open kitchen is like her: nourishing and nurturing. It reminds you of the food your mother used to make—if your mother came from Italy, Greece, Morocco, Spain, Provence, or any other part of the Old Country Joyce happens to draw from that day.

Every day that Joyce goes to work, she goes back to square one to conceive, buy food for, and create the dishes for the day's menu. For me, it all starts with the bread.

At Square One, the whole-grain breads—herbed, spiced, studded, and stuffed with ingredients from all over the world—arrive warm and comforting, leaving you sure that the courses that follow will be just as wonderful. The bread basket comes with two offerings. One night you may lift the napkin to find a crusty loaf with a rich vein of walloping black pepper sitting next to a softer, whole-wheat sourdough. *Pane integral*, a tawny Italian whole wheat with more chew than saltwater taffy, might appear with a *broa*, a Portuguese corn bread, or a rosemary loaf.

It's not surprising to find surprising ingredients in Joyce Goldstein's breads. "I am intensely interested in the little idiosyncrasies, the nuances of the dish. It helps me explore my own palate to see what parts I like the best." *Exploring* is clearly the operative word to describe the chef's approach to food.

"We take people around the Mediterranean, and we introduce them to an entire point of view on a plate. We try to introduce people to what I call the ethnic flavors of that region."

One night I took a cruise around the Mediterranean in about eighty minutes at Square One, and what a satisfying voyage it was! Starting in Italy I thoroughly enjoyed a porcini and chanterelle soup with garlicky house-made grilled croutons. The earthy soup was thick and chocolate brown, loaded with plump bread crumbs.

Boarding the culinary boat to Spain, I got to my next stop just in time (before they ran out) to enjoy the avocado Catalán. Can a salad be a dessert? It can if you pair buttery soft avocados with a mixture of roasted almonds, capers, and herbs in a sweet-sour dressing of orange zest and sherry vinaigrette. Before leaving the Iberian peninsula we headed west to sample three Portuguese salads: black beans with onion, peppers with cayenne and cilantro, and carrots and celery with a mint vinaigrette. Salads and vegetables are favorites of Goldstein's—she calls them "energy foods."

"I live on salads, assorted vegetables, and vegetable pastas. They're the things I probably eat the most because they taste good and give me energy. There are many nights when I just have a plate of vegetables."

Goldstein is fortunate in that cooking in the Bay Area she has her pick of the finest produce grown in America today. "We're very fussy. We've always gotten good vegetables. We're getting a greater variety than ever before. And we're educating the public because they're eating things here that they can't get at their supermarkets right now. But because of what we're doing here, they'll ask for, and eventually get the same produce we get."

Joyce relishes her role as cook-educator. But there's no pedantry here. The learning is fun, because she makes it that way. That's why vegetables are so important in her cuisine. "Vegetables tell a lot about a culture. If you want to have a Greek meal, you start thinking about artichokes, dill, and the vegetables from that country. It's a nice way to learn about the flavors of different areas of the world."

But I digress. On to North Africa, from which came a Goldstein variation on couscous with vegetables. The grains were moist yet al dente, and served with garbanzos, tender grilled eggplant and peppers, carrots, and zucchini. I've eaten in Moroccan restaurants all over the country, and this was the finest couscous dish I've had. The reason might be in Goldstein's approach to presenting the foods of other nations. She's not simply trying to cook recipes from another country.

"I've taken some liberties with certain cuisines, because sometimes a recipe is dumpy. You hate to say that; its heart is in the right place, but sometimes you have to do something to spark the flavor a little bit."

I returned to the home port of Italy to savor the *bocconcini*: little cubes of mozzarella marinated in olive oil laced with oregano, garlic, and red and black pepper, served with arugula salad and grilled garlic bread.

"How was the scenery on your sojourn?" you ask. The food was beautiful. But you don't go to Square One to look at the decor—here, the food is art. The parkside restaurant is plain inside: white walls, blonde woods, and tile—a canvas for the still lifes from the kitchen.

Goldstein left the world of fine art in the sixties after years of painting professionally and displaying her works at galleries in San Francisco. It was during a two-year study period in Rome just after receiving her Master's of Fine Art from Yale that Goldstein fell in love with cooking, especially Italian. "My background was art history and painting. I was very active with my hands. I like cooking because it allows me to work physically as well as intellectually, and that's very gratifying."

After teaching cooking for a few years, Joyce founded San Francisco's first international culinary academy, the California Street Cooking School. Short stints as a kitchen design consultant and as a food writer for *Rolling Stone* made it clear to her that she'd be happiest *in* a kitchen, not designing or writing about one. So she hired on with Alice Waters at Chez Panisse, where Goldstein was chef and manager of the free-wheeling upstairs cafe. But she needed a venue where she could express her own personality. Thus Square One, a venture whose financial risk was shared by the chef and some long-time friends, opened in May 1984.

Square One offers a range and depth of Mediterranean food that you can't get anyplace else, and at moderate prices. But one thing Goldstein doesn't do is to whipsaw you between countries on the same plate. "We try to introduce people to the ethnic flavors of a particular region. So I won't put Mexican vegetables on a Greek plate. I don't do any mishmash, any cross-cultural stuff on the same plate. I believe that everything should be from the same region. I don't like chaotic dishes."

The flavors of her cooking are real and direct. That is by design. "Ours is not a weak cuisine. We tend to have fairly forward flavors. That's a way people learn about the food. The taste makes an impression. Part of my job is to educate while making sure everyone has fun, good-eating fun. I educate the public about seasoning and cultural tastes by intensifying them a little."

Joyce Goldstein does not purport to offer "the food of Italy," "the food of Greece," or "the food" of any country. She hones in on a region and gives you the taste of a special part of a country you might never visit. "I am intensely interested in regional cuisines, the differences in parts of countries. I'm interested in traditional family recipes as they get passed down from family to family. I'm against codifying recipes, amalgamation, assimilation, or watering things down for the public."

The key to her success seems to be her clear understanding of food and of herself. "I'm high energy. In love with food, both for the sensual experience and as a way of reaching people. Food is communication, very powerful. So the business I'm in allows me to do what I like best: work with young people. I'm sort of a food evangelist and Jewish mother at the same time. I like what I do."

APPETIZERS AND FIRST COURSES

Asparagus with Hazelnut Vinaigrette

SOUPS

Mexican Cauliflower Soup

SIDE DISHES

Moroccan Carrots

PIZZAS AND BREADS

Focaccia with Caramelized Onions, Gorgonzola Cheese, and Walnuts

VINCENT GUERITHAULT

Vincent Guerithault on Camelback, Phoenix, Arizona

Southwestern-Provençal Cooking

VINCENT GUERITHAULT whistles while he works. Literally. He has every reason to be happy. In the tenth most populous city in America (with more residents than such restaurant-crazy towns as San Francisco, Washington, D.C., Boston, or New Orleans), he's got the only great place to eat in town. Though he may not cater to as sophisticated an audience as might be found in other cities, the numbers are clearly on his side and getting better. Phoenix has grown by nearly 25 percent in the last six years.

His 130-seat restaurant is located on the hottest boulevard in town, right at the confluence of the up-scale part of Phoenix and the wealthy vacation and retirement city of Scottsdale. But it's not just the prime location that ensures a steady stream of well-heeled residents and tourists to Vincent's—it's Guerithault's Southwest-inspired food prepared with solid French technique.

Temperatures stay in the nineties and hundreds, even at night, two to three months a year in Phoenix. And they stay uncomfortably hot another two. But Vincent Guerithault is used to hot places. He apprenticed in Maxim's broiling kitchens in Paris, working eighteen hours a day.

While at Maxim's he met a young chef from Austria who showed a lot of promise. He, too, had come to Maxim's to gain experience and, he hoped, to parlay his association with the world-famous restaurant into a job opportunity in America. It worked, and in a few months, Wolfgang Puck found himself cooking in Indianapolis.

The two had met a few years earlier at the Michelin three-star l'Oustau de Baumanière in Provence and have remained friends for twenty years. So after Vincent had been working in the United States for a few years as sous-chef to Jean Banchet at Le Français in Chicago and as executive chef for a restaurant near Phoenix, he called his old friend to ask his advice on what kind of food to serve in the place he was about to open for himself.

"Wolf told me to do what he had done when he left Ma Maison to open Spago. He said I should do something completely different and to try not to compete with what I was doing before. So I decided to maybe give my menu a little Southwestern touch, maybe 30 to 50 percent of the menu, because nobody had ever done it in Phoenix before. It was a little risky at the beginning, because I didn't know. But then some people thought it was a joke when Wolf said he was going to cook pizzas. And just like at Spago, taking a risk has worked for me." Work it has. Last year Guerithault served a whopping forty-five thousand people, which averages out to nearly 125 a day.

Over the years I've grown fond of certain dishes at Vincent's. My all-time favorite is chilled avocado soup with jalapeño brioche. Not for the faint-hearted, this deceptively spicy soup masks its mayhem with its cool green color and creamy texture.

The perfect accompaniment to Guerithault's soups are his superior croissants. I normally prefer whole-grain breads to those made with white flour, but though Guerithault's whole-wheat rolls are scrumptious, it's hard to stop eating his buttery baby croissants.

Recently Vincent featured another soup that rivals the avocado. This one's served hot. It's a grilled-corn soup with ancho chili cream that tastes just like it sounds. Guerithault perfectly captures the flavor of just-grilled corn, right off the barbecue.

Guerithault brings corn and avocados together in his frequently featured appetizer of avocado and corn cake with cilantro. The moist, tender little cake comes in a tangy goat cheese sauce.

Goat cheese is also featured in the nachos of mild goat cheese with serrano chilies and summer greens. Another combination of French and Southwestern foods is his roasted spicy green chili filled with minced wild mushrooms in a blue cheese sauce. To finish the presentation, the chef paints cacti or mountain silhouettes on the plate with red and green pepper purees, colorful reminders of the restaurant's desert locale.

Another of Vincent's successful cross-cultural creations is his ratatouille tamales with cilantro beurre blanc. The garlic-and-bay-leaf-flavored mélange of eggplant, peppers, onion, zucchini, and tomato works nicely with the wonderfully coarse *masa harina*–based tamales.

I often ask chefs who specialize in regional cuisines if they think they are riding a temporary trend. I've seen a lot of Cajun-Creole places close recently, and the same might happen to Southwestern restaurants. Guerithault is sanguine on the topic. "I plan on keeping the Southwest influence in my food until one day, if it's not as popular, I can always go back to my French background and my classic training. But I don't think that's going to happen. Cajun food may not be as popular in Los Angeles or New York, but it's still very big in New Orleans. So I would think that Southwestern cuisine will be popular here in Phoenix even if it fades in Denver, Seattle, Chicago, or New York two or three years from now. But I always keep the French influence here in my dishes with my sauces, and there are always some French classics on the menu, so I'm not entirely classified as Southwestern. I don't want to be 100 percent Southwestern, whatever that means."

Guerithault uses the French-Southwestern touches in the design of his restaurant as well. The country French, wood-beamed look of Vincent's is reminiscent of a small, cozy restaurant in Provence. But one look at the southwestern art of the walls brings you right back to the desert. On one wall is world-reknowned artist Fritz Scholder's *Outside*, which became the label art for Vincent's exclusive house wine. But it is another example of local landscape art that Guerithault is most proud of: a colorful eight-foot-high tapestry of the desert, its mountains and vistas handwoven by his mother after a visit from her native France.

Vincent continues to refine the menu at his restaurant. Take his recent "heart smart" additions. Two or three daily specials are listed at the bottom of each menu, along with their ingredients, calories, and cholesterol levels. One such recent offering, a salad of corn, jícama, tomato, and assorted greens in a dill, olive oil, grapefruit juice, and sherry wine vinaigrette, has 365 calories and almost no cholesterol. But the forward-thinking Guerithault isn't stopping there.

"I think my next step will be some vegetarian menus. I just got back from France three weeks ago, and at l'Oustau de Baumanière, besides the special menu of the day, they had another menu, all vegetarian. It was a prix fixe vegetarian menu in one of the finest Michelin three-star restaurants in France."

Well, if it's good enough for Raymond Thuilier, it's good enough for Vincent. Besides, good culinary ideas seem to take root in America's fertile soil and grow healthy and strong—perhaps stronger than in the country of origin.

Guerithault has great hopes for the future of cooking in America: "Outside of France, the best French chefs, and I'm not sure I'm in a position to know what 'best' means for everybody, are living in America. Outside of China, the best Chinese chefs are living in America. Similarly, the best of a lot of other countries, outside of their own country, come here and live in America. I wouldn't be surprised if you see in ten or fifteen years an American cuisine in which all those chefs from different countries blend their cuisines together. If that happens, I also wouldn't be surprised if, a few years later, you find the best food in the world right here in America."

SOUPS
Chef Vincent's Cream of Avocado Soup
Vegetable Stock

SIDE DISHES
Shiitake Provençale and Grilled Shiitake Mushrooms

MAIN COURSES
Ratatouille Tamales

RELISHES AND INTERMEZZOS
Avocado-Corn Salsa

JEAN JOHO

The Everest Room, Chicago, Illinois

"I judge a restaurant by what vegetables they give me. I judge a chef by how many vegetables he uses and how he uses them."

I WAS helping my daughter with her algebra one night when we came to the section on set theory. You remember set theory, don't you? No? Personally, I can think of few subjects as relevant to what I am doing today as high school algebra. In fact, it has been a vitally important part of my life ever since I nearly flunked the course in high school. There have been only a few days when, over casual conversation, I haven't discussed quadratic equations or pure imaginary numbers with someone.

So when the subject of set theory came up for my daughter as it had for me twenty years ago, the same blank, lost expression came over my face as it had two decades before. Finally, my daughter figured everything out (by reading the chapter) and explained it to me excitedly:

"See, Daddy, there's these like mutually exclusive sets. Like two things are mutually exclusive if their outcomes can't be the same."

"Huh?" I replied.

"O.K., like boys and girls are mutually exclusive sets. See?"

"Oh," I said. "I think I've got it now. The two things have to be so totally incompatible that both can't exist in the same category, such as the word *like* and using good diction."

"Huh?"

"Never mind, honey. You've helped me a lot," I said, hugging her. For I now knew how set theory applied to certain kinds of restaurants. View restaurants, to be specific.

Good restaurants and view restaurants are usually mutually exclusive. Most restaurateurs lack the integrity to spend the requisite amount of money to hire a good chef who'll use good ingredients, because the owners know you'll come there for the view anyway.

There is at least one exception that I know of, however. The restaurant is called, appropriately enough, the Everest Room, and it sits atop the fortieth floor of One Financial Plaza, with a completely unobstructed view of the south and west sides of Chicago. The owners had the good sense to hire as chef Jean Joho, an Alsatian who trained at the Michelin three-star L'Auberge de l'Ill with Paul Haeberlin. The food is as good as any I've had on terra firma.

Joho began his training early.

"When I was only seven or eight I was putting on chef's clothes. I had a passion to be a chef then, I have it now, and I always will have that passion."

To channel that youthful passion, Joho enrolled at the Lycée Technique Hôtelier Chambre de Commerce et d'Industrie in Strasbourg. He then served his pre-apprenticeship with Haeberlin, followed by culinary and pastry apprenticeships in France before working with chefs in several Michelin two-star restaurants in France and Switzerland.

"Being top chef is not something to be improvised; it must be learned," says Joho. "When someone comes out of school he is still a student, not a professional. Only practice makes a professional."

So Joho practiced. He returned to the kitchen at L'Auberge de l'Ill for a year and a half as Haeberlin's protege, and only after he felt he had learned all there was to learn from a master did he take the momentous step of going out on his own. It was 1984, and he moved to Chicago to reopen Maxim's, but soon made his way to the top, the Everest Room.

The room is magnificent, with picture-window views of the south and west sides of Chicago. The decor is jungle chic, with leopard-spot carpets and murals of tigers, apropos of Mt. Everest's location on the Nepal-Tibet border where, at lower elevations, such wildlife abounds.

Recently I asked Chef Joho to prepare a meatless *menu dégustation* for me. He began with a fondant of cauliflower: nothing more than a brace of long, graceful parfait spoons lying parallel on a plate, each one filled with a rich squiggle of cauliflower mousse. Atop each fondant was a wisp of chervil. The bread basket contained a whole-wheat baguette, crusty country bread, and the tastiest of the bunch: a rye roll, notable for the absence of caraway seeds, so that the deep rich rye flavor came through. It was the best bread assortment I've had, aside from those at Joyce Goldstein's Square One in San Francisco.

Joho uses plates that would make a Japanese *kaiseki* chef proud. There are oval ones and rectangular ones. A crustless terrine of broccoli and cauliflower dressed with walnut oil came on a rough-textured, square black plate that made the food more enjoyable still. The pure vegetable flavors were unimpeded by crust, sauce, or aspic. But that would figure considering the chef's love for greens.

"With vegetables you can play, you can season: you have all the tastes, you have all the colors," he said, tapping the table for emphasis each time he enthused over another of their qualities.

"There's such a variety. More variety than with meat or fish. A piece of veal is a piece of veal. But no one can say a green pea tastes like a snow pea or yellow beans taste like green beans. We keep fifteen to twenty vegetables in the house at any one time," he said, his voice rising with excitement at the prospect of using them all.

"That's because my concept of vegetables is not just baby carrots. There are some restaurants that put the same vegetables on the plate from January 1 to December 31: two pea pods, a piece of tomato and some parsley. No, I don't want that. I judge a restaurant on what vegetables they give me. And I judge a chef on how many vegetables he uses and how he uses them. [Tapping the table again.] It's the qualification to be a chef."

My next course was an exceedingly light cream of mushroom soup with a sprinkle of wild mushrooms and parsley.

Mushrooms appeared again in the next course, a simple sauté of black trumpet and chanterelle mushrooms in a delightfully sweet shallot reduction on toast points. Joho always goes for the simplest way to present his vegetables.

"There are great vegetable matches that are best cooked simply, like salsify and truffle. It is the most fantastic flavor. You just sauté them. You don't need veal stock or chicken stock. You need nothing else. And when I cook Brussels sprouts, I peel each leaf and then just sauté them for one second. I don't want to let them touch water and lose that immediate flavor. I just sauté them for a second in butter and serve. What else do you need?" he asked rhetorically, his voice falling to a reverent hush. "You *need* nothing else."

Joho served me some red Brussels sprouts as part of his magnificent main-course potpourri of vegetables. The leaves were smoky, vinegary, and sweet. Next to them was *knefla*: buttery, chewy Alsatian spaetzle.

Joho's artichoke hearts were meltingly tender, split and splayed like small fans. There were other little vegetable surprises around the plate: spinach in a tomato *coulis*; beet mousse; a mixture of tomato, vegetables, carrots, and artichokes; and some sprightly ratatouille.

The chef finished with a light cheese course. He served a baked baby potato stuffed with a runny Alsatian Münster, with a sprinkling of caraway seeds.

Joho has plenty of the right raw materials to work with. He surprised visiting chefs Jean-Louis Palladin and Gerard Panguad with the incredible diversity of vegetables he gets from the Chicago produce markets. Many of these are grown locally, but in colder months Joho has no choice but to bring produce in from California. He's too much of traditionalist to serve hydroponically grown produce, asserting, "I like vegetables that come from the ground. *They* have taste."

But the one dish that told me the most about Joho was one he *didn't* serve me. When I was at the Everest Room, he declined to serve me the white asparagus I had heard were being grown in Michigan, because he thought it was too early for them and that they would be bitter. A few nights later another chef served me some white asparagus. Bitter as bile.

MAIN COURSES

Potpourri of Vegetables Pot au Feu-style with Chervil Butter

MADELEINE KAMMAN

Beringer Vineyards, Napa Valley, California

Chef to the Chefs

WHO BROUGHT *nouvelle cuisine* to America?

"I did," says the charming Madeleine Kamman quite matter-of-factly and without a trace of ego. "I was running a cooking school and restaurant called Chez la Mère Madeleine in Boston from 1975 to 1979. While they were doing *nouvelle cuisine* in France, I was cooking the same way in America without knowing very much about what was happening there. When I found out there were such similarities between what we were both doing, then I got interested in what others were doing in France. And nobody else that I knew of was cooking that way yet in America. I found that though I had a different style, I had pretty much the same direction in my cooking as all these French chefs had."

Those days are gone, and Chez la Mère Madeleine is no more. But Madeleine Kamman in the intervening years has grown even more influential, teaching both home cooks and professionals through her cooking schools on both sides of the Atlantic, her books, and her PBS series, and now in her new position as director of the School for American Chefs at the Beringer Vineyards in the Napa Valley.

The Beringer program is a unique school for working chefs, all of whom come on scholarship to polish their professional skills. As Madeleine puts it, "There were so many basic chefs' schools, I decided to offer classes where people can come to learn just what they need to round off their skills in areas like food chemistry and physical reactions, food history, creative work: all the things the schools don't have time to do when they're turning out chefs. We offer ten to twelve programs a year, each lasting two weeks, in which there are 3½ hours of hands-on work [each day] and 3½ hours of lectures by me. They bring the problems they encounter in their professional life, and we work together to get everything clarified."

Madeleine understands the problems chefs face. At Chez la Mère Madeleine (which produced Jimmy Schmidt of the Rattlesnake Club), the lack of ingredients was a significant difficulty. Madeleine recounts, "Getting the right ingredients in the late seventies in Boston was absolutely disheartening and unreal."

Madeleine packed up her cooking pots and headed for Annecy in the French Alps, where, though it's cold, as in Boston, at least there was a tradition of fresh provender throughout the year. There she started a cooking school, but ran afoul of a problem every restaurateur has to deal with sooner or later: the government. The tax structure for her particular kind of business placed a severe limit on the size of her enrollment.

After four years in France, it was back to New England. By this time, 1984, good raw materials were just starting to become available in Boston year round. Unfortunately, her cooking school and restaurant, L'Auberge Madeleine, in Glen, New Hampshire, was three hours from Boston.

A veteran of the restaurant game, Madeleine eventually signed on with Beringer Vineyards. Beringer has long supported the appreciation of food as well as wine. And the chefs who apply from all across the country to work with Kamman will encounter someone who has been at it long enough to have worked out the problems both in and out of the kitchen.

"In the kitchen I am a detailist about food," comments Madeleine. "I like my food to look finished. By 'finished' I mean I want everything to be cooked properly and look good on the plate, but still remain food. Food must look like food. Food has to speak for itself and taste as good as it can taste, even if you don't have the best ingredients. That is the art of the cook."

Vegetables have an essential role to play in her style of cooking. She tends toward a vegetarian diet, preferring "apples, oranges, wonderful veggies, and soups to stuffing myself with all that meat." As a trainer of chefs who will serve all kinds of food, she sees to it that vegetables are correctly prepared and given their due.

"For me, the fresh vegetable puts the color and a large amount of the basic nutrition on the plate. I was always well known for my very good and sparkling clean–looking vegetables. I still absolutely adore my vegetables to look beautiful and to taste nice and crispy and fresh. Of course sometimes they have to be done in the old-fashioned manner, and I don't hesitate to bake them, as in the case of potatoes, with a good stock or heavy cream. That's wonderful.

"But the best thing about vegetables is the color, nutrition, texture, and flexibility they give you in making your plate look like a little painting, without it looking in any way affected. As I say, food must look like food. And I don't like doing extravagant presentations. I'm eating food. So I want people to recognize the vegetables as what they are and have them cooked at the absolute peak of what they should be."

Kamman learned to cook in her aunt's two-star restaurant in the Loire Valley during the war years when there was little time or inclination for affectation, and food was real and hearty. And today her food is cooked and served in a simple and direct fashion.

America knows Kamman best through her cookbooks and PBS cooking classes. She wrote *The Making of a Cook* in 1970 and went on to write *Dinner Against the Clock*, *When French Women Cook*, and *In Madeleine's Kitchen*. *Madeleine Cooks* came out most recently as the companion piece to the TV series. Her latest work is a book on the cooking of the northern French Alps, called *Madeleine Kamman in Savoie*.

In the years to come, however, the American dining public will get to know Madeleine indirectly, through the chefs she polishes at the School for American Chefs.

Kamman is happy to have retired from the fray. She was the first woman chef in America to have her own restaurant. Suppliers treated her like a second-class citizen because they weren't used to dealing with an independent woman in that role. At first Kamman took up the charge in a feminist counterattack, raised her hackles, and clawed back. Then she began to study Mahatma Gandhi's teachings on nonviolence and meditation. By focusing inward, Kamman was better able to deal with the world outside and in.

"Life is hard. I was looking for something to understand life and human relationships better. I found out that if you went the nonviolent way, people would open up a lot more to you. I used to be so angry. Now I've learned to deal with those feelings. When I come home from work each day, I read Gandhi's teachings and meditate for at least a half an hour. It's wonderful."

And so is this woman, who, for the last thirty years, with wit, charm and precision, has helped bring to America the secrets of good cooking.

SOUPS

Curried Green Bean Potage

EMERIL LAGASSE

Commander's Palace, New Orleans, Louisiana

Getting Better All the Time

IT IS notable that Emeril Lagasse became executive chef at Commander's Palace during the restaurant's one-hundredth year. As the Brennan family has lovingly restored Commander's and prepared it for its second century of operation, Lagasse, without abandoning the old, has brought a fresh, new style to the venerable restaurant.

When the Massachusetts native came to Commander's Palace, the Brennan family had been looking for some time for just the right person to lead the restaurant after the departure of Paul Prudhomme, who had left four years earlier to start K-Paul's. Lagasse had the proper training, graduating from Johnson and Wales Culinary School in Providence, Rhode Island, and working in a variety of East Coast restaurants. But he had something more.

As Lagasse tells it: "When I first came to Commander's Palace, Ella Brennan took me back to the kitchen and said, 'What do you think of all this great food?' I said, 'It's just like my mom's.' "

Lagasse's mother is Portuguese, and the saucy seafood and rice dishes of her native land are very similar to those of Creole New Orleans. And like the French Acadians, who were forced by the British to move from their native Nova Scotia to Louisiana, his father is French Canadian. It was these Acadians, or Cajuns, who developed their namesake style of cooking. So it is understandable that Lagasse would feel at home in New Orleans.

But the dark, handsome chef prefers to cook Creole, not Cajun. And he makes the distinction, and the difficulty in learning the intricacies of the cuisine, quite clear. "We have strong, strong Creole influences. We're not a Cajun restaurant. We don't do fiery hot, burning cooking. Our Creole cooking has a lot of taste, substance, and depth. When I came here, with my Portuguese-French-Canadian background and my classical training, I was well prepared. But I didn't just come here and whammo, I knew how to cook Creole. You have to understand the culture before you can understand the food. To create a lot of the stocks, sauces, and soups with all their depth takes hours of cooking. It's not just something you can cook in a sauté pan in two or three minutes.

"All food has background, roots. And unless you know the roots, the culture, the history of that region, whether it's in the South of France, New Orleans, or Seattle, you can't feel comfortable delivering the food of that region. After six years here, I feel comfortable now and so do my customers with the way I deliver my product."

But it wasn't always so. When Lagasse brought his new-fangled ideas about fresh vegetables and lighter sauces to New Orleans, long-time Commander's Palace habitués turned up their noses.

"When I first came here," recalls Lagasse, "I couldn't believe New Orleans' restaurants' attitude toward vegetables. There were very few vegetables on any of the plates, and those that were, like okra, were cooked to death. Mostly, you saw starches like rice, which of course grows here. In my first six months here, I put vegetables on every plate. Customers would send them back saying they were too crunchy, too chewy."

The situation was compounded by the sorry state of produce in New Orleans. Lagasse found that California produce stopped heading eastward at Dallas, and Florida vegetables went north. That left New Orleans, which had few local sources for fresh produce despite its balmy climate, the odd man out. But over time, Lagasse, who describes himself as "a very big vegetable fan," persuaded five or six nearby farmers to grow vegetables, herbs, and lettuces for him. Now he calls the availability of fresh resources for Commander's kitchen "perfect."

"I'm fortunate in that the farmers who grow specifically for me give me vegetables that are not only much sweeter and more colorful but are higher in vitamin count and other nutritional factors than mass-produced vegetables. We get baby kale, all the different squashes including mirliton, which we'll stuff or puree. We get great root vegetables, radishes, even fresh water chestnuts. Eggplant has always been big here, only now we get it organically grown. And since they deliver it fresh three or four days a week, we hardly ever have to store anything.

"Now, the acceptance and excitement people feel toward vegetables is incredible. I do all kinds of combinations of fresh vegetables, like a zucchini, eggplant, and yellow squash lasagne in a light herb sauce. Three of our entrees come with a vegetable timbale of the day. Or I'll feature a vegetable puree or ragout. And in terms of salads, I can do anything. I've got nine different kinds of locally grown lettuce to work with."

Some of Lagasse's finest work is indeed with salads. One features an assortment of local lettuces, cheese tortellini, feta cheese, a *concasse* of tomatoes, and pistachios and black olives in a fresh tarragon vinaigrette. His apple, watercress, and pecan salad is crunchily refreshing, and I think I could drink a bottle of the sesame oil and molasses dressing with which he coats crisp radicchio, curly endive, Belgian endive, croutons, and black olives.

Lagasse tops this salad with a silky goat cheese, which, like the dressing, is housemade. In fact, at Commander's Palace everything is made from scratch, even the Worcestershire sauce and mayonnaise. It's all part of what Lagasse calls "the restaurant's commitment to total quality, total product, total service, and total customer satisfaction."

That's not easy in a restaurant that might serve two hundred lunches and four hundred or more dinners daily. But Lagasse has personalized the service at Commander's Palace to the extent that he's got sixty to eighty regulars that call him before they come to ask him to prepare something special for them. They're free to wander into the open kitchen and chat with the amiable chef. This, along with the layout of the large restaurant, makes the atmosphere both intimate and informal. In fact, you must walk through the kitchen to get to your table, thus reinforcing the homey feel at Commander's Palace. There are nine wonderfully different rooms in all, so that you are always dining in a cozy environment.

Commander's Palace is located in an old gabled Victorian mansion in the Garden District. Far from the French Quarter, where the Creole population lived, the Garden District was formed by what a modern-day expert on demographics might term "yassies"—young Anglo-Saxons—who came to New Orleans right after the Louisiana Purchase to seek their fortune. In the 1880s, Emile Commander turned the big, rambling house into a restaurant for distinguished neighborhood families. Commander's Palace went through a series of owners until 1974, when the Brennan family, sisters Ella and Dottie and brothers Dick and John, bought the place and began to restore and revive the once-proud eating establishment.

After a recent dinner, affable manager Gregory Adragna took me through the grand, turreted gingerbread house. Surrounded by stately oaks, one of which is over two centuries old, Commander's Palace is painted a bright turquoise with white trim. Inside, the Brennans have painstakingly worked to give each dining room its own personality. There's the Garden Room on the top level, whose walls are covered with handmade white treillage. Here guests can relive their childhood by dining in what appears to be a giant treehouse, so close are the oaks. The cherished tables, the equivalent of Spago's tables overlooking Sunset Boulevard, are those next to the room-length picture window perched over the patio.

The more formal downstairs dining room is decorated with bucolic scenes of the Louisiana countryside, all recently painted by a local artist the Brennan family commissioned. The small, bright yellow Coliseum Room is upstairs, adjacent to the shocking-pink Parlor, and the Little Room, a sun porch that seats only fourteen.

But any seat in the house is a good seat when Emeril Lagasse is orchestrating the cooking. Whether it's his svelte cream of celery or cream of eggplant soup with a touch of curry, thyme, and basil; his smoked mushrooms in garlic butter served over angel hair pasta; his Creole succotash with housemade Worcestershire; or his raisin-walnut pilaf, he handles the complex flavors of Creole cooking with originality and a deft, light hand.

My dinner one night began with a surprising but eminently satisfying pizza with a sweet, chewy crust topped with generous pieces of tender fennel, tomato, onion, green pepper, and crispy garlic strands. Actually the pizza was, in a sense, my second course as I had already eaten so much of Commander's addictive buttery garlic French bread topped with herbs.

When the waiter set before me three golden disks sitting in a pool of gold, little did I suspect that I was about to experience one of the most finely wrought vegetable dishes of my life. Vegetable boudins, as the waiter called them, were delightful vegetable patties, with a crunchy bread-crumb coat on the outside, tender and moist vegetables on the in, flavored with a swirl of spices that were familiar yet elusive: a combination of twenty different seasonings including thyme, tarragon, paprika, and red chilies. The pool of gold was a sweet, creamy yellow bell pepper sauce. It was a dish I'll never forget.

A salad of assorted local lettuce with poached pattypan squash and *haricot verts* was dressed with a basil vinaigrette and topped with garlic croutons.

A tangerine sorbet intermezzo preceded the pasta *en papillote:* a parchment bag of bow-tie pasta with olives, mushrooms, sun-dried tomatoes, and black beans, all steamed to perfection.

With talent like this, you might expect Emeril Legasse to be hanging out his own shingle any day now. But he's clearly in no hurry. "I've had a lot of opportunity to do something on my own, and it would be very easy to do. But I really don't need the ego gratification of having my own place right now. I'm working with some wonderful people right now, Dick and Ella Brennan, who have tremendous experience in this business, over forty years. But more importantly, we have a wonderful chemistry. We're all moving in the same direction. That's one big reason I don't need to do something else. I'm very happy with what I'm doing. I touch food every day. I'm involved with food, with the growers. And I'm able to move freely from the kitchen to the dining room. I'm not just trapped in a cage behind the stove.

"I have a personal relationship with the Brennan family. Most people think I'm a Brennan, that's how close we are. It's a unique relationship. To be satisfied and happy, I guess, is the number one priority in life. And I am."

And Lagasse's food shows it. His is a joyful, ebullient style with sass and jazz that also pays homage to the traditions of Creole cuisine. Due in part to his pioneering work, in the six years he has been in New Orleans Lagasse has seen remarkable changes in New Orleanians' attitude toward food.

"The education of our city's diners from the days when chefs would cook the hell out of okra for four hours to today has been remarkable. Whether people are learning from eating out, through reading food magazines, looking at chefs on TV or in their cooking videos, I think people are just getting more and more aware of what good food is and how they should eat. There are some good chefs around town that have made a strong statement and are beginning to educate the public as to what good food is."

Lagasse is one of those chefs. And the public is responding by coming in droves to Commander's Palace to enjoy his cooking. Each year sets a record for more diners than the last, in spite of the oil industry woes that have beset New Orleans's economy. Yet Lagasse isn't ever satisfied. "I try to do what I do as a chef from my heart. I think what I do has soul, because it comes from my heart. It's real. And I'm going to get up tomorrow morning and try to do a little better than I did today. I live by that. And I can see, after being here six years, the progress we've made with that philosophy. We're very fortunate. But we work very hard at what we do. And we'll try a little harder every day than we did the day before to continue to refine, improve, and satisfy our customers."

None of what Lagasse has been able to achieve at Commander's Palace could have happened without the care and concern of the Brennan family. With their lovely Victorian mansion right next door to the restaurant, the Brennans are never far away from the action. The numerous and well-trained waitstaff provides a level of service that seems preserved from a quieter, less harried time. Cooks and service personnel alike are ever vigilant, for they know there is almost always a "B.O.D." (Brennan on duty) to make sure things are running right. The night I was there, doyenne Ella Brennan, in a fire-engine red dress, sat perched on a stool in the kitchen, watching the plates coming off the line. You can bet everyone who dined at Commander's Palace that night left happy. I know I did.

SOUPS

Cream of Eggplant Soup

PASTA AND RISOTTO DISHES

Fresh Fettuccine with Spring Asparagus and Goat Cheese

ROLAND LICCIONI

Le Français, Wheeling, Illinois

A Chef with His Head on His Shoulders

IT'S A long way from Vietnam to Chicago, but some chefs travel like seeds in the wind, landing in fertile ground and growing to their maturity with proper nurturing. Such is the story of Roland Liccioni, born in Vietnam, trained in France and London, and now a firmly rooted perennial in Chicago. In the beginning, the nurturing was provided by Carlos and Deborah Nieto, who operate Carlos' Restaurant in Highland Park, Illinois. They gave Liccioni the respect and the room necessary for him to grow.

Liccioni began his training at an early age in France. "I grew up in France. I went to cooking school there. The way I learned is very hard training. It's a lot of sacrifice, especially when you start young, like I did when I was thirteen. Once you're on the top, it's hard to stay there. I know one day I won't be there anymore, and I want to teach the people who work for me all I know about hard work and dedication."

Liccioni recently took over the legendary Le Français from Jean Banchet. Judging by the tour de force dinner he prepared me, Liccioni will carry on Banchet's proud tradition without missing a beat.

For starters he sent me a complimentary appetizer of two small puff-pastry cups, one filled with avocado and tomato, the other deep-fried and filled with garlic- and dill-scented eggplant. On the same plate were nutty green niçoise olives and housemade cumin-thyme wheat wafers bearing a floral imprint.

The menu had a number of vegetable dishes, including fresh green and white asparagus, fresh mushroom pâté with truffle sauce, and hot cream of artichoke and asparagus soup. But I had asked the chef ahead of time if he would be kind enough to prepare a vegetarian *menu dégustation.*

Liccioni started me off with a fresh baby artichoke bottom filled with mushroom duxelles and surrounded by a fan of artichoke leaves. The tender artichoke bottom sat on a plate divided down the center with cauliflower *coulis* on the left side and tomato *coulis* on the right.

Ravioli of truffles and mushrooms were stuffed with seven kinds of mushrooms, which ran the flavor gamut from over the meadow to through the woods. They shared a rich dark truffle sauce with tiny cubes of tofu. Asian ingredients frequently find their way into Liccioni's cooking, but only as accents. The Eastern influences of his childhood show up in flavorings like lemon grass or ginger, but Liccioni is a French-trained chef who favors the copper sauté pan over the steel wok. A green and white asparagus and truffle fettuccine with black trumpet mushrooms and tomatoes in beurre blanc sauce followed the ravioli.

Now that Liccioni's at Le Français, he does the food shopping, as he did for Carlos', going to the produce market every morning before dawn to make sure that he has the best available. It hasn't been easy for him, because he is used to a higher standard than many of his American purveyors are able to provide.

"It's difficult to get the vegetables I want. But it's important to me because I want my customers to have vegetables whether they eat them or not. I think I have given some knowledge to the purveyors. Now they know what I want, and they try to grow it for me. There are some good local growers now, but during the winter I have to buy what's shipped in from California and Florida."

When I dined at Le Français there was plenty of sweet white Illinois corn, and local *haricots verts* that Roland combined with yellow Florida pear tomatoes and California lettuces to make a terrific salad. The best part was the truffle-hazelnut vinaigrette, featuring the chef's own herbs.

"I like to grow things in my garden. Especially herbs. Fresh herbs are very healthful. They make everything smell good, smell fresh. I grow Vietnamese cilantro: it's kind of a spicy herb. I also grow rosemary, chervil, lemon thyme, opal basil, and green basil. I couldn't imagine meat or fish without these herbs."

Those herbs also found their way into the array of vegetables that the waiter revealed when he lifted the silver dome carrying my main course.

The plate featured such delights as soft smoky lentils; marinated and grilled red, yellow, and green peppers; a hollowed-out beet stuffed with minced vegetables with a tangy Vietnamese cilantro-sesame oil dressing; a bundle of green beans tied with a rosemary twig; creamed potatoes with couscous curry; and a white-on-white creation of what looked like albino snail shells (they turned out to be Chef Liccioni's home-grown Chinese artichokes, or *crosnes*) on a bed of wild rice. The centerpiece of the platter was a rustic-looking, tangy ratatouille framed with paper-thin slices of zucchini.

Each vegetable had its own sauce and its own homegrown herbal sprig. The hollowed beet sat in a zucchini puree, and in a neat red-green turnabout, the green beans came with a beet sauce. An aromatic fennel sauce surrounded the lentils.

I can see why London's Roux brothers were sorry to see Roland Liccioni leave the Waterside Inn and La Gavroche. But after his long apprenticeship, which began at École Hotelière de Biarritz at the age of thirteen and included Rothschild's Parisian Bofinger Restaurant and Mandion Patissier in Biarritz, he felt he was ready to head up his own kitchen.

Now, after seven years at Carlos', during which time Liccioni has become something of a local celebrity, he is finally running his own place. The loyal and appreciative audience that grew addicted to his unique cooking style at Carlos' now experiences Liccioni's unique way with food at Chicago's premiere French restaurant.

SOUPS

Artichoke and Asparagus Soup

BRUCE MARDER

West Beach Cafe, Rebecca's, DC 3, Broadway Deli, Los Angeles, California

"Food tastes good before you do anything to it."

THE BERBER land the natives call *al-Mamlaka al-Maghrebia* drew a young Californian to camp on one of its uninhibited Mediterranean beaches in the early seventies. There, free from the pressures of USC Dental School, he had time to think and to dream. Under the starry Moroccan skies, a lovely French woman from Holland would prepare dinner right on the beach. Often the food was Indian, and as it simmered over an open fire, the fragrance of hypnotic spices wove their spell on young Bruce Marder. The buttery stuffed breads, the rich complex curries, the bewildering assortment of lentil dishes with exotic seasonings not only awakened his taste—they caught hold of his imagination, exciting him with the possibilities of the food *he* could create. Marder wouldn't be going back to dental school. He wanted to cook like this for a living.

The crowd at Venice Beach in Los Angeles was a little like the one in Morocco in the seventies. So when Bruce Marder found a shuttered restaurant only a block from the beach, called Casablanca, it must have seemed like kismet. So he painted the cinder-block walls white, put in black chairs and booths, decorated the restaurant simply with the art of his patrons, and renamed the place West Beach Cafe. He was back at the beach. Only this time, he did the cooking.

Early on, Mexican food seemed a natural. With its emphasis on spices, beans, and fresh-made unleavened breads, it was a lot like the first food Marder had learned to cook: Indian. And unlike most of the top chefs in Los Angeles, Bruce was born here and had grown up with Mexican food. But soon the menu was expanded to include dishes from all the places Marder had traveled while he roamed the world in search of himself. Today, Italian, Greek, Indian, and Moroccan make their appearances on the weekly changing menu.

Although the dishes may lack a single national identity, Marder feels they all have one thing in common: they focus on basic flavors. The witty, enthusiastic chef with a vestigial beard (that he admits was a lot longer in his younger days) gets excited talking about his cooking. "I like the concept that food tastes good before you do anything to it. A raw carrot tastes good. Therefore we cook things quickly using olive oil and a lot of herbs, not much cream."

Marder knows his olive oils like a sommelier does wines. Depending on the dish, he'll use one of several varieties of extra-virgin olive oils from Italy, including the peerless Badia a Coltibuono, or perhaps a Greek or French virgin oil. When Marder does use cream it's to provide richness but not to mask flavor. That's why I think he serves some of the finest vegetable soups in the country. Whether they are asparagus, spinach, broccoli, or carrot, the silky vegetable soups at West Beach taste like vegetables, not cream.

Salads are also exceptional at West Beach. One of the best is the Caesar, in a garlicky dressing with Worcestershire and lemon.

Marder finds a way to make vegetables a part of every dish at West Beach, which is not surprising: he was a vegetarian for a few years, and the cuisines he features come from countries where vegetables are an integral part of the food and not simply a garnish.

"Every plate has a theme. Vegetables are an essential part of that theme. A lot of restaurants think the same vegetables will go well with every entree. I don't, so I put different vegetables with each dish. I mate the vegetables to the meat."

While Marder is a master at finding just the right vegetables to play off his meats, he respects vegetables enough to know he can present strong dishes featuring them alone. A recent Greek appetizer featured dolmas (grape leaves stuffed with rice and raisins), pungent olives, feta cheese, and baklava-sized buttery spanakopita (spinach pie). I've not had better in a Greek restaurant. Once, Marder got a hold of some huge slablike, meaty-tasting mushrooms—chicken-of-the-woods, I think. They were cut in rectangles, marinated in garlic and herbs, and then grilled quickly over high heat. They were as flavorful, chewy, and satisfying as any meat dish.

It is in his rendering of Italian dishes that Marder brings out the best in vegetables, whether it's an Italian antipasto selection of marinated and grilled peppers, eggplant, zucchini, and tomatoes, or a puff pastry herb pizza. Most notable are his pasta dishes. There is almost always a meatless pasta dish at West Beach. Recently I tied into a grand spaghetti and chanterelle platter with a julienne of onions, carrots, celery, garlic, zucchini, and *haricots verts*. Seasoned with fresh basil and Italian parsley, all the ingredients were tossed with Italian extra-virgin olive oil and finished with Locatelli romano cheese.

At dinner one week later I enjoyed a dish of *orecchiette* in a vegetable sauce. The pasta was coated with a perfectly balanced sauce of carrots, red and green bell peppers, zucchini, onions, garlic, and fresh basil.

That same week I tucked my napkin into my collar for one of Marder's tomato-sauced pastas: spaghettini and sautéed cauliflower flowerettes in a tomato *coulis* with garlic, white onions, parsley, and Greek virgin olive oil, topped by Rocca Parmesan cheese.

The Mexican food Marder started with is occasionally featured at West Beach. But cross the street and you cross the border into his second restaurant, Rebecca's, which serves Mexican food exclusively and is very different in appearance from the Cafe.

The first restaurant couldn't be simpler, unless you took down the art. At Rebecca's every time you look around you see something exotic. The restaurant is a product of the well-known Los Angeles architect, Frank Gehry. While some say Marder feels the design (featuring flying crocodiles, an octopus made of beads, and a black velvet tropical fish mural) overshadows the food, he needn't worry—the food is too good to let the decor distract anyone.

Antojitos (appetizers) at Rebecca's include *jalapeños rellenos*, which, though hot, are more smoke than fire. The stuffing of three different cheeses keeps the fire 50 percent controlled, though total containment is both impossible and undesirable.

A plate of three *antojitos* is a nice way to compose a main course at Rebecca's. Along with the rellenos I'd suggest the spinach and Monterey jack enchiladas in a spicy sauce, and the fried corn quesadillas.

For those who can't abide the hot stuff, there are roasted potatoes, and roasted mushrooms in gentled garlic. The *plato de legumbres* gives you a whole pantheon of green stuff.

Marder's approach to Mexican *bebidas*, or drinks, is as refreshing as his approach to Mexican food. In addition to the usual offerings are such tropical refreshments as fresh juices made from papaya, mango, and melon.

Marder never cooked in France, although he did train at the Dumas Père School of French Cuisine in Chicago before signing on at the posh Beverly Hills Hotel. "I think I have a wider range of background in food than just French food. I think I've been involved in more cuisines than many chefs, because I like to know where food came from and why it is the way it is."

His third restaurant features American food. It's DC 3, located at Santa Monica airport, only about three miles, as the Piper Cub flies, from Marder's two Venice establishments.

"Harmless Dinners," proclaims the menu cover (taken from the Ed Ruscha painting that hangs in the restaurant's entrance), and it is difficult to imagine anyone being offended by what's listed inside. There are eight vegetable preparations, including baked banana squash, artichoke hearts, asparagus, and string beans. Potatoes have their own section of the menu, listing eight different versions: French fried, baked, mashed, with rosemary, pancakes, gratinéed, boiled, and with onions.

If I had to choose one dish that best represents what Marder can do with vegetables it would be his stuffed red and green peppers. Oozing a judicious amount of tangy cheese and sitting in their respective pools of pesto and tomato sauce, they are a triumph.

Marder's artist friend Charles Arnoldi designed the interior of the vast (fourteen-thousand-square-foot) space. It is a combination of the familiar and the other-worldly, divided into several rooms. You enter through what appears to be an alien pod—an eight-foot-high gray orb whose polished inner walls reveal millions of fossilized trilobites—into a room of warm wood latticework, marble, and glass. The striking marble light fixtures resemble the letter T in a hurry. There are other surfaces that are cast in shapes and made of materials that must be from Planet Xoron-12.

What his next venture will look like hasn't been decided. But when Marder and partner Michel Richard open the Broadway Deli in the renovated Santa Monica Mall this year, you can bet that the food will be first rate. How could it not be, with two of Los Angeles's finest chefs in the kitchen? It'll be like Gehrig and Ruth, Burns and Allen, Laurel and Hardy.

Says Marder: "It's an international California deli; translating old-fashioned things into a new, lighter, cleaner concept—*international* meaning it's not just a Jewish deli. We want to lighten the food so that the most important thing is that it tastes good."

When I asked Marder what he'd do after Broadway Deli he laughed: "Maybe open a vegetarian restaurant. I could do one easily. I'd have to research it. But actually, I'm going to relax after the Deli opens. I'll probably just take care of things and make sure they're running correctly. I'm a responsible guy. I want to make sure my partners get paid back. In another five or ten years, I'll think about something else. Sooner or later, I know I will, because I'm a foodie. I live and breathe the restaurant business."

PASTA AND RISOTTO DISHES
Vegetable Ravioli in a Tomato Coulis with Fresh Herbs and Romano Cheese

TONY MAY

San Domenico, Sandro's, La Camelia, New York, New York

Changing the Perception

CHANGING THE perception of how Americans view Italian cuisine is Tony May's mission in life. Over the years, he's opened four restaurants in New York, each with a different kind of Italian food. If May ever combined the best of all four under one roof, he might well have the finest Italian restaurant in America. He might already.

May opened La Camelia in 1980, to change the perception that Americans held about the food of his native land. At that time most of us thought of Italian food as heavy, and liberally sauced with tomatoes and garlic. But Sicilian food, on which this Italian-American cuisine was based, is only one of many Italian cuisines. At La Camelia, May introduced us to a classical style of Italian cooking prepared by professional chefs.

Sandro's, on the other hand, which opened in 1984, is a trattoria, featuring food as close to rustic Italian home cooking as you're likely to find in this country. To drive the point home, Sandro's features a great communal "chef's table," where some of the restaurant's habitués as well as unaccompanied newcomers sit together, family style.

In 1986, with the opening of Palio, May brought us *alta cucina*, the

Italian version of haute cuisine, which has no ties to the French. Palio was the most elegant restaurant to open in New York since the Four Seasons, and the food soars. Though May recently sold Palio to his chef Andrea Hellrigl, his stamp will remain.

Then, in 1988, May, in conjunction with the world-famous San Domenico restaurant in Imola, Italy, opened San Domenico in New York to bring to the United States the cooking of Italy's aristocratic families.

May explains: "In France, the aristocratic food went to the populace immediately after the Revolution. In Italy we never had a revolution, so these cooks stayed in the aristocratic families. They're still there now. So the cooking at San Domenico is the aristocratic cuisine of Italy brought to a professional level. The food is much lighter, much more diversified. A tremendous amount of ingenuity went into it. At San Domenico we serve this aristocratic food, taking into consideration the many regions where these families lived and their particular cuisines."

If there is one misperception about Italian cooking May has sought to clear up through each of his restaurants it is this: that Italian food is heavy and basically unhealthful.

"Italian food is tremendously healthful. It always has been. We've always cooked with vegetable oils. We eat salads probably more than anyone else. And vegetables have always been a part of our cooking, from *carciofi alla romana* to eggplant parmigiana to *torta pasqualina* to pizza. It's always a part of our diet. The vegetables, herbs, and spices so prominent in our cuisine give a definite taste to the food. For example we've just introduced an artichoke timbale with cheese fondue as an appetizer. We do a lot of stuffed pasta with vegetables. Because of our reliance on vegetables, Italian food is probably one of the most healthful in the world."

The food at May's four restaurants bears him out. At the elegant La Camelia, Chef Carlo De Gaudenzi offers as starters refreshing soups like chilled cream of tomato, pureed fava beans and chicory, and onion. Your appetite whetted but not dulled, you may then select from pastas such as tortelloni San Remese, little hat-shaped pasta filled with spinach and cheese; or *vermicellini alle erbe fresche*, impossibly thin filaments of pasta with an explosion of fresh herbal flavors and al dente greens. There is also penne with arugula and tomatoes, as well as *paglia e fieno pastorella*, green and white fresh pasta with cheese, peas, and tomato. For heartier appetites, you can tie into *gnocchi di patate al verde*, among the lightest potato dumplings you'll ever taste.

Chef Carlo always makes sure there is a variety of fresh Italian vegetable preparations available, such as spinach in a lemon sauce or a savory stew of mushrooms, zucchini, and eggplant.

For an introduction to the robust foods of the Roman kitchen, sit at Sandro Fioriti's long communal table in the center of the room, and make a meal of his appetizers. If you miss Fioriti's *carciofi alla giudia*, you haven't really eaten at Sandro's. The chef deep-fries small artichokes in olive oil until the petals turn crisp and the heart becomes meltingly soft. You can eat the whole thing. But don't stop with just one appetizer and go directly to the pastas. Pause for the fried ricotta with tomato sauce. The Italian *bufala*-milk ricotta, like the artichoke heart, melts in your mouth in contrast to the crisp bread-crumb coating. Or try the grilled smoked *mozzarella di bufala*.

The pasta dishes may include spaghettini with fruit sauce (choose between melon or lemon), or the same pasta in a sauce of black olives from Liguria, sun-dried tomatoes, and a healthy dose of garlic. Sandro also makes fine light gnocchi of potatoes and spinach, served in a rustic tomato sauce.

There's always a fresh vegetable selection on the menu. My favorite is the mixed grill of radicchio, mushrooms, eggplant, zucchini, and endive. And Sandro prides himself on his salad cart with crisp greens and his own house-flavored selection of olive oils and vinegars.

The decor matches the food, with hand-painted dishes and a terra-cotta floor, creating a homey atmosphere.

San Domenico in Imola, near Bologna, is one of Italy's most reknowned restaurants. So it was no small feat on the part of May to get owner Gianluigi Morini, and more important, chef Valentino Marcattilii, to create a clone of the original San Domenico in New York. One of the best soups I've ever tasted was prepared by Marcattilii from smoked tomatoes and red bell peppers with tiny garlicky croutons. The soup had a deep, rustic outdoor flavor as if it had come right out of a campfire cookpot. He also makes *passato di fagioli*, a bean and barley soup. His pastas include a chewy, almost crunchy, *garganelli*: handmade spinach, carrot, and egg pasta shaped like quills, with porcini sauce. Other pastas include a fresh spaghetti *alla chitarra* with tomato and basil, the noodles made on a guitarlike instrument to give them a rough-hewn look and texture.

May's restaurants give us three different versions of May's definition of good food: "People can no longer eat the way they used to. We live in a different world. It's a necessity of life for people to change their diets. People bring about a change in their diets as they advance as a society, as they become more cultured with food. That's where real Italian cooking comes in, in its many forms. Food is supposed to taste wonderful, but, equally important, it must leave you light and satisfied, not weighed down and worn out. The cooking of the Mediterranean, which was so nutritious thousands of years ago, is the same today. It is the food for modern man just as it was for the ancients."

May champions Italian cuisine on the teaching level also. As a founder of the Gruppo Ristoratori Italiani and board member of the Culinary Institute of America, May strives to get more Italian cooking classes taught in American culinary institutions. As he puts it: "I would say there's hardly an American restaurant menu without a pasta dish. There's hardly an American cook who today does not cook pasta. The problem is they still don't know how to do it! I hope they learn, for God's sake," he laughs.

If Tony May has anything to say about it, they will.

MICHAEL McCARTY

Michael's: Santa Monica, California, New York, New York
Adirondacks: Denver, Colorado, and Washington, D.C.

Going for the Zip

WHEN TWENTY-five-year-old Michael McCarty opened his revolutionary namesake restaurant in 1979 he said brashly: "We're going to do something totally different here. There's lots of room for creativity at Michael's, and my specials will simply blow your socks off."

He was right. But it's not just the specials that remove your hose at Michael's—it's most every dish. Behind the braggadocio of the all-American Michael is a classically French-trained chef who has paid his dues in kitchens and classrooms from Paris to Cornell. Michael revitalized the stagnant Los Angeles dining scene, gave it a direction and an identity. As much as anyone, he created what is today known worldwide as California cuisine.

As a boy growing up in Mt. Kisco, New York, Michael recalls that entertaining was a way of life for the McCartys, who were perfectionists "paying attention to the smallest detail when people would come over." According to McCarty, "the entire environment from the food to the table settings to the service had to be absolutely perfect." One night, his parents took him to dinner at a French restaurant in Manhattan, where the crisp linens, fine crystal, attentive waiters, hospitable host, and great food all combined to make a lasting impression on him. Michael recalls, "At that moment I knew this was what I wanted to do with the rest of my life." He was fifteen.

Fresh out of high school, he enrolled simultaneously at the Cordon Bleu, the Academie du Vin, and the Ècole Hôtelière. Then Michael and two other chefs put their education to immediate use by opening a twenty-two-seat bistro on the Ile St.-Louis. After a year and a half of cooking, entertaining, and eating, Michael returned to the United States. He took Cornell's summer Hotel Management Program and moved to Evergreen, Colorado, where he taught French cooking in French and became part owner in a restaurant. Then he said, "What next?"

Next, it was time to fulfill his destiny on a grand scale. He moved to Los Angeles to open his own place. As he recalls, "I decided to create the best restaurant in Southern California." Michael's fulfilled its owner's prophecy for the first few years it was open, and it's never really lost its preeminent position—it simply shares the winner's circle now with other restaurants (many of which were created by chefs inspired by Michael's example).

It was Michael McCarty who first put serious art on restaurant walls. How serious? His collection of Diebenkorns, Hockneys, Stellas, Johnses, and Robert Graham sculptures, which he installed ten years ago for $150,000, is probably worth several million by now. By doing so, he started a trend emulated by Trumps, Spago, West Beach Cafe, and 72 Market Street, to name a few locals. In 1979, when most twenty-five-year-olds were trying to find a way to convince a bank to give them a car loan, McCarty wheedled almost a quarter of a million dollars out of a banker who was interested in good food. Yet even after going through all that dough and kicking in most of what he had, McCarty ran out of money and had to design (and build) many of the furnishings himself. So how could he justify spending 150 large on art?

"Some people called it excess," says McCarty. "I just called it attention to detail."

Today, at Michael's, as in his father's home years ago, every detail must be perfect. "My restaurant isn't just a place to get great food and wine. It's a total environment of different elements to which I've added my personal touch." That translates into a sand-colored Bauhaus exterior with some nautical connotations, and an interior featuring the famous art on its pastel dining room walls. But for me, the only place to dine at Michael's is in the loveliest outdoor dining area this side of Le Pré Catalan, featuring a profusion of greenery, the Graham sculptures, and gentle waterfalls.

You are served by Ralph Lauren–Polo-dressed waiters in pink shirts and khakis who have set your table with heavy Christofle sterling and thirteen-inch white Villeroy and Boch china. McCarty went to Europe personally to special-order the plates, to create "a stage for the food to emphasize its colors and textures." And to complete the service, only fine West Virginia leaded crystal will do. As McCarty puts it, "dining at Michael's should be an event to be anticipated for the total environment you'll experience here. Otherwise, why not stay at home and cook a nice meal for yourself and your family? Here, a nice meal is simply not enough."

But in the restaurant business, china, silver, gardens, and art are not enough. You go to a restaurant for the food. And here, as in every other category, Michael seeks, and usually attains, perfection. He starts with the finest seasonal ingredients from growers in nearby counties and faraway continents. Then, as Michael says, he "goes for the zip." This "zip" is not something you just shake out of a bottle, or everyone would have it. The "zip" was forged in sweltering Parisian kitchens and in bleary-eyed study sessions at respected teaching institutions on both sides of the Atlantic. It is the culmination of many years of hard work by a very dedicated man and an incredibly adroit chef.

You'll find the zip in the heavenly asparagus *feuilletée* with a beurre blanc of champagne vinegar and Gavalan white wine. And you'll find it in salads such as radicchio, Belgian endive, watercress, and warm Montrachet goat cheese with walnut oil and Pommery mustard vinaigrette; or, my favorite, sautéed shiitake, oyster, and chanterelle mushrooms and pine nuts in a sherry wine vinegar and walnut oil dressing on baby lettuces.

You probably won't find a better cream soup than his nutmeg-scented white corn chowder with colorful peppers, and his asparagus soup tastes of vegetables, herbs, and smoothness—never (thank God), like warm vegetable-flavored cream. His mushroom soup with Tio Pepe sherry and chives is a wonder.

McCarty believes that "vegetables are absolutely imperative for taste, texture, and color as well as for health and nutrition." That is evident in his platter of hand-rolled spaghettini in a Chardonnay cream sauce flavored with opal basil and chèvre, strewn with grilled strips of sweet yellow and red peppers and asparagus. Baby California vegetables in a beurre blanc infused with fresh herbs makes the perfect accompaniment to the pasta.

Who cooks at Michael's? The question is: "who hasn't?" Without McCarty having to make so much as a phone call, the opening foursome of Ken Frank, Jonathan Waxman, Mark Peel, and Bill Pflug came from around the country to be a part of the most eagerly anticipated restaurant in recent memory. As they departed Michael's to open their own places, their positions were ably filled by the likes of Roy Yamaguchi, Gordon Naccarato, and Zach Bruell. They too have gone, but the quality of Michael's food did not leave with any of them, for, like a top college football coach, McCarty coaches his freshmen into stars of the magnitude of those who have graduated and moved on.

Over the years the food at Michael's has remained consistently good, due to McCarty's sure, steady hand at the helm. Michael is forthright about his contribution. After Jonathan Waxman left, he said: "Nobody sets up a fine restaurant to be dependent on one single individual. No one at Michael's is indispensable, except maybe me," he deadpanned.

It's true. He made the restaurant happen, he makes the food happen, and he makes the room happen. There are few hosts who can welcome you as warmly as Michael McCarty. His boisterous, back-clapping, hand-pumping, wise-cracking greeting sets the tone for a dinner that, like the owner, will be long remembered.

In addition to Michaels in Los Angeles and now New York, McCarty also runs Adirondacks in Denver and Washington, D.C. Adirondacks were at first opened as Rattlesnake Clubs, including one in Detroit, in partnership with Jimmy Schmidt, who now, after the break-up of the partnership, runs the sole surviving 'Snake in the Motor City. McCarty supplanted the name with Adirondacks when he opened the Washington, D.C., location because he got tired of explaining "that the restaurants aren't private clubs and they don't serve rattlesnakes."

When both Adirondacks opened, the food was very much Jimmy Schmidt's. Now, however, close your eyes in Denver and Washington, and you're eating at Michael's in Santa Monica or New York. That would make sense. How can you improve on perfection?

Both of these restaurants are located in classic buildings that are important parts of America's past. The Denver restaurant is in the old Tivoli Brewery, complete with its original, giant copper beer vats. In Washington, Adirondacks occupies the former Presidential waiting room at Union Station. The room has been lovingly restored to its former glory. All McCarty added were a few Hockneys and Diebenkorns and some Saarinen sofas and chairs.

Michael is focusing his attention for his biggest project yet. He will open the 175-room luxury Santa Monica Beach Hotel on the current site of the venerable Sand and Sea Club on the Gold Coast. The $50 million beachfront hotel, slated for a 1991 opening, will include, of course, a fine restaurant. McCarty was able to persuade the city to select his project over that of ten other developers by showing how his hotel would benefit the public sector as well as the private. He pledged to provide $1 million annually to the city's beach maintenance fund, to improve beach access by adding one hundred public parking spaces, and, most significantly, to build next to the hotel the Santa Monica Beach Community Center. It will cost him $6 million to construct the sixteen-thousand-square-foot public facility that will be an arts and environmental center, featuring a community meeting hall and a moderately priced beach cafe.

Says McCarty: "We'll build an international artistic landmark by the ocean. The hotel and community center project should serve as a model for governments and developers alike, showing that it is possible to build a public-private joint venture that will work on every level: socially, economically, philosophically, and financially."

There are doubting Thomases who say it's too big a project. That it can't be done. That the public and private marriage can't work. The kid from Mt. Kisco heard the same stories from banker after banker who wouldn't come across when he was a brash twenty-five-year-old looking for a chunk of change to start Michael's. I wonder if they're still in banking?

SALADS

Wild-Mushroom Salad

FERDINAND METZ

Culinary Institute of America, Hyde Park, New York

The Chef-Maker

"Where the demand for quality is constant and the expectations of an increasingly palate-wise society become more and more uncompromising."
—*from a Culinary Institute of America brochure*

WHAT DO Marcel Desaulniers, Dean Fearing, Susan Feniger, Larry Forgione, Bradley Ogden, Brendan Walsh, Jasper White, and Roy Yamaguchi have in common (other than the fact that they're all chefs, of course)? They're all graduates of the Culinary Institute of America, one of the few schools in America that has a one-year waiting list. But perhaps equally important, there are an additional twenty thousand chefs cooking in fine, lesser-known kitchens who also attended the CIA. Though we may not often eat at An American Place or City, we may frequently dine on food prepared by Institute graduates at other restaurants, as well as Marriots, Hyatts, and similarly food-conscious hotels around the country. If you've noticed, the food in such establishments is better, fresher, and more interesting today than it was ten years ago.

It was about that time Ferdinand Metz came to the Culinary Institute from his position as head of research and development at Heinz in Pittsburgh. Metz had all the right culinary credentials to head up the CIA. But the Institute, which had been through four presidents in five years, needed a sharp businessman to put their house in order. When they saw "M.B.A." on Metz's resumé, they knew he was their man.

The Culinary Institute of America is no longer the little New Haven storefront that started with sixteen students in 1946. It's a big business with an annual budget of $32 million, an eighty-three acre tree-studded campus that slopes gently down to the Hudson, and an enrollment of two thousand. The growth experienced by the Culinary Institute can be traced directly to an increased interest on the part of the American dining public in better food and better nutrition during Metz's stewardship.

As the slender, mustachioed president with the wry smile puts it: "At the Culinary Institute we are in a position not just to contribute to the gastronomic pleasures of our clientele, but to contribute to the quality of life. We need to not just react to what the dining customer wants, but to *pro*act. Sometimes you need to take the lead and help educate, and I think our nutritional program allows us to do just that."

Metz foresaw the need for an emphasis on nutrition years ago. And that is one of the reasons why he is so valuable to the Institute. He came to the CIA during what he terms an "unfortunate period," punctuated by unrest and executive turnover, and managed with his culinary and managerial skills to turn things around. "There are two ways to create change. One is where you simply react to the marketplace. The other is where you do a little bit of thinking ahead of time and realize you have to take the lead as well. That requires you to have convictions and the knowledge that you feel comfortable in your beliefs and philosophy."

To that end Metz spearheaded the drive to build a three-million-dollar Nutrition Center, a third of which was funded by a challenge grant from the General Foods Foundation. The eleven-thousand-square-foot facility, which opens this year, will contain a state-of-the-art kitchen specially designed for nutritional cooking, the eighty-seat St. Andrews Restaurant, an outdoor plaza cum dining area, a nutrition resource center and classroom, a bakeshop, and offices.

According to Metz, "We've taken nutrition out of the classroom and put it on the plate. In other words, we've taken a non-traditional approach to nutrition teaching, not just offering a classroom theoretical setting, which hardly ever forces you to prove to what degrees your theories are right and to which degree the public accepts those theories. We've taken nutrition and actually serve it in a public restaurant day in and day out, supported by paying patrons, which means the food has to be very realistic. That is crucial. So we don't just talk about food that is nutritious—we're cooking it. This has led us to seek a better balance in the food we serve and to rely on things like extractions of herbs and spices to provide flavor instead of extra fats. Our type of nutritional cooking forces you to go back to the basics and develop a much greater awareness. You have to become more innovative, much sharper. It's a great challenge. We've discovered some very basic fundamental principles as far as nutritional cooking is concerned, and have implemented very successfully a program that will be the one to emulate in the future. That's why we're building the Nutrition Center. We recognize the evidence of growing dietary concerns among Americans who are providing new challenges to the foodservice industry."

Metz is just the man to meet those challenges. Raised in Germany in a hotelier's family, he apprenticed in Europe and worked in a number of stateside restaurants, the most formative of which was Henri Soule's legendary La Pavilion, where Metz worked under Pierre Franey. After a stint at the Plaza in New York, he went to work for Heinz, where he remained for fifteen years before being tapped for the top job at Culinary. During these twenty-odd years he attained the coveted position of Master Chef (one of only thirty-five in America) and has been intimately involved with the Culinary Olympics, most recently guiding our team (as team manager) to the World Championship in the hot food category.

Metz is aware that restaurant food must satisfy today's nutritionally conscious diner from the equally important aspects of taste and health. This is why he insists on the correct preparation of vegetables in the two-year curriculum at the CIA.

Says Metz, "Good cooking by definition means there's a balance. That goes back to the nutritional concept of a balanced meal. The people who make the news today are cognizant of the many things one can do with vegetables. We oftentimes caution our students that it's not the person who grills a piece of tuna who succeeds, but the person who can make something out of nothing. That is not to say that vegetables are nothing, but they are underrated and certainly not utilized to their fullest extent. They're not appreciated enough. So whenever you cook with vegetables you better be a good cook. You're going to have to provide something in the form of skill and innovation in order to bring off a vegetable dish correctly. We've therefore always paid a lot of attention to vegetables. The way we teach how to prepare vegetable dishes illustrates our overall philosophy on food and flavor: Let food be what it is. Try to enhance it, not overpower it."

That enhancement process takes nearly two years to develop at the Institute, where ninety chefs and instructors (fourteen of whom number among America's thirty-five Master Chefs) from twenty countries utilize nineteen commercially equipped production kitchens to conduct courses in baking, charcuterie, garde manger, American, Oriental, and international cuisines, with related courses in wine, table service, purchasing, stewarding, beverage control systems, sanitation, safety, and, of course, nutrition. Midway through the course of study, students go on a six-month salaried externship program to work in restaurants and hotels around the country (the establishments must make at least 51 percent of their dishes from scratch), producing food for paying customers. This practical experience plus the rigorous course of study at Culinary assures that CIA graduates receive from five to six job offers upon graduation.

The forty-nine-year-old Metz knows why: "I've done two apprenticeships in Europe and have been involved in food for over thirty years. And I can honestly say the exposure our students get here is second to none. I say that knowing what schools in Europe are and knowing schools here. It's just a fact. If we teach international cooking here, and the menu goes to a different country every day, we have the kind of people who have not just read about these menus in a cookbook and are trying to duplicate them. We have people who were born in those countries. The authentic presentation and teaching is preserved here. For example, if we teach Oriental cooking, it's by an Oriental chef. Having been able to attract almost 50 percent of all certified Master Chefs, that also makes a statement. If you look at our course of study, including such subject matter as charcuterie, sausage making, Experimental Kitchen, fish kitchen program, St. Andrew's Nutrition Restaurant, you'll find those things are not present on any other curriculum in the country. Our game is quality, it's as simple as that."

The practical food-serving experience a student is expected to gain at the Culinary Institute is reinforced by the fact that almost every dish prepared there (contributing to some four thousand meals served there each day) is eaten by students and staff or by paying customers. To feed the public, the Institute maintains four restaurants, one of which (the Escoffier Room) bears a three-star rating from the *New York Times*. Whether you dine in the northern Italian Caterina de Medici Room, the American Bounty (which focused on the glory of our country's cuisine long before it came into vogue), or the nutritionally oriented St. Andrew's, you are eating food prepared by the students of the Institute.

The one kitchen where food is not prepared for consumption is the Experimental Kitchen. As Metz describes it, "The kitchen conducts very simple experiments and finds out astonishing things about certain long-held traditions regarding how food is prepared. Oftentimes what we teach verifies those traditions, but now the students have the culinary background for it, such as when we teach whether to salt eggplant or not. When you know the reasons for why you do something you can make judgments that will impact on how you prepare food. Sometimes you're better off to think for a minute rather than just starting to cook wildly. In the Experimental Kitchen we try to teach that a combination of thought along with a good cooking sense gets you further, and faster."

Over and above Metz's culinary and managerial skills is a third quality that does not appear on any curriculum vitae but which may be his most important characteristic. Metz clearly feels he is a man on a mission—a mission to influence the way we eat.

"One has to work here to understand what this place is all about. It's very exciting to be here, primarily because what we do has a present and a future impact on how America eats. Those are big words, I realize. But having over twenty thousand alumni and graduates out in the field, it makes it easier to understand why I just said what I did. We take our job very seriously. We know what we say and what we teach matters. If you look at the scientific evidence that a significantly reduced fat intake can help make an impact on the reduction of the two biggest killers in America, cancer and heart disease, then our mission as chefs becomes so much more important. So we don't just come into the office and say 'Hi!' and at the end of the day leave having made no impact. I think that's why we take our work here so seriously. It's come a long way, but we've got a long way to go, and I'm committed to bringing about the necessary changes in how America cooks and eats."

One final thing that Metz imparts to his students is the sort of practical life experience necessary to succeed as a chef. Like the time he gave up what he says was a "wonderful" $250-a-week country club job with free room and board (back in the sixties when $250 was $250) to take a job for $50 a week to work at Le Pavilion under Henri Soule. Taking the right risk when it can further your career can be a smart move for the chef who aspires to greatness. It worked for Metz at Le Pavilion, and it worked for him again when he left the security of fifteen years with Heinz to take the helm of the struggling Institute.

"I tell my students that learning never stops. And even though I had a good job, you have to continue growing. Today I would kick myself if I had not made that decision to go to Le Pavilion. It would have been a lack in my culinary education. Even though it was a tremendous financial sacrifice, I teach my students you have to make those kinds of sacrifices if you want to succeed. And I want my students to succeed."

Succeed they have, in no small measure due to the guidance of Ferdinand Metz, who is now guiding the destiny of America's future culinary greats.

MAIN COURSES
Haricots Verts a la Repaire

MARK MILITELLO

Marks Place, Miami, Florida

The Spontaneous Chef

FOR YEARS restaurateurs in Florida have named their places Le, La, or Au something and cooked rich, heavy French food to please the people from back east who came south to vacation or retire. Do people really want to each rich, heavy foods in Miami's sultry climate? Maybe not, judging by the crowds at Marks Place in North Miami. The casual white stucco and blond wood restaurant seats only ninety. But they served an astounding three hundred dinners both nights I was there, and I overheard the maître d' good-naturedly say: "If anyone comes in here without a reservation looking for a table, just laugh at him."

Why has Marks Place become so popular in just four years? I think it's because chef-owner Mark Militello, who comes from the same part of the country as many of his vacationing or transplanted customers, took a look at Miami's muggy climate and decided the food being served in a lot of the restaurants was totally inappropriate to the region.

He decided to develop a cuisine for south Florida, based on the 365-day growing season that not only supplies salad greens, tomatoes, and other vegetables year round, but also an unusual variety of vegetables and tropical fruits that grow nowhere else in America. It was the fresh local, often exotic, produce of south Florida that formed the basis for Mark Militello's unique cuisine, on which he has built a national reputation.

"There are a lot of great products grown here. I try to emphasize them, because they're local, available, and fresher than anything else you can get. Of course, if there's something I want to use from another part of the country I don't limit myself. But we like to pick up things with a tropical influence, or things that were brought to south Florida through South America or the Caribbean. It kind of lends a little more of an interesting flair to this whole new emergence of American cuisine."

Every chef has a signature dish. It's hard to pick Mark Militello's, because he has a brand-new menu for both lunch and dinner every day. "I'm spontaneous. I find that I become bored very easily. I'm not the kind of chef who does a dish forty or fifty times, each time trying to perfect every little thing. I find a lot more excitement in being spontaneous. I enjoy that approach. It keeps me interested, and it keeps the people I work with interested."

If I had to pick one dish that is most representative of Militello's south Florida cuisine, I would pick his fresh hearts of palm marinated in tangerine and orange juice and citrus honey. The sweet citrusy juices play perfectly off the coconutty taste of the palm hearts. You can taste the Florida sunshine in every bite. Fresh hearts of palm are about as similar to the mushy canned variety as a French kiss is to an air kiss.

Mark doesn't have to go exotic to feature Florida's bounty. One of my favorite dishes is composed of ingredients you could find in most any good produce market in winter. But what Militello does with them is something you find only in great restaurants. Fresh salsify pancakes are layered with grilled portobello, oyster, tree oyster, brown hedgehog, yellow-footed chanterelle, shiitake, and crab cluster mushrooms, and joined on the plate by two quenelles of goat cheese; mixed Florida greens including cress, curly endive, radicchio, and edible flowers; and a spicy red and yellow Florida tomato salsa.

The yellow Florida tomato is also featured in a dish of herb-studded crepes stuffed with an engaging mixture of ricotta and spinach, on a plate that looks painted half red and half yellow with two colors of tomato *coulis*. How to eat the *coulis* that remained after we dispatched the crepes? With the house-made rosemary-thyme onion focaccia and soft Jewish rye bread.

Militello is as accomplished with pasta as he is with bread. One night, possibly in a philosophical quandry over whether spaghetti was brought from Italy to China by Marco Polo, or vice-versa, the chef gave us a basil fettuccine tossed with gingery julienne zucchini, yellow squash, napa cabbage, carrots, and snow peas in a robust red tomato sauce. On top was some baby bok choy from the chef's own garden, sautéed in soy sauce.

Militello is fond of Asian touches, and some of the fruits he uses to personalize his cooking are Asian in origin. Near the southern tip of Florida around the town of Homestead there is a narrow belt of land that is just right for growing certain fruits that cannot prosper anywhere else in the continental United States. The climate there is so similar to that of Vietnam, India, the Philippines, and Thailand, that food brokers selling to ethnic markets across the country contract with farmers here to grow exotic fruits just like their customers used to eat back home.

The brokers even make seed stock available, so that the smooth, creamy mangoes the Thais and Filipinos like can be grown right next to the stringier Indian varieties. Bengal lychees with a complex taste of curry and rosewater grow cheek by jowl with their cousins, the more straightforward-tasting longans preferred by the Chinese. The tropical climate is perfect for Central and South American flora as well, and Militello makes full use of this exotica in fruit salads, garnishes, ice creams, marinades, sauces, and chutneys.

In the summer it's lychees, longans, sugar apples, and prized varieties of mangoes like the Kitts and the Kents that set Militello's mind racing with new ideas for his warm weather menus. During the winter *jaboticabas* and *muntinjias* show up at Marks Place. The former is sometimes called a Brazilian grape. The fruit grows right on the bark of a tree, however, and is rounder, harder, and blacker than grapes. The taste? How about rice pudding? *Muntinjias*, or strawberry fruit, are small red berries that hang from the branches of a tree. They remind me of cherry Kool-Aid, which I drank by the pitcher growing up in Houston.

On the vegetable side of the ledger Mark is fond of using Cuban root vegetables like *boniata*, or Cuban sweet potato, and *malanga*, with a similar, starchy taste. Together with the exotic fruits and Seminole-grown palm hearts, the Cuban root vegetables help form a cuisine that could only happen in this part of America.

But Militello doesn't pack his menu with unusual flora in an attempt to shock you into thinking he's good just because he's different. Twenty-five years of working in kitchens interspersed with training programs at the Morrisville, New York, Culinary Hotel Program, and Florida International University's Hotel School, gave him a firm base for experimenting.

One of his favorite dishes is grilled Florida vegetables: butternut squash, corn, yellow squash, radicchio, green onions, and mirliton squash, bathed in an olive oil vinaigrette.

Another special dish is bulgur spiced with chilies and cooked in a flavorful vegetable stock with carrots and zucchini, and served with a hazelnut vinaigrette with pieces of raspberry and hazelnuts.

The love affair between south Florida and Mark Militello seems to be deepening. "I've been very fortunate to have the support of the community. I try to get involved with local charities, like raising money for the food bank, to put back into the community what I can to show how much I appreciate their support." The crowds at Marks Place show that the feeling is mutual.

APPETIZERS AND
FIRST COURSES

Grilled Mozzarella in Romaine with a Sun-dried Vinaigrette

MAIN COURSES

Salsify Pancakes with Grilled Portobello Mushrooms

MARK MILLER

Coyote Cafe, Santa Fe, New Mexico

Resurrecting a Great Cuisine

"FOOD IS an important part of history. It's not just something you buy at the grocery store, stick in a plastic bag, microwave, and eat while watching TV. Unfortunately, that seems to be the synopsis for twentieth-century food."

If Mark Miller has his way, the story will turn out quite different. The chef, who was one of the pioneers of California cuisine in his Berkeley days, has settled in Santa Fe, New Mexico, where out of the rich history of the Southwest, he is attempting to create a new cuisine.

Mark Miller didn't start out to be a chef. The Boston native studied Chinese art history and Japanese anthropology at the University of California in Berkeley. He taught there, too, until he decided to pursue another career. The lure of the kitchen overtook him; first he started working at Williams-Sonoma, one of America's top kitchenware stores. Soon after, he caught on with Alice Waters at Chez Panisse, just about the time the restaurant was achieving its national reputation for using fresh California produce. He figures he and Alice must have created over five hundred dishes in the three years he worked there.

In 1979, his own culinary voice well developed, Mark set out on his own at the Fourth Street Grill. The place took off immediately. People crammed into the seventy-five-seat restaurant at the astounding rate of three thousand a week to taste the sharp, direct flavors of the Mediterranean and the tropics. It was Miller's opportunity to bring together food and anthropology. The food was a big departure from the kind he had cooked at Chez Panisse. He used no cream, butter, or salt at Fourth Street, preferring to leave the natural flavors of vegetables, herbs, and spices undisguised. Heady with success, Miller opened Santa Fe Bar and Grill the next year. It was here that he began a flirtation with the foods of the American Southwest that turned into a full-blown love affair.

Even though Miller soon sold his interest at the Santa Fe Bar and Grill, the experience of working with the flavors of the Southwest had made a lasting impression on him. In 1984, he pulled out of Fourth Street and moved to Santa Fe to research, develop, and open Coyote Cafe.

Not since the opening of Michael's in Santa Monica had a restaurant been awaited with such anticipation. That it took three delay-filled years to open did not dissuade newspapers and magazines from doing stories on the restaurant, or the restaurant from running ads in the Yellow Pages, even though a location for the place had not been found.

The Coyote Cafe opened in March 1987 on the site of the former Greyhound Bus Terminal just off the Plaza in downtown Santa Fe. It was worth waiting for. Such dishes as ravioli made with corn flour in a sauce of ancho chilies and Texas goat cheese; roasted tomato and chili soup; cactus salad; and chili corn bread. For Miller, the dishes were a triumph not only in their execution, but in the raw materials that went into them. He had carefully researched the foods of the Hopi, Navajo, and Pima Indians, the Spanish, the Mexicans, and the Americans who had settled in and around our country's oldest city, and found inspirations from the recipes and the raw materials of the region.

"The Indians have hundreds of dishes with corn, and they're all different. In America we have corn on the cob, sautéed corn, baked corn, and that's about it.

I'm trying to help promote a culinary education, to sort of extend people's creativity within tradition. I don't think we need to create new dishes. We should reenact some of the older dishes that were already there and have a better understanding of them."

Miller's research found that in general, the food in New Mexico in 1889 was much better than in 1989, because the ingredients were so much more varied. He found that the Indians who populated New Mexico knew two to three hundred kinds of wild herbs, spices, and edible plants long before the arrival of the Spanish in the sixteenth and seventeenth centuries. Many products were brought up from Mexico as well, further enhancing what Miller discovered was an extremely rich and varied cuisine. His "culinary archeology" of the region further revealed that the Indians there raised dozens of kinds of corn and beans, and that the Pima tribe had over one hundred recipes for piñon nuts. He blames today's food industry for homogenizing that cuisine, so that all we know of its many diverse ingredients are the ten or so things that move fastest off supermarket shelves or on Mexican restaurant menus.

But Miller's out to change that, by introducing his customers not only to a vast array of locally grown produce, but to many different varieties of the same item, often served on one plate.

"We try to get growers to grow things especially for us. I'm particularly interested in the old, indigenous varieties of corns, squashes, chilies, and beans that were used in the Southwest. Vegetables have held a very prominent place in this cuisine for a long time. Unfortunately they've been downgraded to plate accompaniments. Or the generic varieties of vegetables are such that people don't understand them. Or they're not cooked correctly. Or they're not cultivated and picked fresh enough and brought out on the plate properly. What we try to do is highlight the flavors. For instance, squash blossoms. We have particular varieties of squash blossoms that are grown for us. They're picked in the morning before the sun comes up, and they're brought to the restaurant within a three-hour period. They're used that day. So, the delicacy, the nuances—whether it's deep fried and filled with three kinds of cheese and herbs or used in a ravioli—the flavor intensity of the variety of the vegetable actually comes through as the main thing."

Such a philosophy takes great cooperation between grower and chef. And Miller gets that cooperation from people like Elizabeth Berry, who devotes two of her five hundred acres ninety miles north of Santa Fe exclusively to Coyote. Says Miller: "Elizabeth grows all our lettuces, squash blossoms, Mexican herbs like oregano, arugula, wild sage, and ten varieties of corn. We try to highlight her and our other growers on the menu to bring attention to what they're doing." He also works with Indian groups to raise specialty produce such as pre-twentieth-century varieties.

Often Mark will take one product and prepare it in several ways so the public can become aware of possibilities they might never have dreamed of. For example, he'll hold a pumpkin festival during which he'll bring out different varieties of pumpkin and show people how the vegetable can be sautéed, or used in soups, ravioli, sauces, breads, and, of course, pies. Explains Miller, "We're trying to show people the range and versatility of vegetables, how they've been used everywhere from twentieth-century California cuisine to old peasant dishes that have been done in Europe for hundreds of years. We'll do that with pumpkin, onions, corn—all the things that grow here. Sometimes we'll try to do something with different varieties of the same vegetables. We'll serve them side by side. When people become aware of the nuances, that can educate the palate."

Of course, people don't come to Coyote Cafe for an education. They come for a good time, as I did on several occasions this summer and fall. One evening I began with Mexican empanadas, flaky turnovers filled with cheese flavored by locally grown herbs from Mexican seed stock. Another in our party chose corn cakes, made from one of Elizabeth Berry's ten varieties of corn. Miller's vegetable-filled tamales made from fresh corn *masa* were exceptionally light and flavorful.

As a main course I ordered a locally grown, fiery green chili stuffed with soothing cheese; a *tamale de elote* (fresh corn *masa*); and a deep-fried *chalupa* (boat) of savory black beans, lettuce, and salsa.

A month or so later, when the cold north winds blew the leaves off the trees and the tourists out of town, dinner began with cheese-filled empanadas, this time colored with *achiote*, made from the red-hued annatto bean. A warming Mexican winter vegetable soup of *nopales* (cactus), tomatoes, and chipotle chilies in a rich vegetable broth took the chill off the night.

For the main course I opted for a platter of vegetable dishes chosen from the entree section of the menu where they had originally been designed as accompaniments: colorful grilled sweet red and yellow bell pepper *rajas*; braised spinach; a puree of white rose potatoes, turnips, and celery root; and a cilantro-cumin-flavored couscous that reminded me of Miller's ethnic dinners at Fourth Street Grill.

Miller designed the playful and festive interior of the Cafe with the help of decorator DeWayne Youts, who operates Umbrello, the definitive Southwest design store in Los Angeles. The first thing that catches your eye is the second-story dining room, with all sorts of angles and curves that take the eye meandering to a huge painting of a Navajo chief's blanket, pink and turquoise hand-painted floors, and a room-length ledge above the bar and kitchen populated by colorful sculptures of spotted jungle cats, howling coyotes, a rooster, and plenty of cactus. Coyotes appear throughout the restaurant, frolicking on frosted-glass panels and on wall hangings.

The coyote was deliberately chosen as a symbol for the Cafe. Explains Miller: "The Coyote Cafe tries to live the true spirit of what the coyote means. The reason it is our logo is that in Southwestern lore and Indian myth the coyote is the trickster, the merrymaker, the prankster. He is the person that gives life its spirit. That's what the coyote does, and I think food should be the spiritual music in our lives. That's the tradition here in the Southwest, and respect for tradition is important to me."

Respecting that tradition has prompted Miller to find a way of reaching more people with the food and cooking techniques of this region. A new venture, called Coyote Kitchens, will combine his archeological and anthropological approach to the foods of the region with the marketing skills he learned at Williams-Sonoma. "Coyote Kitchens is going to create, for the first time, a center for Southwest cuisine. It has two phases, both being done simultaneously next year.

"One will be an eight-thousand-square-foot *mercado*, or market. It will have a cooking school, a produce market, a cafe, reproductions of nineteenth-century furniture, and wood-burning ovens and smokers. In fact, everything you would need to do any sort of Southwest cooking, like tortilla presses and cutters; Mexican molds; and a lot of colorful plates and bowls designed in Santa Fe and made in Mexico will be available. What we want to do is this: If you were going to do a certain dish the way the Hopis did it, then you could go to the store and get the dried beans, dried wild sage, the berries, and what have you, and you could do the dish traditionally. Hopefully we'll be coming out with a line of products and ideas that show you how to use the native ingredients so that we're not just selling you something. We'll eventually be offering classes. There might be a two-month class on corn or beans or Southwest vegetables, meeting once a week, so that we'd be promoting what we're doing as a philosophy. We want to bring people in and teach, trying to resurrect the great cuisine that did exist here.

"The second phase will be a fifteen-thousand-square-foot factory producing products with the Coyote Kitchens label. In a three-year period our USDA-approved kitchen will be producing between one hundred and two hundred products under that label. Then we'll be in a position to open Coyote Kitchens stores in places like Houston, Dallas, Austin, Phoenix, Los Angeles, and as far east as Kansas City and Oklahoma City: places where Southwest cuisine has been an important part of the history."

When I observed that Miller seemed to be making a major commitment to Santa Fe and the Southwest, he agreed—and hedged a bit. "I love this region. I am attracted to the primitive art here and the expressiveness. To me the Latin palate is the most expressive that's still a part of our culture. I love Chinese, having studied art history there a long time, and Japanese, but I don't feel it within myself. I can cook good Thai food, but I can't *create* within that range; it doesn't come naturally. The Latin, primitive, Southwest things are the closest to my own personality, the things I love. There's an unfinished primitivism about it, and I use that in its best aspect at Coyote Cafe. To me the vitality, the enthusiasm, the passion of the flavors are kept intact, and not at the expense of technique or plate decoration. The food shows an intensity for life."

But will the chef who travels to at least one new country each year settle down in Santa Fe?

"I love it here. As a transitional stage for me, it's been great. I think I'll always have a home in Santa Fe. The next phase is to do something in Mexico, to pull some of that intensity up here. Like taking Laura Chenel down there so she can duplicate in California the goat cheese that's been made in Mexico for hundreds of years. I would say that we've only uncovered 1 to 2 percent of the possibilities of the foods from Mexico. The same explosion that happened with European—especially Italian and French—food in the sixties, seventies, and eighties, I think is going to happen with Latin flavors. There's such a range of palate flavors there."

This is why, regardless of where Mark Miller travels, vegetables are paramount. In his cuisine, vegetables, fruits, herbs, and spices combine to create such a satisfying meal that dessert becomes superfluous because the food has touched every part of the palate: sweet, salty, sour, and bitter. Miller finds the composition of the American food palate "pretty warped because people don't have a discriminating sense in terms of delicacy." And, for him, vegetable flavors have that delicacy and subtlety to awaken and, as he puts it, "cultivate a mental 'file for eating' and a philosophy of taste that's important." To bring this about, in his view, vegetables can't just be something put on at the last minute on the side of the plate.

"Vegetables, besides their healthfulness, are a way of cultivating the palate in terms of nuances of flavor. Vegetables can create excitement in terms of varieties, colors, textures, and flavors and train the palate by adding new dimensions to eating. We try to do things with Southwest vegetables, served fresh, cooked properly, and put on the plate in such a variety that will provide the customer with more satisfaction because of the incredible range of flavors. Then people can eat less. Often they overeat because the flavors of the foods themselves aren't satisfying.

"We're still not relating to vegetables as part of a history, as part of the culture, the ecosystem of the land and how we use it. We must create a culture in which there is a respect for tradition and a depth. Then we can understand who we are in terms of our place in history."

SOUPS

Roasted Tomato and Chili Soup

PIZZAS AND BREADS

Chili Corn Bread

RELISHES AND INTERMEZZOS

Roasted Fresh Pimiento and Saffron Butter

Salsa Verde

GORDON NACCARATO

Gordon's, Aspen, Colorado

High-Altitude Cuisine

RULE NO. 1: Never open a serious restaurant in a resort area. The business is too seasonal to make a go of it.

Rule No. 2: If you have the talent of Gordon Naccarato, you can disregard Rule No. 1.

But you're used to disregarding rules if your culinary training came at the hands of Michael McCarty.

"Michael's was quite an amazing experience. Michael himself was barely twenty-five or maybe twenty-six years old when he started. His restaurant will be ten years old already, this spring. It was just a special time and place, where it all kind of happened—a wonderful coming together of some very talented people. The enthusiasm there, the crackling excitement every day that you were working with some of the most creative and talented people who were all so young. It was just so much fun to go to work there, and that enthusiasm carried me into my restaurant."

Gordon Naccarato got all his training on the job at Michael's. And his mentors read like *Who's Who in American Cooking*. There was Jonathan Waxman, whom Naccarato says was his "biggest influence." There were Mark Peel and Nancy Silverton, who went on to Spago and now operate Campanile in Los Angeles. Roy Yamaguchi went from Michael's to eventually open 385 North and is now starring as Roy's in Honolulu. Ken Frank went to La Toque. Kazuto went to Chinois. Bill Pflug worked at Michael's too. So without having to bounce around from one restaurant to another, Naccarato came in contact with many different chefs, many different styles of cooking.

When Naccarato, like his mentors, was ready to try his wings, he headed to Aspen in 1983. He called his cooking "high-altitude cuisine."

" 'High-altitude cuisine' really means nothing. It was just to free our hand so we could do whatever we wanted to. Our menu is written daily on a word processor, and we do whatever is fresh, starting from scratch every day. I think people still like labels, and I resist being labeled, because when you are, you can't stray too far from what you're supposed to be, or people get upset. I don't like being pigeonholed like that. You see, Wolfgang never lets himself get pigeonholed. Now, when he gets interested in another cuisine, say Chinese food, he opens Chinois. I can't really do that here, so if I'm interested in Chinese food, I can do it that night because I keep my image and my menu flexible."

A recent menu featured an Italian canape, quesadillas, goat cheese rellenos, a Chinese dish, a Southwestern tostada, a Vietnamese salad, and a fennel salad. The selections here are eclectic, the influences ethnic. And the flavors are bold.

"My own personal tastes tend to run toward ethnic foods. I think it's a reaction to not being in a big city where you can get all kinds of good food from all over the world. The only ethnic food up here is Chinese, and it's pretty bad. I think missing those strong, assertive flavors makes my menu go off in ethnic directions."

They say food tastes better at higher elevations. But are there enough tourists to support a sophisticated restaurant like Gordon's in the small town of Aspen?

"The market is sufficient to support us here. Our clientele travel the world. They're much like the people in New York or Los Angeles. People like Rupert Murdoch. Richard Nixon was here. We've had everyone from Supreme Court justices to actors. We get everybody, from soup to nuts."

The menu goes from soup to nuts as well. One night the soup offering was *pistou*, a Provençal vegetable soup with garlic, basil, and Parmesan. There was pasta reminiscent of Michael's, with julienne vegetables, fresh herbs, and wild mushrooms gathered locally. You could order potato pancakes, a charred Anaheim chili–goat cheese relleno coated in blue cornmeal and deep-fried, salads such as fennel-parsley, antipasto, a garden salad featuring Doug and Janelle's edible flowers, and a wild rice salad in a lemon-cilantro vinaigrette with avocado and almonds.

Local growers in good weather and Federal Express in bad provide Gordon's with produce as fine and fresh as any in the country's. And that's important to Naccarato because of the kind of food he cooks.

"I think cooking is moving in the direction of a more healthful approach. We went from horrible food in the U.S. to a preoccupation with *nouvelle cuisine* and how beautiful we could make food look. Now we not only want it beautiful and delicious, we want it healthful, too. That's why in 75 percent of our dishes at Gordon's there's no cream or butter. We don't put butter on the table with the bread. Let's face it, whether you're a vegetarian or not, you need to eat your vegetables."

The restaurant is done up in contemporary L.A. style with local art and is strategically located on the top floor of a shopping complex, giving all one hundred seats a view of the mountains.

One of my favorite dishes here is a grilled vegetable lasagne in a Montrachet goat cheese sauce spiked with garlic, shallots, and thyme. The roasted sweet red pepper provides the perfect foil for the grilled radicchio with its bracing, charred bitterness; the sauce keeps the strong flavors in balance, holding the dish together.

Naccarato has a thing for wild rice. His twice-fried rice salad with napa cabbage, almonds, and water chestnuts is as good as any wild rice salad I've ever had. Or, if you like your wild rice salad cold, you can have it with avocado and roasted red pepper in a cilantro vinaigrette.

Gordon's schedule is tied to the ski season, and each menu features a countdown showing the number of days before it officially opens at Thanksgiving each year. All through the winter the restaurant is busy serving pretty much whatever Naccarato is inspired to do, weather permitting. (Sometimes the snow keeps the white truffles from making it up the mountain.) Along about mid-April, when the snow starts to melt away, so do the visitors, and Gordon takes *his* vacation. He reopens from around the summer solstice to the autumnal equinox to catch the summer crowd trying to escape the heat, and then takes one more break till just before Thanksgiving. What does he do during all that downhill down time?

"We're fortunate enough to live where other people vacation, so we like to vacation where other people live. It's a better balance, I think, to go for two weeks to New York City, and then, when it's starting to get to you, to come back to a place like this. We travel and eat a lot when we're not open. We need to recharge the creative batteries."

I think Gordon is on to something here. And so do other people. In *Food and Wine*'s first-ever list of the ten best new chefs in America, Gordon Naccarato made the cut. But he's not happy just to rest on his sitzmark.

He has now reduced his role at the Aspen restaurant and plans to devote most of his time to a new restaurant he is opening in Los Angeles.

SALADS

Wild Rice Salad with Cilantro Vinaigrette

PASTA AND RISOTTO DISHES

Grilled Vegetable Lasagne

PATRICK O'CONNELL

Inn at Little Washington, Washington, Virginia

The Balance

I'M A city boy. Always have been. Sure, I was born in little Yakima, Washington, but all I remember of small town life was living next to Sno Boy Apple Orchards. After that it was Seattle, Houston, and New York. So it was difficult to go to college at the University of Virginia in bucolic Charlottesville, Virginia—or Hookville, as we called it. The only thing that made it bearable was the frequent jaunts up U.S. 29 to civilized Washington and points northeast.

Route 29 is a winding four-lane blacktop that goes through some of the most beautiful country in America: green rolling wooded hills that turn red and gold in autumn and white with dogwood in spring, backed up by the ever–Blue Ridge Mountains. But I didn't notice that when I was attending the University of Virginia. Then all I wanted was to escape the countryside for Washington, D.C.

Today I am a city dweller looking for some time in the country. Nowadays when I travel the green Shenandoah foothills, I'm headed toward a place about twenty minutes west of Warrenton (about an hour and a half southwest of Washington) on a road to nowhere. This out-of-the-way spot is called the Inn at Little Washington, and travel through these parts without a stop here is unthinkable.

The Inn at Little Washington has garnered almost every dining and hotel award. Supreme Court justices, members of Congress, ambassadors, and media personalities dine there. Craig Claiborne, who celebrated his sixty-fifth birthday there, later proclaimed that the Inn at Little Washington was "of all the inns and restaurants in the United States . . . the greatest."

The remarkable success of the Inn is due to Patrick O'Connell, who learned to cook from a Julia Child cookbook and some Chinese cooking classes. In spite of the fact that the young chef once nearly made his friends ill by combining Chinese sauces with French sauces, he cooked constantly, hosting eight to twelve friends virtually every night. The chef credits one person for helping him on his culinary path.

"I really was leaning more toward becoming a waiter than a chef, so I decided to go to Europe just to leave it all behind for a while and to see the world. Before I left I bought some land out here near the Inn and went to see an astrologer just to get things clear. She was an old one, now dead, who astonished me with specific dates of when my grandparents had arrived in this country from Ireland and events that had taken place in my family life. She said food was everywhere in the chart and asked if that had any meaning for me. I was twenty and told her I'd worked in restaurants since I was fifteen. 'Have you thought about it as a career?' she asked. I told her I'd never considered it but that I was going to France and that I'd heard the food was great there. A year later I got back and went right into cooking again."

O'Connell saw an ad in the *Washington Post* for a position as a chef at the Embassy of Iran. He got the job, and one week later cooked for a party of six hundred. "The country was in quite a state of flux at that time, and so was the royal family. It was a baptism by fire." After stints at the Italian and French embassies, he and partner Reinhold Lynch moved to O'Connell's land in Virginia and started a catering company. With O'Connell's government contacts and developing skill, the company quickly took off. He and Lynch found themselves driving daily to Washington with a car full of food. Then they picked up potted plants and flowers—sometimes even dancing girls—to make an event memorable. And there were the props, the candelabrum, the silver, and, of course, the long drive back to the country near sunup. A few hours later, it would start all over again. Before everything got out of hand, with massive tents and outdoor lighting, O'Connell decided it was time for people to come to him. He wanted to open a place close to his wonderful old farm at the base of the highest mountain in the area, with a river running through the property, and a big organic garden. So he looked close by and located just the right spot only twenty minutes from home: Washington, Virginia, a near-ghost town that had been bypassed by the main traffic artery.

O'Connell and Lynch took a lease option on an old barn of a garage for $200 a month. They chose the building because it could ultimately be developed into a guesthouse, and because there was room for an outdoor garden area, which they considered essential. The little restaurant was a humble place, with rustic tables lit by overhead basket lamps. But the food attracted the attention of D.C. food critics, and it wasn't long before the restaurant took off. The largely self-taught cook was wowing savvy diners and successfully competing with big city chefs with pedigrees as long as (and as French as) a prize-winning poodle's. The diners' only complaint seemed to be having to drive back to the city stuffed and perhaps a little snockered. How nice it would be, many suggested, to be able to bed down right there at the Inn, instead of settling for some roadside motel or persevering over a road that seemed to have acquired many new curves in just the past three hours.

"Whatever I do, I put myself completely into it. For us it was enough just to open the restaurant. A hotel would take a major commitment. So we spent two years touring the greatest hotels in the world, spending three days in each one, doing a kind of a study—measuring bathrooms, looking at laundry rooms and kitchens. We found that when we finally opened the guest rooms it was a tremendous energy drain and adjustment, unbalancing our restaurant for a while until it was totally integrated. Even for ten little overnight rooms, to do them at the same level as the way we were trying to do the restaurant was a major challenge. I see my job as that of a conductor of a large orchestra, wanting everything to come off at the same level of integration, which is very difficult."

When I visited O'Connell he was excited about doing a big party for the London designer responsible for decorating his European-style country inn. There are eight guest rooms plus two suites, and Joyce Conwy-Evans (who had never set foot in Little Washington) did all the designs by telephone and mail, working from plans by Warrenton architect Albert P. Hickley. Each room is done in a different style, from country French to Oriental, with antique furnishings and rich fabrics. In the restaurant the basket lamps are now swathed in silk. The rustic tables are now covered with damask tablecloths, and the walls and ceiling are festooned with silk floral William Morris prints. The floral theme is echoed in a profusion of fresh arrangements from nearby growers encouraged by O'Connell.

Local involvement is important at the Inn, which now contributes enough in taxes to nearly cover the annual budget of the tiny (220 population) once-dying town. A painter who lives nearby can magically change one surface into another, turning metal or plaster into many varieties of wood and marble. And O'Connell has encouraged his neighbors to grow superior fruits and vegetables and to gather wild mushrooms for the restaurant. The local *merkles* (morels), which I remember from my college days, are a special treat. One thing that isn't home grown is most of the staff, which began with O'Connell, Lynch, and a fourteen-year-old kitchen boy and has now grown to over forty. In the beginning, the chef got a long cord for the phone and took reservations while cooking. After some good reviews they hired as many people from the community as possible, but, as O'Connell puts it, "within about two years we ran through the local talent, who were quite sincere, but simply lacked the professional skills." Now O'Connell uses the services of a recruitment specialist who, working exclusively for the Inn, combs the country's cooking and hotel schools looking for just the right personnel for the kitchen, waitstaff, and guest room operations.

What O'Connell is trying to give his guests is a complete experience of enjoyment and, beyond that, a sense of balance and well-being. "The average chef and the average customer these days see only fragmentary parts of the dining experience. They see a plate of food, or ambiance. They don't experience it as a unified whole. So there is a uniqueness about what we are doing. We're going for a complete experience, food being integrated into that. We want the whole to end up having a quality of renewal for the guest. We want to show that life isn't that bad—that it has its highs as well as its lows. So, if we've been successful, people leave with a sense of well-being."

O'Connell was providing that complete experience back in his catering days, creating food and an environment in which to enjoy it. It was then that he noticed what an effect someone who really cared could have. "I used to observe people when I was catering. After they'd been to a great party, they just had a different aura about who they were in relationship to all of mankind. Very uplifting. I think a good restaurant can provide that, seldom does, and should more often."

The chef observed how out of balance many people eat, taking a sugar substitute with their coffee and then loading up on a rich dessert. "People are doing awful things in their own combinations to justify their excesses. I don't think people have learned too much about health at all and are instead being programmed by their doctors without any idea how to balance all the factors. The fact is, though, people don't want to be bloated, overfed, oversauced, and overcreamed. They do have a taste for a dish which leaves them feeling healthy afterwards. This is a start."

Health and balance have always been important to O'Connell. Before his catering business took off he supplied most of the natural foods stores in the District with his home-grown organic produce taken from the rich, loamy Virginia soil, and is now making plans for a larger herb and vegetable garden to supply the restaurant.

"Vegetables are definitely an essential component in my cooking," observes the lanky chef with the easy smile. "I like to sense a plate as something appetizing, nourishing, and healthy just by looking at it. I feel if it doesn't look healthy in some way, then it's not entirely appetizing. In the composition of the plate, we're always looking desperately for color and more and more for crunch. The flavors of meat and poultry are one-dimensional and don't necessarily combine well with each other. Vegetables complete the composition of the plate because they are so incredibly versatile."

Seasonal soups are great starters in the Blue Ridge, where nights are chilly about eight months a year. The cream of sweet red bell pepper soup with Sambuca is a warmer, as is my favorite, a light cream soup made from watercress harvested right next to the nearby Rappahannock River and topped with crème fraîche. As other vegetables come into season this last soup may change, depending on the green being used.

O'Connell's greaseless tempura vegetables are so light they seem to hover a few millimeters above the plate. The assortment ranges from eggplant to zucchini to parsley, all cloaked in a golden batter with an elusive, smoky taste.

Wild rice used to be pretty uncommon. I remember having to mail order it from a Native American in Minnesota who hand parched it. But now that it's grown in California, you see it in many restaurants, especially during chillier months. Yet I know of no preparation as savory as O'Connell's. His wild rice with roasted pecan halves has a sweet, maplely undercurrent.

Whether it's a dish of dry-cooked wild mushrooms with a bacony flavor or a two-toned vegetable mousse like creamy spinach atop gingered carrot, O'Connell has a deft hand with vegetable preparations. In the bread category, you could make a meal of his crusty rye coated in crunchy salt (which I did one day, when the chef insisted I take a loaf because it was just out of the oven and it was still a few hours before dinner).

Balance is a word that crops up frequently in O'Connell's conversation. "I usually feel slightly out of balance. That is the thrust and the momentum, and I think it's a price I'm paying as a part of this culture and as a part of being a developing person.

"No day resembles the last; there's no predictability. I can't even envision where I'll be five years from now. There's a tremendous surge of energy with that, and I happily will pay that price for the thrill of living life to the fullest. I try to create a framework of balance, a mood of balance, an aura in the restaurant of wonderful, composed balance. That offsets how I may be feeling at the time."

Though O'Connell could easily raise prices, his prix fixe menu keeps them reasonable. Though he could charge double for his almost impossible-to-get rooms, he keeps the tariff in line with your average Hyatt. His values are not based on money and prestige.

"What I hope to do, incrementally, is give people a physical and mental high when they eat. We're aimed in the direction of a natural orientation, ultimately becoming a sort of spa without people knowing it. I like the trust that people put in me to give them pure, wonderful, healthy food. And I won't betray that trust. You know, the chef's hat was originally the priest's hat, from the period when priests were the only people authorized to handle tribal food. It was a sacred rite, a very important responsibility. It's a shame that has disappeared. Today's chefs are promoters standing behind Holiday Inn buffet tables. I think it would be nice to get chefs to realize that they not only had the arduous task of preparing food, but were entrusted with nurturing people's well-being.

"We won't let the tragedy of success affect us. If you lose control of yourself, if you lose your sense of who you are, the success syndrome, that can take the ability to please away from you. Particularly if the media controls you, or if you're manipulated in some way. A friend said we located here very carefully, on high ground, with plenty of space between us and them. That's the way we'll keep it."

BRADLEY OGDEN

Lark Creek Inn, Larkspur, California

Thank God He's a Country Boy

DURING HIS last four weeks at Campton Place, I asked Bradley Ogden if he didn't feel a little like a lame duck President. His move to the Lark Creek Inn just across the Golden Gate Bridge from San Francisco was public knowledge, yet he still had to fulfill his obligations until the new place opened.

"I actually feel more like a retiring Cabinet member being wooed by any number of companies," he replied.

"Why are they wooing you?" I asked, densely.

Ogden was patient with me. "They're calling me all day long to get my business for the new restaurant."

And who wouldn't want Bradley Ogden's business? Not just for the money, but to know that your product was going to be used by one of the most talented chefs in America.

Now that the Lark Creek Inn has opened, those purveyors whose products Ogden uses are fortunate indeed, for he's going to need a steady supply. With the kind of crowds Ogden has been attracting, there is very little room at the Inn.

But the chief supplier to the Lark Creek Inn is Ogden himself. From his five-acre farm in the Napa Valley comes a daily supply of just-picked, organically grown fruits and vegetables raised with the chef's expert touch. The produce is superior because Ogden knows what he's doing. Back in the alternative-lifestyle days of the seventies, Ogden was a vegetarian who studied nutrition while working at a natural-foods store and raising organic produce for a living. Today vegetables remain important to him.

"When we go to create menus, we look at all the food groups. The vegetables are as important as anything else on the plate. Without consciously choosing what vegetables you're going to use, you end up with a plate that doesn't make sense. The vegetables complement and enhance the dish, and, let's face it, they're just an essential part of good cooking and well-balanced menus."

Ogden has chosen well-heeled Marin County as the location for his first restaurant: the perfect environment for the town-and-country food served at the Lark Creek Inn.

"I want to create casual, American country food. We have wood-burning ovens, outdoor dining, stone floors, an enclosed porch area, and French doors on one side. It's a very casual environment, moderately priced. The menu is market-oriented, and it's going to change a lot—maybe daily. The philosophy and quality of ingredients is the same as at Campton, but the presentation is different, more country."

Campton Place, with coffered ceilings, peach-fabric-colored walls, and marble everywhere from the arched windows to the entry, is as stately a room as you will find in San Francisco. Yet despite the formal elegance of those surroundings, while he was there Ogden cooked straightforward, robust dishes drawn from America's heartland.

That kind of food seems better suited to the Inn, a stately yellow 1888 Victorian manor surrounded by towering redwoods. You know the food is going to be honest and hearty the moment the waiter brings the bread basket, which holds an assortment of breads like sourdough whole-wheat and rye, poppy seed twist, lemon-tarragon biscuits, and dill rolls. Another night it might be buckwheat dill muffins, Boston Brown bread, and yeast rolls made with sweet potatoes.

There are rarely more than eight or nine appetizers or entrees, but Ogden puts such care into the preparation of those few dishes that you can't go wrong. Soon after the Inn opened I relished his herbed flatbread with yellow tomatoes, eggplant, dry-aged Jack cheese, and olive pesto. The soft, chewy pizza was just the right vehicle for the blend of assertive yet complementary flavors. A summer vegetable stew—just-cooked peas, white corn, carrots, zucchini, green beans, leeks, golden beets, fennel, and squashes in a tomato broth with broad homemade black-pepper noodles—quickly took the chill off of the Northern California night.

Another warming starter was the red garlic and potato soup with turnip greens and croutons. The thick green chowder with a smoky undercurrent was rich and comforting.

The freshest, lightest beginning to any meal has to be Ogden's home-grown tomato salad with goat cheese mozzarella and scallion and basil dressing. The fresh yellow and red tomatoes are as sweet as apples and contrast nicely with the local cheese and Ogden's masterful dressing. For an entree, what could be better than an assortment of grilled Napa County vegetables such as white corn, leeks, wild mushrooms, green beans, and broccoli?

Lunch under the giant skylight at the Lark Creek Inn is a bright, sunny affair. A berry and melon salad with fresh potted cheese is a cool California refresher. Another stunning starter is the roasted and chilled artichoke with garlic mayonnaise. For main courses the chef offers ravioli filled with Swiss chard, or a plate of Blue Lake beans with corn in a saffron vinaigrette.

Though Ogden's food tends to get a bit fancier at night, he promises to keep it moderately priced. The quality of ingredients and portion size are the same as at Campton Place, but the prices are one half what they were in San Francisco, with most appetizers around $5 to $6 and no main course over $18.50.

After training at the Culinary Institute of America (where he was voted Most Likely to Succeed) and years of restaurant experience, Ogden is clearly in his element in the well-appointed, rambling old house on the creek in Larkspur. It's the perfect setting for the Midwestern-raised Ogden, who's really a country boy at heart. He's even doffed the stiff white toque he wore at Campton Place.

The Lark Creek Inn may not be the last stop for Bradley Ogden. He would like someday to have a real inn-restaurant combination à la Patrick O'Connell. But until then he'll concentrate on cooking in the redwoods according to a standard that he feels is essential to the success of any chef:

"The difference between a rather average cook and a chef is that the chef is never really satisfied with what he is serving. He is constantly striving to achieve the high expectation he has set for himself. He is seeking to develop his palate and to enhance the skills of his palate through cooking, travel, and just being open. By keeping yourself open to what's new out there or who's doing something a little bit better, you strive for perfection. I'm always looking to improve on what I do."

JEAN-LOUIS PALLADIN

Jean-Louis, Washington, D.C.

A Generous Guy

I'M IN Disneyland! There's no mistaking it. Look at the food on the plate in front of me. There are the ears, the big eyes—the form is perfect. If that's not Mickey Mouse's face I'll eat it. Hell, I'll eat it anyway. It's not every day you get a salad of crisp, pepper-encrusted shiitake and girolle mushrooms with bitter greens and two pools of cèpe vinaigrette. The chef is doing this intentionally, I say. "A coincidence," you say. "His imagination," you whisper to a friend as you shake your head reprovingly. (Did I just hear a chorus of "tsks?") Okay, okay. Then what was that dish that looked like an ocean-going mine a couple of courses back? You know, the baby green pumpkin filled with sweet potato mousseline into which the chef had stuck asparagus spears, detonator style. And what about the house-made purple potato chips? I'm not making these things up. Jean-Louis Palladin is. For in truth I am in the Magic Kingdom of Food at the chef's namesake restaurant in Washington, D.C.

Whether customers will continue to flock to Jean-Louis Palladin's Watergate Restaurant as they have to Walt Disney's theme parks, only time will tell. Certainly the forty-two-seat restaurant has a somewhat smaller capacity. And with only one location and high prices, it is fair to say that more people visit Disneyland/World in one hour than will eat at the Watergate restaurant in one year. But having been to both, I can honestly say that three hours at Jean-Louis is as much fun as six at Disneyland, where the food isn't nearly as good and the lines are a lot longer.

It was probably a risk for Walt to open Disneyland. Though he had a great reputation with cartoons and nature films, moving from one venue to another can be dangerous. Yet consider the risk of Jean-Louis Palladin, who at thirty-three left his successful Gascony restaurant, La Table des Cordeliers, to come to America. He left at a time when his reputation was peaking. He had just received his second star from the Michelin guide (the youngest chef ever to do so) when he left it all to come to a country he had never visited, whose language he did not speak.

But it was time for a change. And America was ready for a change in French cooking. "At first I prepared dishes I knew Americans would understand. But my job, when I arrived, was to make people aware of what we needed. We needed a lot of vegetables." Jean-Louis, shaking his curly endive mane, lamented the days when it was virtually impossible to get leeks or turnips in America. Now he has a northern Virginia farmer who grows to his specifications. "My first worry after coming here was where I was going to get vegetables. It was difficult in the first years when I arrived, because I needed to import everything. Now my importation is only about 5 percent—truffles, olive and peanut oils, and sea salt. Now it's easy to work."

Palladin started as a kitchen helper at the age of twelve in his hometown of Condom, France. After formal training at a hotel school in Toulouse and stints at Monte Carlo's famed Hôtel de Paris and the Plaza Athénée in Paris, he opened La Table in 1968. Now many think he's the finest chef in America.

Palladin feels the only reason he is able to create the dishes he does is because of his years and years of training in classical French cuisine. Once these techniques have been established, the chef feels, you can have the freedom to innovate. Take, for example, a palate-cleansing coconut sorbet I had recently at Jean-Louis. You won't find it in classical French cookbooks, but unless you've had years of training preparing simpler sorbets, you might not be able to make it smooth as mousse, or it might not occur to you to weave in tender shreds of chewy, toasted coconut for contrapuntal texture and a second, longer-lasting flavor.

Palladin's staff-customer ratio is also part of his restaurant philosophy. "We have twelve in the kitchen and twelve in the dining room for our forty-two-seat restaurant. Because what we're doing here is not only about food, it's an evening we give to the people. The customer is all-important to me. It's a gift we can give to him, to be nice to him, to give him what he wants.

"I don't like to make this a restaurant where you are stuffy. The people are coming to have fun, eat good food, drink good wine. So give it to them, be nice to them. It all comes back."

This is no pop psychology for Jean-Louis. He knows that when you're charging $38 to $90 for a prix fixe dinner ($150 for the truffle menu available November through March) that the customer should be treated like a king. The small size of the restaurant allows the chef to cook for each diner in a personal way, and he often sends out extra treats or a bottle of wine with his compliments. At these prices, you might think, he can afford to say and do nice things. But if the restaurant charged enough to make a profit, most people couldn't afford to go there at all.

An article in *The Wall Street Journal* showed how Jean-Louis' cream-of-the-crop ingredients add up so that by the time the flowers, china, and administrative costs are all factored in, the restaurant only breaks even. The hotel's general manager calls the restaurant a "loss leader," whose reputation serves to draw clients to stay at the parent Watergate Hotel.

It was Watergate owner Nicholas Salgo who convinced Palladin to leave his Michelin stars and come to cook at the hotel. He made Jean-Louis a generous offer and named the restaurant after him. And Palladin has taken it seriously. The name "Jean-Louis" is printed on each plate—the chef's own signature. With his name on the line each and every time a dish emerges from the kitchen, Jean-Louis is going to make damn sure he doesn't disappoint.

My dinner at Jean-Louis began with an appetizer of steamed baby cauliflower (which Palladin calls the "forgotten vegetable") in a pale green sauce of pureed broccoli and Parmesan. Next came the green baby pumpkin filled with sweet-potato mousseline, followed by the mushroom salad, which really was a sight to behold. The two "ears" were pools of tawny cèpe vinaigrette striated with thin lines of white and green vegetable purees. The "forehead" was a mound of crispy, peppery mushrooms, and the "face" was a round bunch of greens dressed with more of the vinaigrette.

Palladin kept the sight gags going with a "vegeburger" of grilled tomatoes and onions, whose pink color, meaty texture, and charbroiled taste reminded me a little of a Big Mac. Of course, McDonald's doesn't serve their burgers on a bun of sautéed straw potatoes nor do they use a sauce of fresh fava beans, but you get the point.

Palladin is obviously in love with vegetables. "You can do so many things with them. With meat and fish, okay, you have variety, but it can be boring. Vegetables are never boring. I love to do a beautiful plate with vegetables, different colors, different designs—fantastic. Vegetables are hard work, so you need to take care about what you are doing. In a few seconds you can lose the color. Then it's all over."

The chef's love of greens came from his boyhood. His Italian father and Spanish mother had a big field near the town of Condom, where they grew "all the vegetables you could ask for. My mother didn't have to go to the market, because the ground was very prolific. Everything grew like crazy. That's why I love vegetables.

Besides, unlike meat or fish, vegetables make your mind work a lot of different ways. And I love that challenge!"

That is obvious to anyone who has tasted his eggplant soup or his cream of fresh onions: dishes that flow from the chef's classical training combined with his uncanny sense for what will work. Palladin calls it "instinctive cuisine," which, in his own words, is his "unique sensibility for food combining southwestern French roots, classical French training, and an insatiable desire to experiment."

In the fall of 1986, everything came full circle for Jean-Louis. The two-star chef who left the big leagues to work in the minors had achieved reknown in France for what he was doing in America. His moment of triumph came when Palladin was asked to return to Paris to cook a dinner for his French confrères. He boldly chose to use all American ingredients. It was judged a great success by the most exacting group of diners he has ever cooked for.

RELISHES AND INTERMEZZOS
Fresh Vegetable Granites

. . . .

CINDY PAWLCYN

Mustards Grill, Fog City Diner, Tra Vigne, Rio Grill, BIX, Roti, San Francisco Bay Area

"Nobody wants garbage."

DETERMINED BY age thirteen to be a great chef, Cindy Pawlcyn started working part-time at a kitchenware and catering company near her home in Minneapolis. She took cooking classes at school while working nights at the store. From there it was on to a local trade school, and then the University of Wisconsin's Hotel Restaurant School, where she worked her way through college by—what else—cooking at a local restaurant.

To master classical technique, Cindy studied at La Varenne and the Cordon Bleu in Paris. And by the time she finished Ken Hom's Hong Kong cooking school, she could cook just about anything.

She got her first job at the Pump Room in Chicago, where she caught the attention of Bill Upson and Bill Higgins of the Second City's legendary Lettuce Entertain You restaurant chain. They saw the food-loving Northern California market as prime territory for a series of restaurants along the lines of the mini-empire of theme establishments they had helped build in Chicago. And they saw Cindy Pawlcyn as the chef who could make it all happen.

Cindy opened the first link in the chain, Mustards Grill, in the Napa Valley in 1983. There she set in place the modus operandi for what would be a line of wildly successful Bay Area restaurants. The food was made from the freshest ingredients possible, the produce often from her own garden. The bread was home baked, the food wasn't expensive, a person could compose a great meal from a variety of categories, and meal times weren't fixed—you could come in just about any time and get fed.

The concept was long overdue. People were tired of the typically stiff restaurant format with its typically stiff tariff. Most restaurants tried to bend their customers to fit the restaurants' schedule and way of doing things. At Mustards, it was the other way around.

Around four o'clock there one recent afternoon, the blackboard menu over the open kitchen window offered a wide array of choices, even for a vegetarian. I started with soup, a cream of tomato with dill. Then I went on to a red and gold beet salad with celery root and bitter greens in a humble let's-just-pick-up-the-flavors-of-the-salad vinaigrette. The salad shimmered with color and tasted as if the ingredients had just been taken from the rich Napa Valley soil.

Cindy Pawlcyn's cooking philosophy is: Grow it organically without death-dealing poisons; get it out of the ground and onto the table as quickly as possible; and don't mess it up with a lot of other flavors.

"Nobody wants to eat garbage," the innocent-looking, yet worldly wise chef later told me on the restaurant patio overlooking its 4½-acre organic garden. "That's why we grow as much of our produce and herbs as possible and bake a lot of our own bread. I feel that you should be dealing with things that are in season, so here with the land and the greenhouses I extend my season as long as possible. And with produce so fresh, my cooking is straightforward and more flavorful. Besides, it doesn't make good business sense to buy all this stuff out of season. It's just too expensive and too taxing to try to bring it in and hope it's fresh and flavorful, even if it looks good. So I'd rather can or freeze from my own garden so I have things to play with. I make homemade vinegars, too. It's fun."

At lunch I had been served a little dish of Pawlcyn's own pickles with a mild garlic bite. Next I ordered a cornmeal crepe with crème fraîche, and a colorful slaw of napa cabbage (what else?), red cabbage, and carrots in a light, unsweetened caraway dressing. Somewhere along the line I got a hunk of house-baked rosemary bread that had been grilled over apple wood.

A few nights later the kitchen offered to prepare me a meal of their own design, and I gladly agreed. But I must admit that when the waitress brought me a plate of radicchio, marinated and then grilled over madrone and oak, in a sauce of blue cheese, prunes, and Oriental sweet-spiced pecans, I didn't know what to think.

Amazingly, it all worked. The cheese flowed sensuously, the prunes were a perfect sweet and chewy balance, the radicchio was nicely bitter, and the pecans were a crunchy, spicy surprise. It was probably the single most inventive dish I can remember having, and one of the most satisfying.

A little bowl of creamy mashed potatoes was next, then a few slices of vine-ripened tomatoes marinated in ancho chili powder and tangerine juice.

The kitchen pressed the grill into action once more for whole shiitake mushrooms, whose woodsy–smoky flavor was nicely enhanced by a delightful mandarin orange–cilantro butter.

I was getting a little full by now, though the portions had all been small. But the kitchen encouraged me to try just *one* more dish: a poblano chili (grilled, of course) stuffed with Sonoma goat cheese and almonds, served with Brussels sprouts, Belgium endive, and crunchy, juicy snow peas. It was a delight.

There seemed to be such an emphasis on vegetables at Mustards that I asked Cindy about their role in her cooking. "Vegetables are very important because you can't enjoy an entree without them. Even if you do a great fish, if the vegetables are lousy, the whole dish is no good. I do a lot of vegetable-stock soups, and I use a great deal of vegetables in the meat stocks as well. And I use vegetables like radicchio to provide bitter contrasts. Or I'll use tubers and aromatics like celery root and fennel to round out other dishes. Vegetables do quite a lot for my cuisine."

It's no wonder Mustards is packed from 11:30 in the morning straight through till closing. So Pawlcyn and company were inspired to create the same type of place in Carmel. Going into its sixth year of operation now, the Rio Grill has captured the hearts of residents and tourists alike with its Southwestern-inspired dishes. The Carmel area has several small local farmers who provide the flavorful and fresh fruits and vegetables for the simple and inventive dishes on the Grill's blackboard menu, which, like Mustard's, changes daily.

But the pinnacle of Cindy Pawlcyn's achievement in the grill genre has to be Fog City Diner in San Francisco. It took her nine months to create a menu to match the sleek, stainless steel railroad car–style building. You can sit at sophisticated booths right out of a William Powell movie or at counter stools out of a William Bendix TV show; the choice is yours.

The menu is divided up into breads, bowls, small plates, sandwiches, salads, large plates, and sides and condiments. And the most expensive item on the menu is under $15 (unless you count the "World Famous 'Don't Worry' wristwatch" for $29.95).

You could almost make a meal of breads alone by ordering the buttery leek and basil loaf and the Cheddar cheese and herb bun. You can also order long jalapeño corn sticks, sweet and chockfull of whole kernels, with a warming bite at the finish.

If you opt for straight diner fare there are onion rings, French fries, and malts aplenty. Or you can start with the "unintimidating mixed greens with vinaigrette," or one of the soups.

I tend to choose the dishes that Cindy calls "small plates," but which, if ordered with the bread course, could constitute a meal. My favorite is the stuffed grilled pasilla chili with avocado salsa. The triangular jade-black chili comes nicely charred, filled with a mixture of asiago, Jarlsberg, and Monterey jack cheeses. The chunky avocado salsa provides a warming bite and helps cut the richness of the cheese. My other favorite is the quesadilla with hazelnuts and ancho chili salsa.

There are so many choices on the menu that it's hard to decide what to eat. Three solutions: (1): Go with a lot of people and have everyone order something different; (2) Go a lot of times; (3) Eschew the sexy booths on the side and the triangular tables in the peristyle end of the diner (from which, on a clear night, you can see the top of the Bay Bridge), and sit at the counter at the business end of the diner. There, smack up against the kitchen, you can smell everything even if you can't taste it.

I asked Cindy to describe her cooking, which seems to cut across all international boundaries. "It's kind of an amalgamation. I can get the products of other countries, but being in America, the foods taste different, and Americans aren't used to the way some other countries' foods taste. And since my sense is different from, say, a Thai cook's, I don't make my Thai dishes as hot as they would. I might personally like the heat, but I have to sell the dish, not eat it. One thing, though, I don't mix styles of different countries. I'm true to the technique of the country where the dish originates."

Speaking of foods inspired by other countries, recently Cindy struck up a new partnership and enlisted one of Florida's finest chefs, Michael Chiarello, to open a "Tuscalifornia" restaurant called Tra Vigne in her adopted hometown of St. Helena. This is the first departure from her grill-style restaurants in appearance, but the food is just as accessible as at her less-formal establishments.

Tra Vigne is a huge stone building with fifteen-foot ceilings. Pawlcyn and her partner put in large windows, refinished the walls and painted them in subtle olive shades, and installed similarly colored chairs and banquettes to make the surroundings resemble a Tuscan villa.

Olives figure prominently at Tra Vigne, and not just as a design element. Each table is issued a plate of complimentary local olives, house-baked bread, and fruity olive oil. You are encouraged, as in Tuscany, to dip the anise-studded bread in the oil.

The menu is similar to those of the grills, with selections available from the oven (i.e., pizzas and calzones), antipasti, pastas, and *piatti del giorno*. No item is over $14.

One evening my wife and I began by splitting the *antipasto misto*, which consisted of mint-and-basil-marinated red and yellow peppers, potato salad Abruzzi, home-cured olives, fresh teleme cheese, grilled eggplant, white beans with tender strings of onion, and the local capers—almost as big as kidney beans and bursting with vinegary goodness.

Chiarello and Pawlcyn are proud of their local ingredients, including their breads and produce. They bake and grow their own because they can control the quality and ensure maximum freshness. This philosophy carries over into the thin, chewy-crusted pizza topped with rich, dark red tomato sauce, soft eggplant, olives, corn, and wild mushrooms. Other pizza selections are porcini with scallions and fontina, as well as a pizza *sfogliatina*, featuring a cracker-thin crust with caramelized onions and Gorgonzola.

An unusual appetizer is called *spiedini al forno*, bite-size pieces of oven-roasted focaccia bread, roasted red and yellow peppers, teleme cheese, and red onion—alternating on a skewer shish-kabob style, grilled, then drizzled with balsamic vinaigrette.

Salads from the restaurant's garden include the *insalata del nordico*, featuring arugula, escarole, fennel, pecorino cheese, and white beans.

The pasta dishes are as finely wrought as the appetizers and pizzas. Whether one opts for the pasta ribbons with grilled artichoke, red mustard, and chervil, or the pasta tubes with English peas in a black pepper cream sauce, all the house-made shapes come out chewy and substantial. My favorite this night was the *ravioli di magro al burro*, half-moon-shaped pockets filled with homemade ricotta, spinach, and red chard in sage butter. Like most of Pawlcyn's restaurants, Tra Vigne is open from lunch through closing.

The exception is BIX, Pawlcyn's recent move away from the grill concept toward more formal restaurants evocative of the Pump Room, where she first cooked professionally. But even here, the food is relatively inexpensive, cooked to order, and available in small courses as well as entrees. The former assay office, then supper club on San Francisco's Gold Street has thirty-foot ceilings, hanging lamps, dark wood-paneled walls, and Corinthian columns, and is the perfect setting for Pawlcyn's rendition of between-the-wars American food, featuring dishes such as an updated Waldorf salad (with Oriental candied walnuts and blue goat cheese), hearts of romaine vinaigrette, potatoes O'Brien, and creamed spinach with a golden bread crumb crust.

You'd think that with five restaurants to manage Cindy would have her hands full, but she has just opened Roti at the Hotel Griffon in San Francisco. The clubby, comfortable room furnished in warm wood tones is the perfect setting for roasted and grilled dishes such as plump leeks with potato shreds and chervil, smoky eggplant with goat cheese and artichoke hearts, and a ragout of eggplant and tomato in beurre blanc. My favorite dish is potatoes 7th arrondisement: new potatoes mashed in the pan and cooked in butter and garlic.

Pawlcyn seems to thrive on the nomadic life, traveling from restaurant to restaurant, checking on the operation here and there, working with the mostly self-sufficient staffs. She has great front and back office support from her partners and can concentrate mostly on creating and perfecting the food for her phenomenally successful and trend-setting restaurant empire.

SOUPS

Cold Beet and Buttermilk Soup

SIDE DISHES

Fennel with Tomato

GEORGES PERRIER

Le Bec-Fin, Philadelphia, Pennsylvania

Good Taste

MANY PEOPLE asked me during the course of the writing of this book where I had my best meal. That's an impossible question to answer. The food was great at every restaurant I chose to include or I wouldn't have written about it and the chef who prepared it. Accordingly, you won't find much in the way of criticism in this work. I'm not a restaurant critic: I'm a celebrant of good food and those who prepare it.

There were many "best meals" and "best restaurants" during the three-year period I researched this book, but a meal prepared by Georges Perrier one night in Philadelphia was one of the finest classical French dinners I experienced. There were no flaws or lapses, just a continuous stream of superlative dishes, served in a lovely room by an attentive and knowledgeable staff.

All this on a night Georges Perrier was half-dead with the flu. That's true greatness.

It figures, though. Perrier is known as an exacting boss. He misses nothing. One of his staff told me conspiratorially, "He has a thousand eyes."

It seems as though he has a thousand staff as well. For a restaurant that serves fifty-five, he has a staff of forty-two. They're everywhere when you need them, and yet they're nowhere when intimacy or solitary reflection is desired.

The magnificent dining room is oval, with towering, Scalamadre-damask-covered walls, three huge leaded crystal chandeliers, and seventeenth-century portraits of Louis XIV and Catherine the Great by Rigaud. Balloon-back chestnut chairs covered in tapestry, silver candelabra and flatware, Limoges china, Cristal d'Arc stemware, and discreet flower arrangements create a dining environment from another, less harried time.

Each plate Perrier designs is as well-thought-out as the design of the room. This was evident from my first dish, a salad of mâche, artichoke hearts, *haricots verts*, and black truffles in a mustard vinaigrette. The mâche sat in a loose clump in the center of the plate, with alternating green beans and artichoke slices fanning out like spokes. A judicious coating of vinaigrette brought the salad to life. And truffles, little *bâtons* of gorgeous, fragrant black truffles, were everywhere.

The second course of my prix fixe dinner was tender ravioli, one green, one orange, and one yellow, stuffed with sautéed chanterelles and served in a tawny thyme and chive butter sauce.

Perrier is a master saucier who has trained with some of France's best: Jacques Picard at L'Oustau de Baumanière and Guy Thyvard at La Pyramide. His skill was apparent in the wondrous *feuilletée* of green and white asparagus in a lemon *neige*. A disc of feathery puff pastry was the hub of a wheel whose spokes were alternating green and baby white asparagus. But the sauce! The sauce of concentrated vegetables, butter, and lemon was extraordinary.

More truffles and more *feuilletage* were paired in a puff pastry turnover filled with morels, spinach, and carrots. It was served in a cream sauce with slivers of truffle echoing in a lower octave the deep notes of the morels.

A stunning cheese board offered a perfectly ripe Dolmen Bougon; a dry, sharp goat cheese; a semisoft Tourre; a dry, two-year-old Gouda; and a Saint André. The cheeses were served with Perrier's own whole-wheat fruit-and-nut bread, studded with nubbins of dried apricots, raisins, figs, and pistachios.

Dinner at Le Bec-Fin is costly, although the food is worth the expense. A satisfying alternative is the $27 prix fixe lunch, which one day featured puff pastry with white asparagus in hollandaise sauce, a light lentil soup with celery, mushroom ravioli with herb sauce, a puff pastry filled with goat cheese, and fresh pasta with vegetables and wild mushrooms.

Why the emphasis on vegetables for Perrier?

"They're the most important thing in cooking," he told me. "You can't do a thing without them. There's no color, no presentation, no care without vegetables. I've been interested in them all my life. Now, finally, the last few years, people are starting to care about them. There are small farmers in my area who bring me great produce, like the baby white asparagus. Now people are beginning to care."

Consideration is one of the reasons Perrier came to the United States in the first place. Though he is grateful for his training in the French apprenticeship system, he rails against the "industrial" approach there.

"There's no respect for the people in the kitchen there, though it's better now thanks to the reforms of Paul Bocuse. Young people are treated like something on an assembly line. I came here in 1967 to do things a little differently. I hope it's working," he laughed.

Perrier has a reputation for being a stickler for detail. His drive for quality is relentless, but most of his crew have been with him for years and speak reverentially about him with a combination of fear and love. They understand what Perrier is trying to do, and they want to be a part of it. That's why his customers are so loyal. Everything at Le Bec-Fin is in good taste.

SOUPS

Cream of Celery Root Soup

SIDE DISHES

Pommes Paillasson

MAIN COURSES

Gratin des Capucins

RICHARD PERRY

Richard Perry Restaurant, St. Louis, Missouri

"Nobody told me I couldn't do it."

"FACT OF the matter is," the stately Richard Perry told me in a matter-of-fact, Midwestern way, "I'm basically untrained. I think that ignorance on my part has led me to do things I may not have done had I had some sort of traditional background. We've made a lot of mistakes, but we've broken a lot of new ground, too. We do things that nobody else does just because we want to. Nobody told me I couldn't do it."

Right from the start Perry did things a little differently. He loved the food of his Illinois farm boyhood but couldn't find it in any restaurant. He enjoyed the dining experience only a fine restaurant can provide, and though he had no previous experience, he decided to create such a place, which would serve the hearty food he remembered so fondly. Lacking any formal training, he went to the Chicago Historical Society to research turn-of-the-century recipes from famous St. Louis eateries like Tony Faust and the Planter's Hotel. And he sniffed around St. Louis to learn more about the substantial food of its boardinghouses—food, he would find, like the kind he had enjoyed as a youth.

Drawing on all this information, in 1971 he opened the Jefferson Street Boardinghouse, in an old red brick building that used to serve that function, and faithfully reproduced the dishes of the Midwest from around the 1900s. Perry offered one prix fixe dinner nightly, just as you might have found at a boardinghouse of the era. Everything was made from scratch, including dressings, sauces, vegetable butters, and relishes. And instead of serving tasteless white rolls from a commercial outfit, he baked his own whole-grain breads and biscuits, based on recipes of a bygone era. Jefferson Street Boardinghouse created an instant sensation. People flocked to Perry's restaurant for the honest, hearty tastes veteran restaurateurs hadn't been interested in providing.

But as often happens with instant sensations, business eventually fell off. The lack-of-experience sword cut both ways. People never knew what to expect when they came to eat the dealer's-choice menu, and, with no options, they were wary of coming.

As Yogi Bhajan, my spiritual teacher, once told me, "You can either learn from a teacher or from the hands of time. The lesson is the same; the second way just hurts more and takes longer." Time was his teacher, and Perry did learn. He renamed the restaurant after himself and abandoned the boardinghouse concept altogether, adopting a regular menu that changed weekly, save for a few popular items. It worked.

Fortunately, St. Louis isn't new York, where the fickle dining public is unforgiving and one mistake can take a restaurateur over the precipice. When the conservative St. Louis dining public heard what Richard Perry was doing, they came back. What they found was a chef who, with a few years of restaurant experience under his belt, was confident enough in his own growing skills as a chef to give up reproducing earlier dishes in favor of lighter renditions and new creations based on the robust flavors of America's heartland.

Now that Richard Perry Restaurant is nearly twenty years old, it has moved to a new location in a beautifully restored, landmark St. Louis hotel. The growing pains are over. The restaurant is a St. Louis institution.

"The basic thinking behind the restaurant hasn't changed a lot since we opened," commented Perry. "We got started by thinking restaurant cooking had neglected American food. And it really had back in the early seventies. There was no restaurant I knew of where you could get good American home cooking in a fine restaurant setting. That's still what we do. We present the produce from our area; we cook according to the tastes dictated by old recipes. That's not to say we recreate old recipes, because we don't do that. We did that for quite a long time. We recreated turn of the century recipes, but we've kind of moved on to adapt our new thinking about cooking to the flavors of our area."

Those flavors include white corn from the Ozarks, which goes into Perry's corn chowder along with earthy ingredients like parsley root and celery. Many of the tastes Perry presents are re-creations of foods he enjoyed from his childhood.

"I can remember as a kid sitting down to a meal, we always had a platter of at least a half a dozen raw vegetables, carrots, radishes, celery, green onions, and so forth. I think it's an extension of the Pennsylvania Dutch seven sweets, seven sours concept. We'd always have pickles, relishes, and olives on our table." So it's not unusual to find Perry's juicy cucumbers marinated in red wine vinegar, sugar, onions, and olive oil on the table to pique your appetite.

The Midwest is farm country. A cuisine faithful to the region would have to feature lots of farm-fresh produce. But what can a Midwest restaurant that wants to feature local produce do in the frigid winter months?

"We're concerned about using the produce of our area," Perry told me. "We watch the seasons and, in winter, instead of bringing in a lot of produce from California, which by the way, isn't the sole producer of all the fine vegetables grown in the U.S., we'll use winter vegetables and fruits. Plus we'll serve a lot of summery things we preserved when they were in season. We have so many wonderful things available to us. The only time the availability of fresh produce really gets to be a problem is in March, when the winter root vegetables aren't really plentiful and the spring things haven't arrived yet. But midwinter is just a wonderful time for produce."

Wonderful indeed. When cold weather settles over St. Louis, Perry features dishes like green risotto cooked in a hearty root-vegetable stock of turnips, carrots, onions, and leeks. This time of year, frigid Midwesterners find that Perry's wild-mushroom chili's just the thing to throw off winter's chill. Stewed with kidney beans, garlic, onion, and chili powder, the dish fortifies you for the cold like antifreeze does a car. Equally sustaining are Perry's sautéed butternut squash with rosemary and his creamy celery root and apple puree.

Regardless of the season, Perry relies heavily on locally grown fruits and vegetables. In doing so, he has encouraged a burgeoning cottage industry that has benefitted other restaurateurs as well. "We've always presented vegetables as an integral part of the meal. Main courses are always served with at least three different fresh vegetables, and we feature salads as a regular part of meals, too. I think we're really concerned about the production of good food. We try to encourage local people to grow things for us. We've always had the philosophy that we really don't ask the price. I think it's necessary for us to commit to taking production regardless of price, so as to encourage things to be grown. We have many, many incidences where we have been the sole outlet for a farmer. Then they go on to find other people who are interested and develop a customer base. In general, I think we have taken the lead over the last fifteen to twenty years in developing new products and reviving some old, forgotten ones."

Fresh fruits also are important to Perry's cuisine, both on his dinner menu, where you'll find parsnip-apple pancakes with such accompaniments as pear butter fragrant with orange and nutmeg, and at brunch, where he offers at least six different fresh-fruit courses on the sit-down menu. Perry was serving fresh fruit years ago, at a time when the only fruit you'd find at most St. Louis restaurants was canned fruit cocktail in heavy syrup with that insult to nature, maraschino cherries. A recent brunch offered a green and red old-fashioned apple compote in white wine; an open-faced pear pie; honey-glazed ruby-red grapefruit; crème brulée with Concord grapes; sliced oranges with Grand Marnier; and fresh orange and grapefruit juices, the last of which came swirled with grenadine.

Over the years Perry has moved from the heavier fare of his youth to the lighter dishes favored by today's diner. At lunch he offers a completely vegetarian three-course meal that's extremely popular. In addition, there are creamless soups like tomato bouillon and clear celery, as well as cold spinach-fettuccine salad with Pommery mustard dressing and seasonal vegetables.

Faithful to his location in America's breadbasket, Perry bakes his own bread. Though the new restaurant doesn't have baking facilities, he kept the old site to serve as a commissary, where such treats as parsley biscuits and cracked-wheat bread are baked fresh every day. At Sunday brunch, you may never get past the heaping basket of cinammon buns, fruit muffins, streusel cake, and scones with Perry's own marmalade and apple butter.

But the new location in St. Louis's venerable Majestic Hotel has something the old Richard Perry Restaurant didn't have: a central downtown location where residents and tourists alike can more easily enjoy Perry's food. The L-shaped dining room has seventeen-foot ceilings, Oriental carpets, Victorian lamps, comfortable green velvet booths, and rich cherry wood paneling. The sheer lacy curtains are pulled back to open the room up to the street.

Richard Perry has overcome the odds in his desire to present regional food of unstinting quality and integrity. He is living proof that taking chances and trying new approaches can pay off.

As Perry good-naturedly puts it: "Every time I think I've made every mistake possible I find that it's not true. I can always find new and inventive ways to make mistakes."

SIDE DISHES

Celery Root and Apple Puree

Sauteed Butternut Squash with Rosemary

MAIN COURSES

Wild-Mushroom Chili

PASTA AND RISOTTO DISHES

Green Risotto

RELISHES AND INTERMEZZOS

Marinated Onions and Cucumbers

ALFRED PORTALE

Gotham Bar and Grill, New York, New York

Call Him Mr. Touchdown

THE HEISMAN Trophy, awarded annually to the nation's outstanding college football player, is no guarantee to success in the pros. But such winners are always highly prized by professional teams because they know that former Heisman recipients like Roger Staubach, O.J. Simpson, and Tony Dorsett all went on to distinguished careers in the pro ranks. Sometimes all it takes is one key player to turn a franchise around and put a losing team on the road to success.

When restaurant owners Jerome Kretchmer, Jeff Bliss, and Richard and

Robert Rathe were looking for a chef to turn the Gotham Bar and Grill around they drafted Alfred Portale. The young chef had received the cooking equivalent of the Heisman, graduating first in the 1981 class at the Culinary Institute of America, and had polished his skills in the kitchens of some of France's top chefs: Michel Guerard, the Troisgros brothers, and Jacques Maximin.

Despite his impeccable credentials, the move was a risk. Portale had never run his own kitchen. And time was short. Once New Yorkers turn their backs on a restaurant, it's difficult to change their minds. Meanwhile, the expenses keep on mounting.

But only a few months after Alfred Portale began as chef there, Gotham was awarded a coveted three-star rating from the *New York Times*. And Portale was selected for the chef's equivalent of the all-star team: Wolfgang Puck's Meals on Wheels, which calls on some of the finest cooks in America to prepare food for a $150-a-plate buffet benefit to feed elderly shut-ins.

Portale purged the menu of its eclethnic grab bag of dishes and gave it a consistent voice with food attuned to the way Manhattanites like to eat. And he mated his food well to the Knickerbockerish design of the restaurant. Set on several levels, the restaurant occupies a Greenwich Village warehouse with thirty-foot-high ceilings converted to half-resemble an outdoor cafe. And in the middle of it all is a down-sized replica of the Statue of Liberty complete with glowing torch.

"In New York there are important factors to consider," observed the lean, brown-haired chef with the slow, shy smile. "People in the city expect a lot of style in the presentation. They want some drama to their food in terms of how it's styled and in terms of the interaction of ingredients. At the same time, they want to be able to readily identify flavors and accents without a lot of masking. So while the food has to be sophisticated, it can't be silly or contrived. Add to that their time constraints and a growing concern for nutrition and you've got some idea of what kind of food to create to satisfy our young, professional audience."

Portale packs his menu with as many as sixteen appetizers rich in vegetables and grains to give his hurried diners plenty of nourishing choices even if they don't have time for a second course. His first and second courses, following the current trend in France, are close to the same size, but the second courses have more things happening on the plate. Meat, fish, or fowl will generally be served with no less than three vegetables. For example, quail might come with celery root, marinated shiitake mushrooms, and mixed greens in a sherry vinegar-walnut oil dressing. Pheasant is often served with sauteed endive, leek, and roasted red pepper, as well as wonderfully light little mushroom quenelles.

Portale insists on a different vegetable selection for each dish. "The way I incorporate vegetables here, each dish has very specific vegetables. Never do we put the same vegetables on every plate. I use vegetables inherent to the preparation in some way, either because of the tradition or history of the dish or because of the blend of flavors, textures, or colors. And I don't just use vegetables. I do just as much with beans, lentils, couscous, and rice. A lot of thought goes into the creation of the garnish, and I hate to use that term, because it's not the meat that's difficult to prepare, it's the accompaniment, the preparation of vegetables that I find challenging. I get much more inspiration taking what's seasonal and creating something unusual. With meat and fish I have only a few things to work with, but there's a much greater range of choices with vegetables."

Gotham's habitues often tell Portale that they order certain second courses just to get wonderful accompaniments like eggplant caviar custard, deep-fried shallots, or grilled leeks with white beans and blue cheese. But to experience the widest possible range of Portale's vegetable preparations you need only scan the first-course portion of the menu. The chef encourages his customers to order two to three appetizers to experience a variety of his many interpretations highlighting the market's best seasonal ingredients.

One night some friends joined me at the Gotham for a meal of first courses. We began with one of Portale's personal favorites, eight-tomato salad. Working with a variety of local growers, Portale is able to get his hands on Jersey beefsteaks, Golden Jubilees, persimmon, cherry, pear, currant, green, and white tomatoes. The colorful melange is arranged in the center of a large white plate. Fanning outward are tender green beans and slices of raw fennel sprinkled with a few pungent olives. On the side, Portale serves a creamy mustard-garlic dressing. The chef's penne with wild mushrooms followed, the pasta in a light thyme-and-bay-scented cream sauce loaded with morels, shiitakes, and chanterelles and liberally sprinkled with toasted pine nuts and tiny cubes of fresh tomato.

Another first course featured goat cheese blended with cream cheese with a brilliant vein of tomato running through the center, sitting in a vibrant tomato coulis. Later we were served a tomato that had acquired a smoky flavor after being slow-baked for six hours with garlic, thyme, and olive oil. All through our meal the waiter supplied us with basket after basket of chewy seven-grain muffins and sourdough rolls.

Portale has a personal interest in eating healthfully. He has the look of a runner and works out daily. "I'm very conscious about nutrition. Working in a restaurant, there's a lot of inherent dietary problems. Tasting, nibbling, smelling food all day, you're less likely to sit down and eat a good meal. Then there's the pressure. When I eat, I tend to go for low-fat, low-caloric, nutritious foods including whole-grain breads, vegetables, beans, grains, and salads."

While the chef might prefer to eat a simple salad, he knows that his audience wants some excitement in the presentation of the foods he serves them. Presentation is never sacrificed for the sake of nutrition at Gotham, however. One night Portale sculpted a salad that rose from the plate like a volcanic peak. It was a swirl of warm asparagus, green beans, and beets, paired with chilled salad greens and savoy cabbage in a lemon, orange, and olive oil dressing. Portale, who credits Michel Guerard with "inspiring such drama," knows that his audience appreciates food that looks glamorous and is full of vitality. Seeing such a creation I mused over the probable success Portale would have achieved had he pursued his original career goals and become a jewelry designer.

I followed the salad with a good-sized platter of goat cheese ravioli with tomatoes, thyme, and garlic. The pillows of pungent cheese came in a chive-flecked garlic butter sauce brightened with fresh savory, thyme, and black pepper.

There was a time when the raw materials needed to create Portale's kind of food weren't available at prices a large-volume, medium-priced restaurant could afford.

"Since I came to New York there has been an explosion of cottage industries supplying odd varieties of produce, baby vegetables, wild mushrooms, and so on. There has been an amazing change. I can remember when you couldn't get wild mushrooms grown in America. A couple of months a year you could import them from France, but some of the time they'd be rotten. You couldn't get *haricots verts* or baby lettuce, even from California. Now there's an abundance of this stuff, and the prices, though high, are much less now than before. Now I'm able to cook year round with the abundance of fresh vegetables my customers appreciate."

The Greenwich Village location draws local artists and designers as well as a significant uptown crowd. So Portale will soon open a big project midtown.

"Don't expect $100 a plate prix fixe dinners there, though," says Portale. "I still want to keep it casual and accessible." It is a winning formula.

PASTA AND RISOTTO DISHES
Wild Mushroom Pasta

WOLFGANG PUCK

Spago, Chinois, Los Angeles, California, and Postrio, San Francisco, California

"I don't try to imitate."

MANY HAVE theorized as to why Wolfgang Puck has become the best-known chef in America. Some attribute his success to the highly visible personalities frequenting his Los Angeles restaurants. Others point to his personal appearances on such programs as "Good Morning America." Still others note that his frozen desserts and pizzas, which are available across much of America, offer Puck the kind of exposure few other chefs can hope for.

But none of these things would be possible without (1) Puck's sheer talent as a chef; and (2) Barbara Lazaroff, who helped develop and design Puck's restaurants and who, behind the scenes and in the dining room, continues to create the magic and the excitement necessary to keep them on top.

It's easy to get caught up in the hype and the hoopla that swirl around Puck and cling to him like cotton candy on a stick. And it's facile to oversimplify his food and talk about duck sausage pizza, grilled meats, and great salads as though those things were all he makes. But people would not continue to flock to Spago, Chinois, and Postrio if his food were not formidable.

No one cooks like Wolfgang Puck, with his disarmingly casual, yet elegant style. But the levels of taste that underpin and weave throughout the simplest dish show a culinary intelligence that comes along only rarely.

While it is true that Puck has tailored his menus to an audience that pays as much, if not more, attention to each other as they do to the food on the plate, they must be sufficiently satisfied at the table, or they would find another hangout. Some might argue that the food Puck serves up is only as good as it has to be. I maintain that it is good enough to satisfy a range from the least discriminating diner (who may not know or care how good it is) to the most exacting.

Those who come to Puck's places for the food find the kind of cooking a chef can create only after he has refined his technique so well that what appears to be simple is simply devastating. For example, Philippe's chopped vegetable salad appears to be nothing more than a mélange of rough-cut vegetables with croutons and some dressing. But what vegetables they are! Lovingly grown by Tom Chino at his organic San Diego County Ranch, carefully harvested and packed at the peak of freshness, these carrots, onions, celery, tomatoes, and peppers, dressed with a spicy, herby vinaigrette and accompanied with crunchy whole-wheat Parmesan toasts, are such a perfect alternative to the usual plate of leafy greens that you wonder why no one else had thought of it.

As Puck puts it: "If you have good stuff to start with, you can spoil it if you make it too complicated. But as long as you keep it simple enough it should be good without being complicated. And I think most of the chefs today try to impress the customer, to make things complicated. People really don't like that. Most people would really rather have something simple and really good than have something elaborate. But if you have nice vegetables you shouldn't have to make it really so elaborate in its presentation, because it should look good the way it is. Simply."

Another reason for Puck's success is his and Lazaroff's originality. Wolfgang observes: "I think everybody tries to imitate everything, and it gets so boring for the customers. You go to one place and they serve one thing. All of a sudden you go to another place and they start having the same thing, and pretty soon everybody has the same thing. Barbara with the design and me with the cooking—we will always be different because we don't try to imitate."

Barbara Lazaroff arrives at Spago and Chinois each night turned out like a tropical butterfly. She glides from table to table, alighting here and there, often pollinating one with newsy nectar from the others. Wherever she goes, she brings with her a natural radiance, easy laughter, and an unaffected charm that could brighten up the mood of someone who'd been passed over for the lead in a Spielberg film. Armed with her own camera, she can make you smile even if you don't feel like it. The forums for Puck's food are based on Barbara Lazaroff's vision. She is responsible for the interior design of warm woods, tiles, and brilliant colors at Spago.

Puck reminisces: "When we thought about Spago at the beginning, it was conceived as a much simpler place. Less expensive than it is. It is what it is really only because of Barbara, who built it really nice and made it into the beautiful child that it is. It would have been a good restaurant, probably with good food, but very simple. People would have come, but it wouldn't have gotten so well known, and a certain crowd that helped build the restaurant and make it famous would have stayed away."

That "certain crowd" is, of course, the scores of celebrities who come to Spago and Chinois to dine and table-hop, helping keep the party atmosphere alive and bringing in hundreds of patrons who fight for what are, after eight years, still the toughest tables to get in town. They come to see the stars and, if they are so attuned, have a taste of heaven.

Since Spago's inception, Wolfgang Puck has seen to it that the food remains at a consistently high level. He achieves this by taking a hand in every aspect of his business despite his ever-expanding worldwide enterprises. Besides Spago, Chinois, and Postrio, there is Spago in Tokyo, Puck's consulting duties for the Dallas-based Rosewood Hotel restaurants, the frozen pizza and dessert line, and two new restaurant ventures. One will be in Malibu, and the other will be the Eureka Brewery in West Los Angeles, where Puck will serve international deli food and homemade beer that he will sell not only at the restaurant but all over the country.

Puck sees to it that quality is maintained by making sure his produce is the best. He still drives the two-hundred-mile round trip to Tom Chino's ranch before dawn twice a week to buy the organic produce that is the foundation for his pizzas, sauces, salads, and vegetable dishes. Here, even before the cooking process, Puck distances himself from many restaurateurs, who buy from large commercial growers and produce houses where the food is grown with synthetic chemicals, sprayed with insecticide, picked half-ripe, and put in cold storage. The love and admiration Wolfgang has for Tom Chino's operation is unabashed.

"Chino Ranch is my favorite place to go. It is the way they treat the things which is so impressive. If you go to the produce market they don't care. They just throw it like it was a sandbag. The way he treats his raspberries and strawberries or tomatoes is amazing. And I think Tom and all these people at the Chino Ranch, they feel like they *make* vegetables. They look at them and they're proud. They are really happy when something is ripe. They cut it carefully off the vine and put it gently in a case. It is so different. They really like what they're doing, and it's really, really wonderful to have people that you know who do such things with food."

With Chino Ranch produce Wolfgang can wow customers with such simple dishes as cold asparagus with lemon-mustard sauce. Chino's fat tender spears are so sweet you hardly need the sauce, but the combination of sour, spicy flavors and asparagus is a perfect marriage.

Chino Ranch isn't Puck's only source for produce. To make sure his salad ingredients are fresh daily, he buys Andrea Crawford's many varieties of organically grown lettuces, as well as her radicchio, arugula, and field greens. His garden salad with fresh goat cheese sautéed in olive oil has become a Spago classic, remaining on the menu by popular demand through the years. Often Andrea's greens will show up in salads featured as specials, such as sautéed wild mushrooms in citrus or wine vinaigrettes. With these kinds of salad ingredients available, Puck sees his task as simply finding the dressing that will best show off the rich and varied flavors of the greens.

Spago is Italian slang for "spaghetti." So it would figure that Puck's pasta dishes would be special. One has been on the menu as long as I can remember: angel hair noodles and broccoli tossed in a creamy goat cheese sauce and seasoned with thyme. Pasta perfect.

Puck has a penchant for spicy food, and this shows up in his pepper-infused pastas. Sometimes he'll incorporate black pepper in the dough, as in the case of a fettuccine with julienne Chino vegetables in a red bell pepper cream sauce. Another recent offering was white-pepper ravioli stuffed with squash. When he's not stuffing raviolis with squash, he'll fill them with something equally delicious, like a sweet pumpkin puree or mushrooms. And Puck has created a new kind of lasagne dish: baked pasta rounds with wild mushrooms and black truffle butter is about as earthy and rich as you can get.

One memorable dish that appears as a special is Puck's wild rice and corn risotto. Wolfgang is fond of Tom Chino's many varieties of corn, and in late summer when the sweet white corn is at its peak, this chewy, hearty dish makes its appearance.

No discussion of Wolfgang Puck's food would be complete without talking about pizza. I'm unclear as to whether Wolfgang got the idea from Alice Waters or vice versa, but the oak-burning pizza oven at Spago is responsible for the craze for a new kind of pizza: a sweet, chewy crust with innovative combinations of vegetables and cheeses.

That same oven bakes country loaves of olive sourdough and six-grain breads. Like the rest of Wolfgang's food, they are some of the best you can find.

One of the lightest and most satisfying ways to eat at Spago is to order a platter of Chino's vegetables and wild mushrooms cooked on the grill to seal in their juices, and served with several dishes of the wonderful sauces Puck serves with his entrees. Wolfgang may have been glad to get away from French food with its heavy emphasis on saucing, but he learned the saucier's craft somewhere between L'Oustau de Baumanière and Maxim's. His grilled eggplant relish, spicy kumquat-ginger sauce, Maui onion–herb vinaigrette, onion marmalade, and black pepper–Cabernet butter should give you some idea of what I'm talking about. For me, his sauces, more than anything else, show his exceptional talent as a chef.

In the early days, people criticized Puck for dividing his time between Spago and Chinois, the Rosewood Hotels, and his other ventures (some of which, like his Wolfgang Puck Charitable Foundation, reap him no financial reward), but the chef feels that the range of his interests is part of the cause of his success. "It's helped a lot that I'm doing so many different things. It kept up the publicity and kept us in the public light. There are great restaurants that open up but they still don't get the publicity we get because we're doing so many different things."

Wolfgang Puck understands human nature and knows how to work with people as well as any politician or psychologist. When it comes to his patrons, there is almost nothing he won't do for them, such as the time he made mock Maryland crab cakes for my party at the drop of a hat. In eight years of dining at Puck's restaurants, I can't remember making a request for a substitution, addition, or subtraction, no matter how peculiar, that wasn't honored—with a genuine smile. As Puck sees it: "I think flexibility is the main thing people want from a restaurant."

Flexibility is the key to working with, training, and keeping good staff as well. Puck knows he can't build this restaurant empire alone, so he spends time making sure his staff is as happy as his customers. When I asked him how he had developed such a good staff, Puck, ever ready with the humorous riposte, replied, "I think it's because I'm so mean to them!"

He went on: "Everybody's really creative if you let them be. Once we sent Mark Peel to Dallas, and his replacement, John Stewart, made a ravioli with truffles. We had never made it before and people really liked it. So we kept it. There are many other kitchens where if somebody would have done that the chef never would have told him, 'You did wonderful. The people really liked it.'

"If you're just one person, you can't do as well as if you have five good people around you. That's why it's good at my restaurants, because we have so many good people. It's important to have good people around because somehow at one point you run almost out of ideas. And then you're drained so much you need time to recuperate before you can get going again. If you have good people around you at that time who are willing to listen and who know what the philosophy is about, what you want, then it is much easier."

As Puck was preparing for his new ventures, some wondered what would happen to Spago and Chinois. Puck thinks what he's doing will enhance them and help him hang on to chefs instead of losing them. In the past five years he has lost Mark Peel, who with his wife (and former Spago pastry chef), Nancy Silverton, now has Campanile restaurant in Los Angeles; Richard Krause, who owns Melrose in New York; and most recently Hiro Sone, who now cooks in the California wine country.

Puck has an answer for everything, and this subject is no exception. "With these new restaurants I'll have a place for some of these great young chefs to go and still stay within the family. It's going to help all the places."

And that's precisely what happened when Puck opened Postrio in San Francisco. He sent husband-and-wife chef team Anne and David Gingrass to open the stunning restaurant with the dramatic sweeping staircases. Postrio has been packed since the day it opened, and Puck hasn't lost a chef in the process.

Wolfgang Puck is used to criticism. "People told me I couldn't open a Chinese restaurant, that I didn't know anything about Chinese food. They think it is the biggest mystery in the world when it is not. It is as simple as any food. You talk about the food. I don't know the names. But I know what I can do with it."

But how many projects, restaurants, frozen food companies, charitable causes, and breweries can one man handle? Wolfgang will put no limits on himself. He's always taken risks: leaving home at fourteen to study cooking; breaking away from Maxim's in Paris to cook at a view restaurant in Indianapolis; relinquishing his post at Patrick Terrail's Ma Maison.

"I left Ma Maison when I was thirty years old. I was thinking at the time: even if Patrick Terrail gives me half of the restaurant, I'm going to wake up one day when I'm fifty-five and say: 'Wow! I have been sitting here for twenty-five years, and

I'm still doing the same things.' It would have been so boring. I think that's the main reason why I do so many things. The only thing I'm scared about is to get stuck somewhere with one thing."

There's little chance of that. Puck, recently slimmed down and getting plenty of exercise despite his breakneck schedule, thrives on his active life and his philosophy of simple food.

"You know, good cooking is like dressing a woman. You don't have to hang ten pounds of jewelry on her to make her look good. It is simple clothes and nice shoes and hair combed somehow nicely that look beautiful and really elegant. And that is what I try to do with my food."

. . . .

STEPHAN PYLES

Routh Street Cafe, Baby Routh: Dallas, Texas,
Goodfellow's, Tejas: Minneapolis, Minnesota

"I want to develop a kind of cooking that did not exist."

AS A boy, Stephan Pyles looked in wonder at the shiny tractor-trailers that rolled into his parents' West Texas truck stop snorting diesel.

"How do ya'll drive those? They look like they're about a hundred feet tall and about a mile long."

"Mom, this guy, he wants some chicken-fried steak, cornbread, and some strawberry cobbler. Yes ma'am, with *lots* of whip cream."

Then the big rigs would slowly pull off into the clear Texas night, their running lights visible for miles. The whine of those eighteen-wheelers on the West Texas highway that ran past Pyles's boyhood home seemed to last for hours. Maybe someday he'd travel far from home. Maybe someday he'd even have his own truck stop. But he was only eight.

"Yes ma'am," he muttered dejectedly, as he shuffled back from the highway, hands in his pockets, head hanging dramatically, "I'm comin'."

Stephan Pyles eventually opened his own place in Texas. But it's unlikely that many truckers go to Routh Street Cafe or its younger sibling, Baby Routh. The atmosphere and the food are as sophisticated as you'll find in Los Angeles or New York. You don't see too many guys with Caterpillar Tractor hats in either the trilevel cafe with its peach walls and white latticework, or in Baby Routh with its pastel walls and oak-tree shaded courtyard. And there's no chicken-fried steak on the menu, either.

But Pyles's cooking *is* Texan. There are many kinds of Texas cooking, from barbecue to Tex-Mex to East Texas Cajun-Creole to just plain Southern. And Stephan uses all of Texas's culinary traditions, creating out of them his own cuisine.

A recent prix fixe dinner at Routh Street Cafe (which just celebrated its fifth anniversary) began with complimentary hors d'oeuvres of melba toast spread with sweet and spicy Texas pecan-ginger butter. While it's unlikely that Mrs. Pyles used much ginger root at the truck stop, those sweet Texas pecans must have found their way into many a pie. And what could better epitomize the merging of Texas and Southwestern styles than Pyles's jalapeño–blue corn muffins?

The meal began with a grilled-vegetable tamale filled with diced mushrooms, baby carrots, black beans, leeks, peppers, and asparagus, then sprinkled with edible flowers and dressed with a smoky green-and-yellow-pepper butter. It was a visual, textural, and flavorful treat.

The chef insisted I try the salad of watercress and mâche with hazelnuts and young red and gold beets in a hazelnut-raspberry vinaigrette. It was unique and refreshing. Even more wonderful was the salad served to my brother: wilted mustard greens, arugula, and julienne smoked carrots, peppers, and onions, with warm, breaded Texas goat cheese in a cranberry-pecan dressing.

While I was putting away course after course, my less-hungry brother snacked on that salad and a plate of select American cheeses with seasonal fruit. Since we were on the cusp of autumn, there were berries and melons as well as apples, pears, and grapes—all grown in Texas—to enjoy with a wonderful selection of goat and blue and Cheddar cheeses made in the U.S.A. You'd be hard put to find anything in Pyles's restaurants that comes from outside our borders unless it's the ground black pepper on the table. And if you give him time, he may find a way to get that grown here, too.

Stephan Pyles is dedicated to using all American foodstuffs. "I'm committed to using indigenous products, ingredients I've grown up with and feel real comfortable with. It's sort of my mission to take ingredients that have always been considered kind of lowly and reinterpret them in a fashion that is innovative and creative. I want to develop a kind of cooking that did not exist by using the ingredients I love and know so well."

Well, there's no doubt that the papaya quesadilla never existed B.P. (Before Pyles). The soft, sensuous fruit was just sweet enough to offset the goat cheese and chewy flour tortilla. The side dish was tiny deep-fried breaded okra from Routh Street Farm.

Pyles insists on the best fresh local produce he can get, and one way of ensuring a steady, reliable supply is his arrangement with the farm (which, despite its name, is not owned by the corporation headed by Pyles and partner John Dayton).

"Vegetables are grown specifically for the restaurant at the farm. We have a financial agreement that we will take everything he grows. And it's all organic. The Texas growing season is long, so we get lettuces year round. And he has greenhouses, so that there's always something being produced, except one month in winter when it's too cold and one month in summer when it's too hot."

Pyles says that it's the availability of good farm-fresh producc that sets his cooking apart from that of others.

"The quality of raw ingredients is essential to the kind of food I produce." Though he is largely self-taught, when he trained with three-star French chefs like Michel Guerard, the Troisgros brothers, and Georges Blanc at Robert Mondavi's Cooking School, Pyles realized how important freshness was to fine cooking. He learned to pay close attention to seasonings and to harvest produce only when it's ready.

Another thing that sets Pyles's cooking apart from that of others is his pioneering use of herbs. He singlehandedly put marigold mint, which tastes like intensified tarragon, on the culinary map. You now see it in restaurants all over the Southwest. His next baby is *hojasanta*, an herb with an aromatic, eucalyptoid, almost mentholated flavor that Mexican chefs make a kind of pesto out of and serve with *chilaquiles* or fish.

Pyles's interest in herbs puts an unexpected and delightful spin on his dishes. While he also uses many different varieties of chilies, he does so more for flavor than for heat. In fact what struck me about meals at the Cafe and Baby Routh is that Pyles's cooking isn't all that spicy—it's herby.

Pyles heads up the cooking at the Routh Street Cafe and relies on chef

Rex Hale at Baby Routh. During our interview he kept one eye on me and the other on closed-circuit TV, monitoring kitchen operations in both facilities. But he has no such hookup to keep an eye on his two new Minneapolis restaurants, so he frequently travels to partner Dayton's hometown to make sure things at Goodfellow's and Tejas are running smoothly.

Both restaurants are located in a huge new glassy indoor shopping mall known as the Conservatory. Goodfellow's occupies the high ground on the third floor, where the food, though earthy, is refined. Tejas is down but not dirty, below street level, serving a brand of Tex-Mex food that is so original it ought to be patented.

Twin rows of translucent golden columns lighted from within guide your way to Goodfellow's. The restaurant overlooks an atrium, from which live piano music flows up into a warm space of pink walls and a ceiling composed of a golden marble square framed by blond wood latticework. The fireplace adds a final cozy note to what must be one of the most beautiful restaurants in the Midwest.

But you can't eat decor, and even Minnesotans aren't going to brave snow blowing at them horizontally to look at pretty pink walls and hear somebody tinkling the ivories. They want heaping platters of good food with plenty of different kinds to choose from. So they come for the inventive, always satisfying food of chef Tim Anderson, served in big Midwestern portions at reasonable prices.

Anderson knows how to use his native local products as skillfully as Pyles does those of the Southwest. Wild rice comes with tangy-sweet cranberries and chunks of pecans, or in soft puffy fritters with a wonderful sweet melon relish. Any farm table would be proud to offer the purple cabbage slaw with thin strands of julienne carrot. Fortunately, Anderson had the good sense to keep away from the sugar bowl when he made it.

Beans are served with an unusual twist at Goodfellow's: lentils are cooked al dente in a *masala* of cumin and garlic, with little chunks of jícama for a nice moist crunch. A grill of marinated and peppered wild mushrooms, fennel, zucchini, and multicolored peppers served with a fruity olive oil is another example of Anderson's robust fare.

As satisfying as all this is, and as filling (each dish I mentioned was listed as a $2 appetizer but by itself could have fed one of the Minnesota Vikings), Anderson's greatest skill may be as a saucier. One evening the menu listed no fewer than eight wonderful and original sauces: smoked pepper–tarragon, honey-mustard, roasted garlic, sweet corn, white Cheddar, tomato-basil, roasted pear, and pepper-barbecue. I can still smell their delightful bouquets. (And that's not just because take-out containers of all eight sauces spilled on the plane back to L.A., permeating my notes with their essences.)

Also notable are Anderson's hearty soups of roasted pumpkin with honey crème and apple-pecan relish, and his white bean soup with red-pepper cream. And Goodfellow's breads, a garlic-and-serrano-pepper-studded blue-corn stick and a whole-wheat loaf, are not to be missed.

Four or five days after eating at Goodfellow's should be a sufficient length of time before you need to eat again. And I can think of no better place for your next meal than Tejas. Salads range from one with jícama and endive in an orange-tequila vinaigrette, to a Southwestern Caesar with cumin-tamarind dressing and cayenne croutons. On days when the windchill factor is low, you'll be warmed by soups like black and white bean with red jalapeño cream, or tortilla soup with Monterey jack, smoked tomatoes, and avocados. There are unusual sandwiches, onion rings and fries, and a hearty three-potato salad. Dinner gets a little fancier, but prices never break the nine-dollar barrier, even for huge entrees.

Pyles credits vegetables for giving his cuisine its personality. "Our dishes just wouldn't be there without the vegetables. For example, we'll get a small Anaheim chili, stuff it with a local goat cheese and some almonds, and we have a chile relleno. In that dish, the produce is the most important thing. The same with our corn soup, or our red-pepper soup. When you think of all the herbs and chilies I use, both dried and fresh, the tomatillos, the jícama—of all the different ingredients that speak for what I do, produce would account for about 75 percent. In some dishes it's 100 percent."

One thing's for sure, Stephan Pyles's cooking is 100 percent enjoyable, whether you taste it in Dallas or Minneapolis. But if those two cities are too far away, don't worry. You're probably already tasting food inspired by Pyles at the restaurants in your part of the country.

SALADS

Arugula and Fried Okra Salad with Roast Corn Vinaigrette

RELISHES AND INTERMEZZOS

Jícama-Melon Relish

JEAN-JACQUES RACHOU

La Côte Basque, New York, New York

"There is only one cooking: the right cooking."

WHEN I first walked to La Côte Basque on a frigid December night a few years ago, I didn't know that Craig Claiborne had once called it "the most beautiful restaurant in America." All I knew was that my formerly substantial northern blood, having been thinned by seventeen years of Southern California sun, was providing me roughly the same protection as last year's anti-freeze might a New York City taxi.

But by the time I had been solicitously ushered to my table next to Bernard Lamotte's luminous mural of the Basque countryside, every bit of my chill had vanished in the warmth and beauty of the room. And I hadn't yet begun to eat one of the most memorable meals of my life.

La Côte Basque was first created by the legendary Henri Soule in 1958. The once-proud restaurant had gone into a slow decline when Jean-Jacques Rachou, successful chef-owner of Le Lavandou, bought La Côte Basque in 1979. Within months he had reopened the restaurant and surpassed its original glory. Now, ten years later, the glory continues. But it was a long, hard road for Rachou from his childhood in Toulouse to success in New York.

Rachou began working in a restaurant in his native Toulouse when he was eleven. Through the years he polished his skills until today he is at the peak of his craft.

Jean-Jacques Rachou was taught in the traditional French manner. And for him, the principles he learned are as inviolable as the law of gravity. "Too many people are changing the basic principles of cooking. There is only one cooking: the right cooking. That's why it takes ten to twelve years to become a *chef de partie*. Some people go to cooking school for one or two years, cook for about six months, and think they're already chefs. But they can't even cook ratatouille. There is only one ratatouille. And there is a principle to making it. It's not just a sauté. It takes about three hours to make it properly, and when it's done right, it should just melt in your mouth. Otherwise it's just some kind of vegetable stew. How they make it now, even in France, is no good anymore. Cooking is a matter of principle. I spent ten or twelve years learning it, and that's the way it is."

Jean-Jacques is dedicated to his profession. As he puts it: "You have to give all of yourself." That means being at the restaurant from 7 A.M. to 2 A.M. if necessary, six days and nights a week. It means overseeing the food not only from preparation to presentation, but also noticing if any is left on the plate when it returns from the dining room. "There's nothing I can't do in this business. I order the food, I cook it, I talk to customers, of course. But I can also fix the plumbing, the electricity, and do the books."

At one time Rachou stayed out of the dining room, leaving those duties to his staff. But he realized that hiding behind the stove kept him cut off from his clientele. So now he makes a point of working the room. "Without that important contact with the customer," he explains, "I would have no idea what they expect."

La Côte Basque does not offer too many vegetable choices on the menu, so I arranged with Rachou for a prix fixe vegetarian dinner.

Rachou started us off with a beautifully balanced, silky cream-of-potato-leek soup. We were next served buttery, woodsy morels that had been sautéed and placed in the middle of a puff pastry boat so light that only the morels, no doubt, kept it from sailing away on the next breeze.

After a lemon sorbet intermezzo and another visit from the chef, who was looking for a progress report, we received plate after plate laden with vegetable delights, including a broccoli timbale, herbed wild rice, and intricate vegetable terrines.

La Côte Basque is expensive. There's just no inexpensive way to produce food of this quality, in a splendid setting one door east of Fifth Avenue, with the kind of staff Rachou employs. Rachou could charge even more for what he serves, but he's concerned with giving value to his customers in the increasingly tight restaurant market. Rising costs prompted Rachou to close Le Lavandou, which predated La Côte Basque. He specialized in seafood there, and when the prices for raw materials skyrocketed, he simply closed the place rather than charging more.

Rachou plans to open a bistro later this year. Instead of five to eight vegetable garnishes, many of them intricately carved, he'll put just one vegetable on each plate. Instead of expensive cuts of meat and ingredients like truffles and foie gras, he'll serve less costly cuts in cassoulets and pots au feu—hearty bistro fare. As Rachou puts it, "I'll go back to the old-fashioned fare, good and tasty."

But there will always be an occasion so special, you must go to one of the most special rooms in America. My wife and I celebrated our anniversary at La Côte Basque recently, and as we sat in the remarkably beautiful dining room, the chef began by serving us a creamy risotto with peas and asparagus spears. The next dish was a woven pastry basket filled with three different kinds of sautéed wild mushrooms. The basket came complete with a pastry top and sat in white and green marbled sauces of beurre blanc and parsley puree. We simultaneously devoured a side dish of Rachou's justly famous potato balloons.

The salad with which we concluded our meal was another special dish: mixed lettuces with goat cheese rolled in pistachios, Belgian endive, *haricots verts*, and beets. The plate was alive with color. The balance of tastes and textures was masterful.

The look of the plate is as important to Rachou as it is to chefs who cook post-nouvelle. But Jean-Jacques will not sacrifice flavor for design. This is where his strict adherence to tradition kicks in. "Too many chefs want to give people what they want to eat. But that doesn't mean it's right. Something may not be cooked long enough, but a chef will put it on the plate because he likes the color. It may look good, but has no flavor. Flavor comes with the right cooking. But on these plates, where is the flavor?"

That's a question that's rarely posed at La Côte Basque, where tradition reigns over trend, and where staff and customers remain loyal to a chef who cares about them.

SIDE DISHES

Ratatouille

THIERRY RAUTUREAU

Rover's, Seattle, Washington

Expected the Unexpected

YOU CAN expect the unexpected from the voluble Thierry Rautureau. When he untied himself from Laurent Quenioux's apron strings at the Seventh Street Bistro in Los Angeles, Rautureau surprised nearly everyone by choosing Seattle as the venue for his first restaurant. Seattle does not appear in bold print on the culinary map, although the region boasts an abundance of some of the finest raw materials in the country.

Thierry further confounded people by taking over Rover's, a restaurant in the Madison Valley district: an area into which, before Thierry arrived on the scene, not too many Seattlites cared to venture.

Rover's, which occupies a building behind a nondescript strip of shops, features a sun-splashed patio overlooking a grassy knoll. Rautureau has created a country-French idyll where, for a few blissful hours, you are as far away from the marginal neighborhood that surrounds it as Seattle is from L. A.

While it's not unusual to find a chef frequently popping into the dining room to share in his customer's enjoyment, it *is* unusual to find him wearing black Bermuda shorts and giving guided taste tours of his herb and flower garden. But the most unexpected thing about Rautureau is his unique way of cooking.

Take, for example, his cold vegetable salad with avocado. Poached home-grown snow peas, *haricots verts*, and splayed baby vegetables with Belgian endive are mounded together, topped with nasturtiums, dressed in a light fennel-scented vinaigrette, and then ringed by meticulously cut wafers of avocado.

Rautureau is fond of doing variations on an ingredient within the same dish. Zucchini flowers stuffed with zucchini mousse and pan-fried goat cheese in a goat cheese sauce are two examples. But Thierry doesn't stop there. He embellishes the zucchini flowers with a garnish of diced tomatoes marinated in balsamic vinaigrette and lamb's quarters. The goat cheese sauté is served with fermented black beans and roasted bell pepper garnished with sweet ciceley and opal basil.

His dishes are complex. But the chef says that shouldn't intimidate anyone from cooking his recipes. "Don't be afraid of cooking anything complex. Nothing is ever *that* complex. Cooking is a very scary thing when you look at it as a foreign thing, but it's actually not that hard. It takes imagination and a good palate, and you have to start developing an idea of what things taste like. Once you get that down, you can create a lot of things, because you know before you create what it's going to taste like."

In both the planning and the execution of his cuisine, fresh herbs play an essential role. He grows over thirty-five varieties in his restaurant garden. That's one of the benefits for a first-time restaurateur starting out in a nonmainstream-restaurant city: rents are cheap, and what might seem a luxury in a city like Los Angeles or New York is easily affordable in Seattle.

And it's a good thing, for Rautureau's cooking requires a cast of hundreds, if not thousands of ingredients. Besides the familiar herbs, Thierry grows lemon verbena, anise basil, tricolor sage, silver thyme, borage, apple mint, and chocolate mint. And he reserves one corner of the garden for edible flowers, which he uses to garnish most every dish. Calendulas, violets, begonias, daisies, roses, nasturtiums, pansies, and geraniums all find their way onto the dishes at Rover's.

"I've learned a lot from having the herb garden," the enthusiastic chef told me during a tasting tour. "I wanted to know more about herbs, which I do, plus it's incredible how many more ideas you get from just having them in your backyard. You start by learning about them, then you use them, and soon you know herbs very well. When you don't have the availability of things it's very hard to conceive of all the things you can do with them. At Rover's, when garnishing a plate or making a sauce, I just go in the back with a pair of scissors and get my herbs. You can't get any fresher than that!"

Rautureau has room for a garden but not a farm. And though he grows some peas and beans, he, like most chefs, relies on outside sources for many of his raw materials. Luckily, Thierry is blessed with an abundance of wonderful, organically grown vegetables and fruits from small, conscientious growers happy to have someone appreciate what they do.

The availability of high-quality produce works perfectly with Rautureau's penchant for vegetables. He says he would like to open a vegetarian restaurant someday. In the meantime he contents himself with originating dishes like ratatouille wrapped in bok choy and oyster mushroom timbales. "I think as far as vegetables are concerned you have a lot of doors that haven't been opened yet. That's what makes it so interesting. We must give more consideration to vegetables. Not enough is being done with them."

There are other benefits for Rautureau in Seattle as well. "I think a restaurant is definitely easier to do in Seattle than in L.A. There's a different mentality. In a city like Los Angeles, many things add to the stress of what is already a pretty stressful business. The payroll costs there are very high. I wouldn't be able to serve there what I can serve here for the same price. I would have to change more because of the overhead and everything else. Plus, in Seattle, people are still at the stage of liking someone for what they do, not for who they are, which is a beautiful thing—something that's getting lost everywhere in the world. Here in Seattle, people care more about what they eat than who made it."

How could you not care for a suave eggplant puree topped with a mélange of crunchy sea vegetables, or a tomato *concasse* in a ginger-vermouth sauce, or pompano and chanterelle mushrooms sautéed in olive oil, garlic, and shallots, then topped with a generous shaving of truffles and served in a bold red burgundy beurre blanc?

Ingredients like these are expensive, and Rautureau's food costs are an astronomical 47 percent of his budget. (Most restaurants peg their food costs at about 33 percent.) Even though the rest of his overhead is low, with only thirteen tables, Rautureau has to strike a balance between the love for what he's doing and the money to make it all happen.

"I think in America people are based on the business side, which means they want to acquire a lot of money for what they do. I still haven't got that part right. I'm still caring a lot about what I'm serving, more than the money side of it. I have a hard time believing that when I want to make money I'm still going to be as good . . . that everything will be as perfect. And I have a hard time with wanting to make money because right now what I'm trying to do is basically satisfy myself. For the last fifteen years I've worked so hard to be able to come into the dining room and have people say: 'That was wonderful.' I have enough pride in what I do to make sure it stays that way."

Not only is Rautureau satisfying himself, he satisfies the forty to fifty devoted diners who have discovered this little hideaway. "I have a lot of loyal customers who keep coming back. In fact, I've never in my life worked in a restaurant where I've seen so many customers return. The clientele is incredibly open. And they come here for one reason: they like the food."

Perhaps the most unexpected thing about Rautureau is his attitude toward success: "People are too greedy, I think. I tell my customers when I want to make money I'll open a big place that seats two hundred people. That's how you make money: pack them in and pack them out. That's not the purpose of my restaurant. I'm not interested in that right now. I'm more interested in the customer saying: 'Wow, I never had this before. This is the best thing I ever had.' "

For serious diners in Seattle, Rautureau may be the best thing they've ever had.

APPETIZERS AND FIRST COURSES

Pan-fried Goat Cheese with Roasted Peppers

Sauteed Wild Mushrooms with Sage Sauce

Filo Pastry Stuffed with Ratatouille and Tofu

Tarragon and Chervil Cream Sauce

Stuffed Zucchini Flowers with Tomato Marinade

LESLEE REIS

Café Provençal, Evanston, Illinois

From Spaghetti to "Franco-American"

SOME PEOPLE are born cooks. Leslee Reis isn't one of them. In fact, she had never cooked in her life when she married at age twenty-one. For her first meal, Leslee got out some spaghetti and tomato sauce. The instructions on the economy-size package said to boil the spaghetti in eight quarts of water. So she proceeded to get out every pot in the kitchen, measure out the water, and start it boiling. Meanwhile, she heated the eight-ounce jar of sauce in the oven. When all the spaghetti was cooked Reis knew she was in trouble.

To cover up her mistake, she dumped most of it in the trash, covered it with newspaper, and somehow stretched the eight ounces of sauce over the still-mammoth two pounds of remaining spaghetti. A little Kraft Parmesan cheese from the green foil can. Et voilà. Dinner was served.

So, for all you home cooks out there who received little or no culinary training at home, take heart. After all, Leslee's mother had only these parting words for her when she left home: "Buy your meat at the butcher shop, dear. Not at the grocery store." Twenty years later, Leslee's Café Provençal was voted the fifth best restaurant in the country in the national Zagat Restaurant Survey. It was a long way from that first spaghetti supper to what Reis calls her "Franco-American" style, which has won the hearts and minds of so many.

Reis and her husband were living in Cambridge, Massachusetts, in the sixties so Leslee could get her Ph.D. in biochemistry from Harvard. But on a trip to Europe she took a course at Le Cordon Bleu. When she returned she took time away from her studies to help her neighbor, Julia Child, by volunteering to wash dishes for a television cooking show called "The French Chef." After working with Child for one season and studying with a Cordon Bleu cooking teacher in nearby Boston, Leslee was hooked.

After graduation, she moved to Evanston, where she took a teaching position at Northwestern. But after a few years, Reis made the change and started a successful catering company. Soon it became clear to her that one of the reasons people wanted her to bring food to them was that there were no fine restaurants on the North Side. So, in 1977, with little planning or money, Leslee Reis opened Café Provençal in a residential hotel on a quiet side street near Lake Michigan.

How did Reis become a fine chef? "My cooking is French based, French inspired. That's where I learned to cook. I think there are only a few great cuisines in the world in terms of style and basics. And a good chef just can't abandon the basics. But I like the countryside style—not real rich with tons of cream—because I think too much cream and heaviness diffuse and dilute flavors. I like flavors to jump right off the plate at you. And living in America, we've had to adapt because we have different tastes, different ingredients here. So I call my cooking 'Franco-American.' "

I sampled some of Reis's Franco-American cooking at her romantic cafe recently and found that millions of Zagat contributors aren't wrong. The flavors are indeed jumping.

I started with the *tarte aux legumes et chèvre*. Preventing the feathery crust from floating up to the wood-beamed ceiling was a mélange of zucchini, yellow squash, and eggplant in a bang-up herby roasted red pepper and tomato sauce with warm goat cheese: Provence on a plate. An honest hunk of nutty whole-grain bread allowed me to mop up every last drop of sauce. A bowl of butternut squash soup topped with crème fraîche and chives was next, a smooth-textured composition the color of butterscotch, with sweet and savory flavors in perfect symmetry. The eggplant-chestnut timbale was also a wonderful combination of flavor, sweet and nutty and earthy all at once.

There's nothing like a beautifully turned-out gratin on a chill night—for example, the mixture of butternut squash, artichokes, and caramelized onions at Café Provençal one blustery night as the wind off Lake Michigan blew a frigid rain horizontal, rendering umbrellas useless. Rich and creamy, with every vegetable getting a chance to harmonize, the gratin was warming and fortifying, making me, if only in my imagination, impervious to the storm.

Whatever the weather, each season brings Reis new ingredients to work with and, therefore, new inspiration. "Seasonality plays a big part in my cooking. When the seasons change, my spirits change. That's why fruits and vegetables are so important to my cooking. I mean, you can get meat and chicken year round. But certain fruits and vegetables come only once a year, so the flavor of that first asparagus or that first basil leaf is inspiring to me. Seasonal produce is the catalyst that excites me about cooking new things in a new way. You see, flavor is the most important thing to me, and when I get new flavors to work with, fresh from the ground, I get excited all over again about cooking in a new way."

Reis brings in as little produce as possible from faraway sources, relying on local suppliers throughout most of the year. In winter she has no choice but to import things like lettuces from the Republic of California.

"We do a limited amount of imported produce because people can get bored if they can get anything anytime. And I'd get bored cooking it, too. Besides, the things that grow closest to you taste the best, because they go from ground to kitchen to plate in the shortest possible time without losing any flavor. For me, vegetables would be the last food group I'd give up, not only because of the seasonal inspiration I get from them, but also because of the incredible variety and contrast of flavors, both in and of themselves and played off against other things like fish and chicken."

In Reis's Provençal cooking style she likes to focus on one strong flavor with a supporting cast usually featuring fresh herbs. But if it doesn't have a distinctive "jump-off-the-plate flavor," chances are you won't see it at Café Provençal.

"Maybe it seems like a foregone conclusion, but for me flavor is very important. I don't put things on a plate just because they're pretty. They have to have flavor value. That's why I don't use edible flowers on my plates. I think that sort of thing is contrived. It's the antithesis of fine cooking because it sacrifices flavor for looks." Then she added, dubiously, "Maybe if I was in a blind tasting test I would think a pansy tastes just as good as a spear of asparagus, but I don't think so."

With Reis, what you see is what you get. The ebullient, outspoken chef leaves no doubt as to where she stands on any issue. And neither does her food. It is simple and direct, without any ambiguity or vagueness. And I think that's the reason her restaurant remains so popular after twelve years. Leslee Reis's culinary talent burns brighter every year, like a cozy fire at Café Provençal.

APPETIZERS AND FIRST COURSES

Provençal Artichoke Hearts

Crispy Tarts with Vegetables

MAIN COURSES

Summer Vegetable Tart

SOUPS

Café Provençal's Avocado-Cucumber Soup with Fresh Oregano

Butternut Squash Soup

SEPPI RENGGLI

The Four Seasons, New York, New York

"It's the only thing that can make me go. It's my life."

SOME MAINTAIN that the so-called "new American cuisine" movement started in the 1980s. But the first Four Seasons menu was based on American food in 1959 by James Beard, Joseph Baum, and first chef Albert Stockli.

For a time the restaurant reigned supreme in the supreme restaurant town in America. Then, in the early seventies, the Four Seasons slumped. Those were the years of the New York restaurant doldrums, when such once-proud institutions as the Forum of the Twelve Caesars, La Fonda del Sol, and the Tavern on the Green either folded or slipped badly. So Paul Kovi and Tom Margittai put chef Seppi Renggli in the kitchen to bring back the glory that was the Four Seasons.

Renggli, who had started at La Fonda, was the right man for the job. He had been supervising the corporation's 130 restaurants up and down the eastern seaboard for eight years. But when Seppi was brought in as executive chef in 1973, the Four Seasons was serving only seventy lunches a day and perhaps a hundred dinners. Those would be good numbers for most places, but not for this gargantuan place, whose waiting area could encompass the serving space of 90 percent of the restaurants in New York. The magnificent Philip Johnson–designed wood-paneled rooms stood nearly empty. The unique chain curtains still danced in the air currents and water filled the white travertine Italian marble pool, but few people passed by Picasso's theatre curtain in the massive hall, or Richard Lippold's sculpture of brass rods in the Grill Room.

Seppi Renggli found thirteen bedraggled cooks badly in need of guidance. The chef decided the restaurant needed a new culinary direction. How did he do it? "We had to start from scratch. We worked *very* hard," he says. And today? The Four Seasons now employs as many people as they used to feed in the bad old days (a small militia of over one hundred serves a whopping 750 meals daily). It is once again the premier restaurant for New York's powerful.

Seppi Renggli is used to working hard. His first two professional jobs were at hotels in England, where he was berated for the way he cooked. But he had been warned by his boss in Switzerland. "He told me that when I went to England that whatever I learned I might as well forget because the English don't know how to cook." Finally Renggli headed for the West Indies and then to South America, but Seppi got his real training in cooking vegetables when he married his Indonesian-born wife. She told this professional chef with twenty years of experience that he always overcooked them.

When Seppi burst on the New York restaurant scene at La Fonda del Sol in 1966, he was ready to show off a culinary skill that included a way of cooking vegetables most New Yorkers had never experienced . . . and some didn't care to. Reviewers said: "Seppi Renggli thinks we are rabbits and serves us uncooked string beans and cauliflower." But Seppi stuck with his guns, and five years later, when crudités and al dente vegetables became popular, he was vindicated.

The significance of Renggli's pioneering use of superior vegetables in remarkable combinations and his development of the Four Seasons' trademarked Spa Cuisine in reviving the once-flagging restaurant cannot be overstated. The new menu was designed to offer the kind of food that would appeal to a new, more health-oriented clientele.

"Years ago, you couldn't put stuffed cabbage on the menu. Nobody would eat it. Now, everybody wants it. The same with pasta. People went to Italian restaurants for it and came here for steak. Now they come here for pasta, and fewer people eat steak. The big boom came in the late seventies. We started making a vegetable platter. We even bought a special platter for it. We served about twenty kinds of vegetables on this platter, served them with a little sea salt, olive oil, lemon, and lots of herbs. It went like crazy. It was a full-time job for two people in the kitchen. But now people want something just as light, but a little more interesting."

Something like a cream of green pea soup with knob celery, or an autumn vegetable soup. If you can't decide, pick the pea soup. You can have the autumn vegetables in a mild curry sauce as an appetizer. That dish qualifies as Spa Cuisine, but you wouldn't guess it by tasting the rich, creamy mixture. If you want to continue in a non-cholesterol-clogging vein, the Spa Cuisine offerings feature dishes like earthy mushroom won ton in clear broth, vegetable ragout braised in savoy cabbage with a velouté of green peas, or a vegetable timbale with peanut sauce, an Indonesian dish no doubt inspired by Mrs. Renggli.

But you don't have to go Spa to go light at the Four Seasons. A recent Pool Room menu featured such an array of vegetables and salads that you could have feasted on them alone. There were raw mushrooms with pepper dressing, baked fennel with Parmesan, Bibb lettuce with watercress salad, sautéed autumn greens, wild rice with pine nuts, sautéed sugar snap peas, and broccoli hollandaise, as well as baked and rösti potatoes. The chef uses a modicum of butter and salt in these preparations, relying on the flavors of the vegetables to carry the dish.

By long-standing tradition, the restaurant changes its menu with the seasons. In the Grill Room, selections change more frequently. And even in the giant Pool Room there is a new flexibility, due to a great leap forward in computer technology.

"My boss bought a laser printer, so we print every menu before lunch and dinner. Now, if the man who gives me the salmon and oysters doesn't come up with what I want by 11 o'clock in the morning, I can tell the steward to just have our office type in something else that is good. In the old days you had to have this big fancy menu ready for the printer almost five weeks before it went into effect. So you could never express the items you only get five or six days a year. You know, if you get from Canada or from Europe yellow asparagus or beautiful melon from Israel, you could never put it on the menu. You could never even put figs on the menu because you may get them one week and not the next. Now I can put on whatever's good. It makes it easy for me, and it gets everyone in the kitchen going."

Getting people going in the kitchen is where Seppi excels. With an operation as large as the Four Seasons, a reliable, well-trained cooking line is indispensable. Of course, hot young chefs fresh out of cooking school sometimes have their own ideas, and Seppi has to retrain them. "I scream in the kitchen if somebody bends over the table and works like this (hunkering over his desk to demonstrate). I say: 'It's not necessary. Sharpen your knife, stand up straight.' I don't want this to be a hard-labor camp. But I'm not a grumpy person; it's just that I want my chefs to respect food. I can tell when a cook touches something if he has a feel for it. If somebody takes a beautiful celery and holds it up in the air and just hacks away at it, I get wild if I catch somebody. I can't stand this. But what really gets me is when a cook comes up to me with *his* idea of a great dish. One guy brought to me a special veal stuffed with foie gras. I said, 'What are you doing? It's like decorating gold with gold.' More or less stupid. So I asked him what he was going to serve it with? He didn't know.

"And it's so surprising considering what's available in the market now. From twenty years ago there's no comparison. When I founded La Fonda I got a little box of jalapeños from Mexico. That was it. I use up that much in one evening now. When you order carrots now, you don't just order carrots. You have five or six varieties. And with peppers, you can get red, green, beautiful ones from Holland or West Coast anchos, Anaheims, every color of jalapeño. Years ago when I'd cook six, eight hundred dinners a night, they'd give me six bunches of chives. And chervil—forget it. But things have changed now, and I expect my chefs to respond."

One chef named Andy started in the bake shop at the Four Seasons, then went to work for Wolfgang Puck at Spago for a year. After a short stint back at the Seasons, he took off for Puerto Rico to work as a sous chef in one of the island's finest resort hotels. Seppi misses him, though, he confesses, he did feel slightly competitive with Andy. But it made the master more involved, and by carpooling together from Rockland County, the two worked out any creative differences. They had to do it in the car, because Mrs. Renggli won't let father and son discuss business at home.

APPETIZERS AND FIRST COURSES

Vegetable Timbales with Peanut Sauce

SOUPS

Puree of Celery Soup

Mushroom Won Ton

MICHEL RICHARD

Citrus, Michel Richard, Broadway Deli, Los Angeles, California

Forever Young

THE THING that makes Michel Richard so delightful is that he never really grew up. He has no pretenses. New discoveries excite him so that he just *has* to tell you about them immediately. Like the time he perfected licorice ice cream and ran over to me during the middle of my salad course and insisted I have some.

A man half his forty years would have considerable difficulty keeping up with Richard's eighteen hours-a-day, six-days-a-week pace. During mealtimes his energy takes him from station to station in the kitchen at Citrus, then out to schmooze with his customers, and once again back to the stove. Like an adventuresome youngster, nothing delights him more than the prospect of experimenting with something different, and if he had his way, he'd do a new menu every day, maybe even twice a day. In fact, if you ask him to, he'll prepare a meal on the spot, his version of improvisational theatre. The stage will be your table as Richard brings forth a succession of dishes whose composition was unknown to the chef until the muse whispered in his ear.

Michel loves to crack jokes and tell stories, and always sees the humor in a situation. He wants everyone to have a good time. So there's nothing serious about Citrus, except, of course, the food. The restaurant is divided into two rooms. The smaller one, with carpeting and contemporary art, is for grown-ups. The larger room is a big outdoor patio (with a retractable canvas cover) full of huge white umbrellas and yellow linens, jungles of foliage and flowers, and a floor-to-ceiling window through which you can watch all the fun in the white tile and stainless steel kitchen.

To start, try the sheer joy of the curiously refreshing chilled melon soup: a colorful array of melon balls in a fruity broth of melon juices, with a hint of pepper and balsamic vinegar. Or ask for the "cheeseburger," made of sautéed onions and tomatoes fused into a patty covered with melted goat cheese, and served on Michel's fabulous whole-grain bread with a brick of his needle-thin French fries.

Sometimes Michel will form shreds of potatoes into a cup, deep-fry them, and then serve the crispy critters filled with his spicy black bean soup. His potato risotto in scallion sauce is made without a grain of rice. Once he surprised me with a bowl of potato-leek soup topped by a crunchy tangle of fried onion rings and surrounded by a pretty border of goat cheese rosettes squeezed through a pastry tip, pointing up one of the most enjoyable aspects of his cooking: the use of varying textures.

Michel is not fond of using cream and butter. He doesn't like the way they mask other flavors. As a professional baker, he cut way down on the amount of cream traditionally used in pastries. And at Citrus, to prepare five hundred to six hundred meals a day, he'll use only about five pounds of butter. Even though his sunny restaurant is so quintessentially California you almost need sunglasses indoors, Michel insists, "I don't want to be called a California restaurant. People laugh when you say that. They don't take you seriously. What I am is a French regional chef. My region is Los Angeles."

When I mentioned that Los Angeles isn't normally thought of as one of France's regions, he rejoined: "If a French chef in Normandy cooks with cream and butter and the ingredients of that region, he's doing the regional cooking of Normandy. So I, as a French chef in Los Angeles using the local ingredients and a light fresh cooking style, am doing the regional cooking of Los Angeles."

The light touch is especially evident in Richard's intense sauces. There's no flour, cream, or butter in them, just reduced herb and vegetable stocks and purees with a touch of vinegar or lemon added at the end to bring out the flavor. Says Richard, "You have such nice sharp flavors and colors with the vegetable sauces. Why dull them down?"

Besides, there is one region in France where olive oil predominates over butter and cream: Provence. And the sunny disposition of Southern California, with its year-round abundance of fresh fruits, vegetables, and herbs, is so similar to that of Southern France that Michel Richard's cooking has a definite Provençal flavor to it.

Richard's restaurant almost didn't happen. Rents in Los Angeles are so scandalous on the West Side that Richard, despite his ten-year reputation as the finest pastry maker in the city and his devoted following of catering clients, almost gave up hope. Then someone persuaded him to have look at an empty dentist's office on Melrose Avenue at Citrus Street, on the decidedly déclassé side east of La Brea, where rents are still within reason for a restaurant of this size.

When Citrus opened in 1987, it was clear that Michel Richard, who brought the first real croissants and best-ever pastries to Los Angeles, knew how to cook. The restaurant was mobbed, and, even in these uncertain times for fine restaurants, it has continued to serve more than five hundred people a day. They come for inventive salads like julienne strips of zucchini, cucumber, carrot, and lemon in a citrus vinaigrette. They come for tomato-zucchini tarts made of the airiest pastry and the zingiest, brightest colors and flavors imaginable. And they come for his marvelous terrines of eggplant and tomato or endive and leek.

Almost from the start reviewers recognized the genius of Michel Richard. And the chef appreciates their support, even though he was a little miffed that they swooped down on him during his first week, proclaiming Citrus the next Spago. "We need the support and help of reviewers," noted the chef. "But I really wish we had a Michelin rating system here. Then a restaurant could slowly establish a good reputation and with one or two stars do well over a number of years. Here, one bad review and someone can be ruined."

Richard needn't worry; no one will stay away from Citrus. The prettiest new restaurant in a long time, with some of the best food and absolutely the finest service in Los Angeles, it should have a long successful run.

Richard wants to make his restaurant an established institution on the Los Angeles dining scene. For Michel is a man who respects tradition. He grew up in a small village in France where everybody knew everybody. People got married and stayed married, because, from their childhood on, they knew everything about the person they were settling down with.

Yet Richard is also a man who does not like to stand still. Soon, he, Citrus partner Marvin Zeidler, and local restaurateur Bruce Marder will be opening Broadway Deli in Santa Monica. With wood-burning ovens to bake fresh bread and the two chefs' light touch, deli food will never be the same.

But in his heart, Michel Richard, the chef with the twinkle in his eye, will always remain the boy of nine who walked into a restaurant and fell in love with the idea of being called "chef."

APPETIZERS AND FIRST COURSES

Onion and Black Olive Tarts

SIDE DISHES

Salsify and Fava Bean Ragout

JACKY ROBERT

Amelio's, San Francisco, California

Cuisine de Pensée

WHEN JACKY Robert entered the dining room at Amelio's for the first time one night, the last seating's customers began to clap. Then every one of them stood to continue the ovation, in a spontaneous response to his well-thought-out, intricately wrought food. Robert has worked hard to deserve this kind of appreciation: his signature dish, woven green and white fetuccine in a basil-goat cheese sauce, took five years to research, practice, and perfect. That kind of careful approach to food results in dishes that rarely disappoint.

Robert describes his approach as *cuisine de pensée*, "cuisine of deep thinking." "I'm a deep thinker. My cuisine is of the same type. I think over and over. My ideas are not just fully formed in a minute and put onto the menu. I have to develop that idea."

Robert was developing his ideas long before Amelio's and long before he came to San Francisco on vacation in 1975 and decided to stay, single-handedly reviving a moribund Ernie's and making it one of the top restaurants in the city again. The Normandy native had gone through the apprenticeship that is de rigeur for aspiring French chefs. When he got to Paris he worked at Prunier-Traktir and at Maxim's, where he got the greatest inspiration of his career working under Michel Bourdin. It was Bourdin, along with René Verdun, who later sponsored Jacky Robert into the prestigious Maîtres de Cuisiniers de France.

When Jacky moved on from Maxim's, they lost a Michelin star. Rankings dropped when he left other restaurants in Switzerland, Italy, Boston, and Ft. Lauderdale. And so it went at Ernie's, which has had trouble maintaining the Himalayan level it attained when Robert was in the kitchen.

So when Chris Shearman was reviving Amelio's, a former speakeasy that opened at the edge of San Francisco's Chinatown in 1926, he made sure Jacky was there to stay. He formed a partnership with the chef in 1985, and the restaurant took off like a rocket. Shearman had lovingly restored the once-shuttered restaurant, whose huge red leather guest books were filled with names like Clark Gable, Lana Turner, William Powell, and Groucho Marx. And Robert cooked and cooked and cooked, seven days a week, because Amelio's was the place for romantic dining where the food outshone even the huge crystal chandeliers.

But now Shearman has sold his half of the business to Jacky, and for the first time in his career, the chef is solely in charge. And for the first time in thirteen years since coming to San Francisco, Jacky takes a day off. In fact, he takes two days off each week. He wants to move away from the daily demands of the restaurant and take some time to think about what he will do the five days Amelio's is open. This allows him to change the menu more frequently, because at last he has the time to plan it instead of just execute it.

"My food has become more sophisticated, and I've made the restaurant more exclusive. I'm accepting fewer reservations so I can control the quality better and think about how well I'm going to cook for the smaller number of people who are here.

"Money is never an issue in my life as long as I have enough to live on and to send my daughter to school. I'm not in this business to make a lot of money. I do this because of my respect for people and my love for food. I do this because of the way I can treat people because I care for them and the food."

That was obvious from a special *menu dégustation* the chef prepared for me after I called ahead and told him what I'd like to have.

Robert told me that night: "It's like a vacation to get a request like this. It brings out my creativity. Especially with vegetables. You just can't find colors and flavors like that anywhere else. Served alone they are marvelous. For example, in my vegetable soups I don't use any chicken stock because I feel it takes away from the flavor of the vegetables. And you must serve vegetables with your main course because meat or fish just don't taste as good without them. There is nothing I'd rather eat or cook than vegetables. In fact, if I could find someone who cooked like I do, I could be a vegetarian."

But there is no one who cooks quite like Jacky Robert. One recent night Robert's opening gambit was a rice-paper "beggar's purse" filled with curried wild rice, raisins, scallions, and carrot shreds. The purse sat in a pool of crème fraîche strewn with tiny mint shreds and was tied with a thread of chive near the top, leaving enough of the bag for the chef to fashion a tiny tea rose.

Wild mushrooms en croûte was a ragout of mushrooms and onions inside a pastry shell. The delightful sweetness obtained by cooking down the onions contrasted nicely with the dark mushroom flavor.

Fresh grape ice came as an intermezzo, followed by an artichoke bottom filled with sautéed beet greens in a light curry sauce, served with hand-braided pasta.

The penultimate course told me a lot about the way Jacky Robert cooks. The dish was called "two kinds of potatoes with black truffle." One kind was potatoes Maxim, which is paper-thin slices cooked crisp-tender in Normandy butter. It was paired with an ethereal potato pancake, set on its edge and sliced. The truffles were sprinkled generously on top, enhancing the flavors of both preparations.

Robert's method of cooking black truffle seals in its mysterious juices, so that when a shaving as thin as a butterfly's wing rests on your food, it flavors the dish from first bite to last. He starts by making a dough of half flour, half rock salt, with just enough water to form it. Then he wraps the truffle in a cabbage leaf and inserts it into the center of the ball of dough. The dough is baked forty-five minutes at a low temperature so that the truffle cooks slowly by steam, with the cabbage leaf holding in all its truffly essence and imparting no flavor of its own. The rock-hard "bread" can either be cut in the kitchen or tableside (by waiters who are smiling but cursing under their breath at its granite-like hardness); the black nugget is then removed, disrobed, and sliced for distribution over pasta, vegetables, or, in my case, two kinds of potatoes.

Last was a cheese course, which I remember by referring to the scale drawing I made in my notes showing the approximate locations of the Loire Valley chèvre with herbs, the Doux de Montagne, the white Vermont Cheddar, and the Brie, as well as their juxtapositions (and approximate angles) to the grapes, lettuces, walnuts, and apple slices with which they shared the plate. Why does Robert go to so much trouble to compose his presentations?

"Food is to be enjoyed now and savored later in the memory. By an artful presentation you can remember easily how the food looked and tasted. So that the mind can more easily remember, I don't make my portions too complicated, generally putting only about three elements on a plate. There are enough courses in a *menu dégustation* that you'll be able to enjoy many tastes and still be satisfied. I try to put myself in the customer's place so that when he leaves the restaurant he is not too full and yet he is satisfied."

Making people satisfied is what satisfies Robert. He has plans to open a coffee shop–bakery, "not for the sake of the money," but to provide a place for his loyal staff to spread their wings. "I want to reward my friends who have worked so hard with me. Maybe it will give them a chance to express themselves."

Jacky Robert's reward is the recognition of his skill as a master chef—something he has worked for all his life. The applause of a grateful dining room is enough for now. But he has hopes that Michelin will one day come to America and rate the restaurants here. "That would motivate me to do even better, if they would come."

And if they don't?

"I'll probably return to Paris one day and open a nonprofit restaurant, just to achieve a three-star rating before I finish my career."

MAIN COURSES

Pipe Organ of Baby Zucchini Stuffed with Vegetable Purees

MICHAEL ROBERTS

Trumps, Los Angeles, California

The Beatles in the Kitchen

I USED to get my car fixed at Trumps. Of course it wasn't called Trumps then. It was called Chet's Auto Repair.

In those days I ate at places named the Source, Zap, Help, the Marathon Meatless Messhall, and the Golden Temple: vegetarian restaurants serving food that was hearty and healthful. But after seven or eight years of this sincere but somewhat uninspired fare I started to long for something a little more interesting. I was raised in an atmosphere of excellent food; *Gourmet* magazine sat on the coffee table next to *Life* in my family's living room, and my mother ran Rudi's, the only fine restaurant in Houston during the fifties and sixties.

In the midst of my health-food doldrums I chanced to pick up a *Gourmet* magazine. Caroline Bates, in her "Specialties de la Maison" section on California restaurants, mentioned Michael's, a restaurant in Santa Monica with some interesting-sounding vegetable dishes. I tried it. I liked it. And I began a whole new way of eating. Over the next few years, I followed Bates's reviews and those of Lois Dwan in the *Los Angeles Times* and was led to La Toque, the West Beach Cafe, and one day in 1981 to Trumps, which I was surprised to learn had opened in my old car repair shop right across from where I used to practice yoga. I decided to go there for lunch.

A basket of corn bread and whole-wheat rolls was the first indication that this restaurant was serious about creating food that was both fun and wholesome. The salad had some of the same healthful lettuces I got at the natural foods market, but what were these other strange-looking greens and reds? And what was that dressing that made it all taste so good? (A walnut oil vinaigrette, I later learned.)

And the tomato soup! The waiter poured the hearty red-orange broth tableside from a white porcelain pitcher into a white porcelain bowl. It was pure tomato delight, with unidentifiable herbal undercurrents that enhanced rather than competed with the dominant flavor of the soup.

But it was the main course that made the deepest impression on me: corn risotto with shiitake mushrooms, peppers, and okra—a rich, creamy dish that had flavor, freshness, and heartiness. The corn was cooked al dente in a vegetable stock enhanced by shallots, cream, pecorino cheese, and butter, and served mounded in the center of the plate with the sautéed mushrooms, red and green peppers, and okra spaced around it like multicolored lava streams.

I recently spoke with Trump's courtly chef about how he arrived at such satisfying dishes.

"The way I put flavors together, to me, is the most fun part of cooking. That's where I get improvised and creative—where I experiment and do things I like. It's also the thing about cooking that intimidates most home cooks, I believe. I have the advantage of cooking for hundreds of people, so I get to try many more things and find the common denominator of how things work, or don't work, more quickly."

That common denominator for Roberts may be such variations on a theme as his corn risotto and his green pea guacamole. In fact, Roberts planned to be a composer until he decided the immediate fulfillment of cooking was better than the long road to uncertain success in music. So he went to Paris to study cooking at L'École Jean-Ferrandi and worked in France and in New York before coming to Los

Angeles, where he eventually was selected for Trumps by a restaurant partnership headed by Doug Delfeld.

Nine years later, Trumps has established itself as one of the city's premier restaurants. Some attribute its success to Waldo Fernandez's naturalistic design (featuring gray stone tables, tile floors, raffia chairs, and lots of greenery), or to the number of stars that come out to go there at night. Others might give credit to the restaurant's innovative high tea with cucumber sandwiches and scones, or the late-night supper, which few Los Angeles restaurants offer. But Michael Roberts is the prime reason anyone comes to Trumps.

No other place offers such imaginatively rethought dishes as potato pancakes with goat cheese and sautéed apples, or, my current favorite, an onion-raclette cheese pie in puff pastry. You can smell the aroma of the sensuous cheese, caramelized onions, and flaky pastry several tables away.

It seems to me that Roberts is so popular because he takes the familiar and adds his own personal, delicious spin, creating food that is accessible and unusual at the same time. I asked him how he would characterize his cooking.

"It's hard for me to put a label on my food," he answered thoughtfully. "There's some sort of personal stance that makes my food *my* food, whether I'm cooking Italian, Japanese, or anything. Like Picasso. He had all these different periods. And no matter what painting you are looking at, you know he did it. I think that's true about everything. It's true about the food I cook. I'm a chef whose personality is strongly in his food. The personality is there whether I'm baking a cake, making soup, cooking vegetarian, whatever. Like the Beatles—after a few notes you know it's them."

Now that Roberts has published his first cookbook, people who have never been to Trumps can experience his genius both as a cook and as a first-rate writer. *Secret Ingredients* features cooking that isn't overly complicated, isn't tied to expensive ingredients, and can be prepared by anyone in a relatively short period of time with foods on hand or readily available.

Roberts plans to do more writing. "I have some other books in the works. I see this as a shift in emphasis. I don't plan to give up my restaurant, but I think I can balance my life out well by cooking and then writing about it. When you work in a restaurant for a period of years, you feel drained, like everything is going out. I think, in a way, taking time out to write this book has improved the quality of my food."

Roberts feels strongly about the importance of vegetables in cooking. "Most flavors occur in the vegetable kingdom. All herbs and spices are in this kingdom. That's what all our food is flavored with. I don't understand where this misconception occurred that vegetarian cooking is bland and boring."

Well, I understand where that misconception came from. From someone who hadn't tasted Michael Roberts's cooking. Now, by sharing his talent with a wide audience through the printed word, Roberts will help to improve the cooking of other people who love food.

SALADS

Braised Belgian Endive and Baby Bok Choy

MAIN COURSES

Corn "Risotto" with Okra and Shiitake Mushrooms

SOUPS

Tomato Stew

ANNE ROSENZWEIG

Arcadia, "21," New York, New York

Just Like Grandma's

NEW YORK is intoxicating. The pulse of the city becomes the pulse in your brain. The winter wind whips the hell out of you, and the summer sun pushes you slowly but inexorably to the pavement. But you don't care. You're in New York! The extremes of pleasure and pain seem to come with the territory. But at some point each day the thrill-inebriated visitor and the pumped-up Knickerbocker alike need to escape. Many New York restaurateurs create soft nests of silk and leather to swaddle the nerve-hammered and weatherbeaten and make them temporarily forget the wondrous monster outside.

Native New Yorker Anne Rosenzweig has taken a different tack with Arcadia. Because she loves the outdoors, she has brought it inside and created a peaceful bower of a dining room. As you contemplate the four-seasons wraparound mural of a harmonious natural world, you are served food that seems to come from that same magical place. Arcadia is the ideal spot to escape from the realities of Manhattan.

Even though Anne never planned to be a restaurateur, she inadvertently had the right training. While doing ethnomusicological research in Africa and Nepal, she became fascinated with the role of food in cultural rituals, and decided to make food her career.

Returning to New York in the early eighties, she couldn't convince hardened chefs that she would be an asset in their kitchens. She was small, female, and untrained. But she was determined. She started as an unpaid kitchen person, the equivalent of an untouchable in the Indian caste system. She was given jobs like lugging huge stockpots and cleaning twenty thousand leagues of squid. But she hung in there, and eventually landed a paying restaurant job at a place in the village called Vanessa. Her grit and determination paid off. She got great reviews as a brunch chef and before long parlayed that good ink into the opportunity to open Arcadia with partner Ken Aretsky.

Her brunch training may be one reason why Rosenzweig's food is so likable. A brunch chef, unbound by the conventions of what people expect at a particular meal, has the freedom to improvise and experiment with combinations that might otherwise seem forced or contrived.

At Arcadia, anything goes. Take, for example, Anne's warm figs stuffed with Gorgonzola and walnuts on autumn greens. The figs had been slit on top at right angles so that they opened like flowers. The fig meat had been scooped out and mixed with lightly toasted walnut pieces, sweet Gorgonzola, crème fraîche, and fresh-ground pepper. The figs were then filled with the mixture, warmed through, and placed on top of various lettuces dressed simply in walnut oil. This is what you eat when you outgrow s'mores. I think if I were stranded on a desert island and I could only eat one dish, this would be it.

On second thought, I would probably choose the grilled leeks in puff pastry with onion *confit* and chive butter.

But no—I think instead I'd choose Rosenzweig's sauté of tart green tomato slices paired with sweet-tasting cherry tomatoes and served in a balsamic basil vinaigrette.

But why choose? At Arcadia, you can have all three. There are so many enticing dishes on the menu here that you'd be cheating yourself by simply ordering one appetizer and one entree. Take the three-dish approach and try, say, the grilled vegetable salad with niçoise mayonnaise, the corn cakes, and the savoy cabbage with kasha. Another tri-partite triumph would be the ragout of lentils, spring vegetable risotto with orzo, and the Gorgonzola-mascarpone fritters.

Good things have always come in threes for Rosenzweig. "Growing up, my sense of food was shaped by the TV dinner. Not that I ate that many TV dinners, but the idea of this three-compartmented meal where you have your protein, and you have your starch, and you have your vegetable stuck with me. And there was something about that that was very satisfying. As you ate it, you looked at it, and ultimately it was a kind of soul-satisfying thing, because everything you needed for complete nourishment was there. And from then on, a meal didn't seem complete unless all three components were there. So when I started developing the menu for Arcadia, vegetables began to take a larger and larger role because of their early importance for me. And I would always try to think of new ways to present them and new vegetables to use."

You won't get reheated corn, pea, and lima bean succotash with reconstituted mashed potatoes and greasy fried chicken at Arcadia. Nor will you get what Anne calls "effete vegetables." "On our menu there are lots of grains and unusual vegetables. We use kasha, toasted barley, Swiss chard: things with guttiness and hardiness. We call it 'grandmother's cooking,' except with more sophisticated ingredients."

Rosenzweig's suppliers are part of her success. "We have a lot of little farmers coming to us and growing unusual things especially for us. And that's important to us because vegetables play an equal role here with the meat and fish. That's pretty unusual compared to a lot of places. There are so many restaurants where you just get a piece of fish on a plate with a sauce. *That's* not a meal," she says emphatically.

Though Anne eschews "toy food," she's not a meat and potatoes kind of "boy food" cook, either. But she mounted up and rode west to "21" last year, and that's serious "boy food" territory, pardner. She almost got bushwhacked by some of the gang at the club, but she aims to clean up the place and bring it back to its former glory.

Her latest project? Oh, it's nothing, really. She's just convinced a bank to give her a tiny space on Madison Avenue for use as an herb garden. Well, if anyone can convince a bank, which the last time I checked was not an eleemosynary institution, to donate a chunk of the most expensive land on earth outside of the DeBeer's diamond mine for an herb garden, it's Anne Rosenzweig.

With such a high-pressure life, you may wonder how Anne relaxes in her leisure time. She kayaks and climbs mountains. These pursuits give her a chance to get back to the nature she loves, and to keep in shape for navigating the rough waters and overcoming the mighty obstacles she thrives on facing every day in Manhattan.

PASTA AND RISOTTO DISHES
Spring Vegetable Risotto with Orzo

. . . .

JIMMY SCHMIDT

The Rattlesnake Club, Detroit, Michigan

Looking for America

AMERICAN COOKING is not homogeneous. We cook the way we speak, with diverse regional distinctions. That is why Jimmy Schmidt and Michael McCarty teamed up to open a different kind of restaurant chain: the Rattlesnake Clubs in Denver, Detroit, and Washington, D.C. Though Schmidt and McCarty have parted amicably, with the latter taking over the easternmost and westernmost links in the chain, the restaurants still celebrate the regional riches of our culinary landscape.

Jimmy Schmidt's restaurant is devoted to the traditions and the bounty of the American Midwest. The Rattlesnake Club is located in a once-abandoned pharmaceutical plant that is listed in the National Register of Historic Places. Schmidt breathed new life into the building, completely restoring it and adding a striking decor. His food, though emphasizing dishes appropriate to its locale, broadens and deepens the flavors of the region.

"The restaurant focuses on Midwestern food with its very earthy flavors.

The food is lighter in style than most of the things that are done in the Midwest, but heavier than you would find in California. There's a lot of forage foods here in Michigan, like cattails, tubers, and ground nuts. It's really a bountiful region, with everything from wild mushrooms to cherries and apples, and even tiny white asparagus. As you go toward northern Michigan there are terrific cottage industries like American Spoon Foods. This is a part of Michigan that provides us with the ingredients we need to define our cuisine. The food here is not as spicy as Denver's, just because of the ingredients and what people like to eat here."

That won't keep Schmidt from sneaking onto the menu a terrine of corn, squash, and peppers lit up with jalapeño, or his masterpiece potato and yam gratin with artichokes and poblano chilies. But it's Jimmy Schmidt's "collision of flavors" approach, combining local with long-distance ingredients at their perfect stage of readiness, that makes his food so intriguing. So Schmidt combines local acorn squash and red onion with New Mexican pine nuts and Southwestern herbs to awaken Midwestern tastes to other influences. He also uses sunny ingredients like papaya, ginger, and hearts of palm in Detroit, but not to make strictly tropical dishes. That would make the restaurant there into a kind of culinary tanning salon for the snowbound, and that is clearly not Schmidt's intent. Instead he uses unusual products to create a dish that you can appreciate on several different levels of taste.

That is why vegetables play such an important role in Schmidt's cooking. "You can take vegetables, restructure them, change their textures, build them high or low. You can do different combinations and textures and even create different chemical reactions, whether they are starches or acids or proteins, allowing them to interact to bring about a change in flavor or color.

"The levels of flavor and texture in vegetable dishes are so much broader and deeper than in meat dishes. You take a tenderloin steak, what can you do with it? You can grill it, you can roast it, maybe brush some sauce on it, or alter the texture a little bit, but it's going to come out pretty much the same. What other dimension can you go to? You're stuck. The only other alternative at this point, and this is what really makes the dish great, is the vegetables. Vegetables give you the control to change the perspective of how you view the main item on the plate. For example, a crisp vegetable will really emphasize the tenderness of the beef. Or use a gratin. You can make a gratin using any root vegetable like potatoes, sweet potatoes—even taro root. Add the herbs, a little cream or cheese, and with all the flavor choices available you can get a whole lot going in that gratin. With the combinations available, that little gratin will out-distance the beef dish any day. The possibilities in vegetable preparations are endless."

I guess so, since you can get beet ice cream and grilled pumpkin at the Rattlesnake Club. And Schmidt insists on using the best organic, unsprayed produce he can find, because "that pesticide has to have some bad effect on us."

Vegetables here are not mere garnish. "We look for the main focus of the dish, and then all the other ingredients we use are really there for structural support, either through contrasting or similar textures and flavors, so that, say, if the peppers are the main focus, then they'll be reinforced by the olive oil and the vinegar and the olives. You build the flavors consciously rather than just throwing things in."

That consciousness about food could not have come without the inspiration and guidance of culinary genius Madeleine Kamman. Schmidt graduated first in his class at her Modern Gourmet School in Massachusetts, simultaneously working to pay the tuition at Kamman's late lamented Chez La Mère Madeleine. From dishwasher he rose to senior chef, a post that acted as a springboard for Schmidt's first solo dive at Detroit's London Chop House. It was there he developed his style of focusing on dishes representative of the area. Schmidt works with dedicated small growers to obtain the best local ingredients, harvested just when their flavors are peaking.

Schmidt sees to it that the main focus of the Rattlesnake Club is to represent the restaurant's locale. "Our overall philosophy is that the foods and the style of the foods should be reflective of the environment. By environment I mean the weather, the place you live, and your own past—what for you are traditional foods. These form the food memories generated from your childhood, and just as you eat turkey at Thanksgiving and barbecue in the summer, you feel comfortable eating certain foods at certain times and in certain places. We try to follow those same concepts, not necessarily using the exact same ingredients, but being true to the region. For example, here in Detroit there are great things available in the winter like seckel and comice pears and icebox pears; there are root vegetables. For me, using those ingredients is a better representation of what I want to do because these are the normal flavors you would experience in this region this time of the year. If I ended up using a kind of lazy susan from which I took ingredients that came in from all over the world at any time of the year, I think that I would fail in my duties as a chef by representing products out of their context."

"Does that mean that if you could get raspberries from Chile in December and you wanted to do a raspberry napoleon that you wouldn't bring them in?" I asked, fiendishly.

"That depends on whether they're great raspberries," he responded squelchingly.

The point is that, while not being a regional food idealogue, Jimmy Schmidt brings out the best in the region in which he cooks, pleasing the diner's palate by varying the chef's palette.

I dined at the Rattlesnake Club one rainy November night with the wind whipping up a froth on the Detroit River. The restaurant occupies a huge corner space, with rosewood lattice windows overlooking the river. The walls and columns are off-white, and the cherry wood floor is inlaid with green marble tiles. The idea was to bring the greens and browns of the Michigan countryside indoors.

The menu, too, features food from Midwestern fields and streams, occasionally spiked with Southwestern flavor notes. My starter was hearts of palm with red onion, acorn squash, and pine nuts. The mélange was coated in a garlicky vinaigrette with plenty of heat from chilies that sure didn't come from Petoskey, Michigan. One of the most enjoyable aspects of the dish was its textural range, from the soft hearts of palm to pliant squash cubes to crisp onions to crunchy roasted pine nuts, with each element remaining distinct yet well integrated.

Equally fascinating was the saffron risotto, filled with pearl onions, asparagus, and wild mushrooms. In another colorful dish, green and white noodles were combined with red, yellow, and orange tomatoes, peppers, fennel, and chives.

Other winning light dishes included the grilled eggplant and peppers with roasted garlic and basil: a juicy, olive-oily dish perfect with the whole-grain dill-and-red-onion rolls served at the Rattlesnake.

A nice alternative to the more formal dining room is the Grill Room just off the bar. For a quick, relatively inexpensive bite there is informal fare like blue cheese won ton, little pizzas, sandwiches, and light entrees.

Schmidt's restaurant is unusual in the States for including a service charge on the check, a practice that does more to improve the level of service in restaurants than any other. People working in such a restaurant have a reliable, verifiable source of income that enables them to do the things that most of us take for granted, like applying for loans and budgeting. The management treats the waitstaff like salaried professionals, and they respond accordingly. It's part of the corporate philosophy at the Rattlesnake Club, which believes that it is paramount to hold on to its employees. In fact, it would have been impossible to open the Rattlesnake Club in Detroit without the considerate attitude toward staff fostered at the Denver Rattlesnake.

Jimmy Schmidt says: "I've got half a dozen guys who have been with me as long as nine years. And there's a lot more I've had from three to six years. They're able to grow on a steady pace because they know there is something else coming up. They feel, 'Maybe if I get everything balanced out here and learn to do this I can go on and take this other position.' It's a good internal challenge. There's a very strong support group of people helping each other. It's a team. And I have the utmost confidence in them."

SIDE DISHES

Potato and Yam Gratin with Artichoke and Peppers

CORY SCHREIBER

Gordon Restaurant, Chicago, Illinois

Take It Lightly

ON MY first visit to Gordon Restaurant, I took refuge from a vicious thunderstorm. Since I was still a little chill of body, I chose pumpkin soup with Frangelico cream for my first course. The waiter poured the burnt umber broth tableside into a bowl centered with a plop of hazelnut-scented cream, and sprinkled it with roasted filbert shards and pumpkin seeds.

My next course arrived: impossibly light, totally greaseless, golden artichoke fritters served with perfect herb-studded béarnaise. A conundrum: How can a fritter, by definition crispy and crunchy, melt in your mouth? It was like cotton candy for grown-ups.

I looked around at the eclectic decor of Gordon: bleached knotty pine walls and ceiling, Persian rugs, billowy tied-back white raw silk drapes, and metal sculptures holding a mixture of dried and fresh flowers. I felt warm and welcomed, by the surroundings as well as the food.

Cory Schreiber was a fresh-faced young chef from Portland, Oregon, when he came to Gordon, a restaurant that had gone through two chefs in two years. He must have felt like a rookie pitcher coming to a ball club that had lost its last two starters to free agency. In his first stint as chef de cuisine all he had to do was satisfy over two hundred diners nightly, seven nights a week, in what is becoming a highly competitive restaurant town. And Schreiber does it, in the relaxed way of a man who has lived in the tumult of the kitchen all his life and takes it in stride.

Around the end of World War I, Schreiber's grandparents started Dan Louis's Oyster House near the Portland docks. When Cory was about knee-high to a mollusk he worked at the restaurant busing tables and shucking oysters. He went on to Westin Hotels' three-year chef apprenticeship program. His last Westin assignment was in Boston, where he jumped ship to work for Lydia Shire at the rival Bostonian Hotel. Four years with Shire culminated in his promotion to executive sous chef at the hotel.

There he came to the attention of Chicago restaura-raconteur Gordon Sinclair. The art collector, film buff, American Institute of Wine and Food power in Chicago, and all-around mover and shaker had heard how quickly young Schreiber was progressing, and hired him as executive chef in April 1988. Cory hit the ground running, employing Shire's tough-love management style to create a tight, family-style kitchen cadre that delivers its high-quality product to the table via a well-trained waitstaff. "My approach is not to take things too seriously. Take it lightly. Laugh at serious situations, and have fun. That way the kitchen crew works with you in a crisis, not against you."

While not a member of the meat-and-potatoes school, Schreiber avoids complicated, played-with, supercalifabricated creations. "I'm trying to make food people can eat with their hands—food that will recall the best memories of simple outdoor cooking, barbecues, and picnics."

Vegetables play a major role in Schreiber's cooking. And the symbiotic relationship the chef has developed with local growers (notably Mike Michaels of twenty-five-acre Ladybug Farms in nearby Spring Grove, Illinois) has given him the raw materials he needs to turn out superior vegetable dishes. The restaurant is so proud of its vegetable side dishes that the waitstaff often introduces them like this: "Good evening. Tonight's special is oven-roasted tomatoes and grilled green beans from Ladybug Farms. With them will be served fresh swordfish."

"I get more excited thinking about how I'm going to use vegetables and sauces than anything else," he told me. "The meat—that's what we sell the plate with. I only regret we can't get fresh, local produce here year round. I mean, how good can produce taste when it's flown in from a thousand miles away?"

One of the chef's ideas is serving the same vegetable prepared three ways: sautéed, grilled, and blanched, for example. "I look to vegetables to provide the color and texture to my dishes. *They* hold the artistic value." And in Schreiber's art, less is often more. For example one of his favorite offerings is artichokes drizzled with truffle oil—nothing more. "It's so simple. It's so obvious that these two flavors are perfectly mated. You don't need to do anything more."

Gordon has some of the best light cooking coming out of the Midwest. The eggplant with duxelles and Maytag blue cheese glaze, for example, is a paradigm. Tender eggplant underlies a sauté of coarsely chopped, not minced, mushrooms coated with pungent Iowa blue. Other interesting side dishes are creamed salsify with Parmesan, chanterelle-orzo pilaf, and spaghetti squash with pumpkin.

The square jawed, brown-haired chef has dreams of eventually returning to his Oregon homeland to open a small inn along the lines of Patrick O'Connell's masterpiece at Little Washington, Virginia. "Chefs in France and the Napa Valley have shown that where there's good wine there's good food. What's coming out of Oregon's vineyards is first-rate. I hope someday my cooking there will match it well." Regardless of where he cooks, Cory Schreiber's first-rate cuisine is not to be missed.

SIDE DISHES

Eggplant and Mushrooms with Blue Cheese Glaze

Braised Leeks with Saffron and Sherry Vinegar

JOHN SEDLAR

St. Estèphe, Manhattan Beach, California

Cactus on the Beach

I FIRST ate at St. Estèphe eight years ago, acting on a tip from Lois Dwan, then restaurant critic for the *Los Angeles Times*. "The chef, John Sedlar," she told me, "was trained by Jean Bertranou at L'Ermitage. I think you'll be pleased with what he does."

She was right. Sedlar surprised and delighted me with such dishes as wild mushrooms in an airy puff pastry shell, and elaborate *mosaiques* of vegetables so well constructed it was difficult to tell where food left off and plate began.

Sedlar hails from Santa Fe, and though he got his first restaurant experience there, he thought he was leaving Southwest food behind when he adiosed New Mexico for what he deemed to be the real Land of Enchantment: Los Angeles. After working for a few years in kitchens around Manhattan Beach, he offered his services free of charge at Chambord in Beverly Hills, so he could learn classical French technique from Maurice Peuget. From there he landed a position at L'Ermitage, where the great Jean Bertranou would make a lasting impression on the young chef. Under Bertranou, Sedlar not only became an accomplished saucier and chef, but acquired an eye for design and presentation that would become his hallmark.

When in 1980 he and partner Steve Garcia, a boyhood chum from Santa Fe, opened St. Estèphe in a Manhattan Beach shopping center, they made a smart move. Both rents and expectations were low, so that the inevitable mistakes young restaurateurs make were less costly then they might have been in Los Angeles. About the time all the bugs got worked out, St. Estèphe began attracting the attention of Caroline Bates of *Gourmet* and Lois Dwan, then of the *Los Angeles Times*. St. Estèphe became a raging success, catering both to the local South Bay crowd and Anglenos alike.

Then, over the course of the next few years, Sedlar slowly began introducing the foods of his native New Mexico through his "menu within a menu." But the dishes weren't the typical green chili enchiladas and sopaipillas of northern New Mexico. Sedlar took the ingredients he was familiar with from his youth and glorified them in artistic creations inspired by Bertranou.

The "menu within a menu" was overwhelmingly popular. It wasn't long before Sedlar made the menu within the menu without. And though on his new menu he kept the selections printed in French and still prepared boudins, mousses, terrines, and quenelles, he started using ingredients of the American Southwest almost exclusively. Mushroom duxelles appeared, stuffed in a green "Big Jim" chili imported from Española, New Mexico, and served in a garlicky sauce of New Mexico–made goat cheese. A salad of red cabbage and Roquefort was served with sage crouton arrows. And there were elegant two-toned soups, such as tomato and green-chili served in one bowl.

Eventually enchiladas, tacos, and tamales worked their way onto the menu. But they weren't your ordinary Tex-Mex, or even New Mex kind. The tamales were filled with juicy morels and black truffles. Enchiladas came stuffed with spinach and French cheeses, served in twin pools of red and green chili sauce.

As the food has slowly evolved at St. Estèphe, so has the decor. The once French-inspired interior has just as gradually given way to a decidedly Southwestern look, with the requisite desert art, cacti, sandy-colored walls, and a warming *horno* fireplace. Recently the restaurant increased its size with an enclosed patio (which can be chilly unless you're directly under the blue warmth of a propane heater or right next to the fireplace) and added other Southwest touches, so it feels like you're dining in the desert even though the ocean is only two blocks away.

Sedlar's menu is now in English. Yet he retains the fanciful shapes (triangular ravioli) and colors (pasta infused with red or green chili powder) and designs (sorrel, chili, or saffron sauce arrows and lightning bolts piped onto the plate) that have made his food unique. His food is some of the most beautiful I have ever seen, though never at the cost of flavor. With all he does to create that look, isn't his food too fussed with or fancy?

"You could say that," nodded Sedlar. "But I'd rather say our food is special. There are over thirty thousand restaurants in Los Angeles, and probably twenty-eight thousand of them are casual cuisine. There are many people that like that kind of food. I just happen to enjoy cooking special-occasion food. And there are people who come to my restaurant as a special-occasion restaurant. I try to make it as special as I possibly can for our customers. The food is not overfussed. The food isn't overhandled. But there is an idea, a conception behind what we do, and we work to make our food visually appealing when we execute our ideas on the plate."

The key to Sedlar's visual style of cooking is, of course, vegetables. His *mosaique* of carrots, turnips, and beets is stunning. But John relies on vegetables for more than just color.

"Vegetables are as equally important as meat in a balanced cuisine, especially now with the nutritional standards that people like to keep. In fact, I'd say vegetables are now taking over. Between salads, main courses, side dishes, and desserts, where you see a lot more fresh fruit, people are eating more vegetables than meat, fish, or dairy. Besides, vegetables beat the entree in flavor hands down. Other than sauces, which are often vegetable based, they're the most interesting flavors on the plate.

"And when you compare the variety of flavors and designs you can get with all the different vegetables you can put on the plate to the few choices you hve between beef, lamb, and poultry, there's no comparison. And I'm not just talking about a side dish of carrots and green beans. A good assortment of eight to ten vegetables beautifully conceived and displayed tastes and looks incredible. If you take it a step further you can create things like a checkerboard of vegetables. Or what I like to do is blanch a tomato, skin it, take a wedge and put some truffles on top, and it looks like a ladybug in the salad. You can create things that look like butterflies. You could replicate an entire garden scene with vegetables or play tic-tac-toe with them. The possibilities are endless."

But, I wondered to Sedlar, were the possibilities of Southwest cuisine endless? And did it have staying power?

"You mean can it become a permanent part of the food landscape?" he asked. "Absolutely. For a long time people didn't realize the good value inherent in Southwestern food. It's a very nutritional cuisine. Like anything else, the bad elements have be be separated out. So if you do away with the big gobs of cheese, and avoid an abundance of pork or too much deep-fat frying, the food has a redeeming value of being healthful as well as exciting and interesting. Especially if you liberalize the parameters a little bit and infuse other influences, which I occasionally do, like with my New Mexico sushi with jalapeños and corn or my potato salsa with fresh herbs or my papaya salsa with tropical fruits. Then the cuisine becomes very diverse."

But not diverse enough for the ever-expanding culinary imagination of John Sedlar. Now that St. Estèphe is running like a sewing machine, he and partner Garcia will open a "new American restaurant." To Sedlar that doesn't mean meat loaf, pot roast, and apple pie. His restaurant will be heavily influenced by the foods of Pacific Rim countries such as China, Thailand, and Japan, but it will include other traditions as well.

"We'll do Caribbean, South American, Eastern and Western European. Our cooking will reflect the diverse ethnic mix of America. That's what I mean by 'new American.' We'll have things on the menu like fresh sauerkraut enchiladas—a little bit of everything."

Sedlar tells me he wrote the menu for the new restaurant ten years ago, but had to wait until everything at St. Estèphe was perfect before he could contemplate opening a new place.

Sedlar sees the restaurant world as moving in ten-year cycles. A batch of places will open up and hold sway for about ten years, during which time the number of copycats grows increasingly larger until the originals are no longer unique. Then a new batch comes along.

Sedlar picked Santa Monica for its more "advanced food audience" as a site for his casual, mainstream restaurant, and other savvy restaurateurs are turning to the same area. This past year saw the opening of Opera, Fennel, and DC3 in Santa Monica. And Bruce Marder (operator of the latter) and Michel Richard are soon to open Broadway Deli on the renovated Santa Monica Mall. When Sedlar's new offering enters the picture, Santa Monica could be the West Hollywood (where Spago, Trumps, and La Toque held sway during the eighties) of the nineties.

APPETIZERS AND FIRST COURSES

New Mexican Sushi

GUNTER SEEGER

Ritz-Carlton Buckhead, Atlanta, Georgia

High-Energy Food

I'M TRAINING my seventeen-year-old daughter, Hari, to appreciate fine food. It hasn't been easy. She used to turn her nose up at the hotsy-totsy places I'd try to take her, pleading with me to drive through Taco Bell (a travesty, an affront to all people who need to eat on a daily basis) and Burger King (whose Whopper without the meat isn't half bad when a vegetarian is slumming). But after weekly doses of John Sedlar's amazing cooking at St. Estèphe, which was on the way home from her doctor, and numerous trips to Spago, our family's special-occasion restaurant, she began to appreciate what I was trying to do. She even created her own dish at Spago, a *feuilletée* of fresh vegetables and wild mushrooms in red bell pepper cream sauce, which she trained a succession of chefs from Mark Peel to Hiro Sone to cook to her exacting specifications. Of course it took no convincing on my part to get her to appreciate their raspberry–chocolate mousse cake. But, despite all the calculating inculcating I've done, I have learned you can keep the junk food away from a teenager, but you can't keep a teenager away from the junk food.

Still, on a recent father-daughter trip to Florida I felt no trepidation about inviting her to join me for a stop-over dinner at Atlanta's Ritz-Carlton Buckhead, where Jean-Louis Palladin had told me I would find one of the finest chefs in America. There I discovered Gunter Seeger, who has devoted his career to changing the food consciousness of the public.

The food exceeded all expectation on my part, but drew from my nacho- and pizza-loving daughter the most surprising of assessments.

"Daddy, *this food* I could eat every day of my life. Like that salad would be easy to make. I could make it. And look, if I could like have Mama make that clear mushroom soup with the little green things and the whatdayacallit mushrooms, we could put it in a thermos. Then, those four *won*derful vegetable purees on the same plate, well, it wouldn't *look* that good by the time I got to school, but it would *taste* fan*tas*tic. Now my *fav*orite was that creamy sweet potato thing in the hollowed-out sweet potato with the black truffles? [My daughter often turns declarative statements into interrogatives.] We could like pack that up, for school, couldn't we? Oh, and the those little flat onions between those layers of sweet pastry, we could put it *rea*lly carefully in one of those Tupperware things, and put the curry sauce with the fresh ginger slices in a little baggie and I could warm the whole thing up in the microwave at school. *God,* I would be *so* skinny if I ate this way every day. Whadya think, Daddy?"

"I think that if you appreciate Gunter Seeger's food so much that you want to eat it every day then all my struggles against the abominable diet you eat have not gone in vain."

In a way, my daughter is representative of the kind of eating mentality the lanky chef rails against in his never-ending struggle to raise the food consciousness of America.

"It's disgusting how people eat in this country. Ninety-eight percent of the public goes to fast food places. They eat hamburgers, french fries, ketchup. Seventy acres of pizza day! I can't believe how people eat. It's not that they're stupid. Too many people in this country just don't think about the food they're eating."

It's a situation familiar to Seeger. In 1976 he opened his first restaurant in Pforzheim, Germany. "At that time, the food was just dead there. There was no good produce, no lightness. No one was growing good produce then and many changes had to come. So I started working with small farmers there to give us nice fresh young vegetables. Then I had to train my employees and my customers. Finally, I succeeded."

The Michelin guide gave Seeger a star, something they are quite circumspect about doing beyond their borders.

In 1984, Seeger was invited to Washington, D.C., to become chef at the Regent Hotel. There he found a situation similar to the one he had found in Germany, despite the pioneering efforts of Jean-Louis Palladin who, four years earlier, "couldn't even find one fresh leek on the eastern seaboard."

"In Washington, I did the same thing as I did in Pforzheim. I wanted to bring freshness, color, and lightness to the table, and that takes first-class produce, the best produce."

He achieved that limited objective in Washington, then headed down to

Atlanta where cooking, like other Southern traditions, has been slow to change. But Gunter Seeger is a dedicated man who usually accomplishes what he sets out to do. He thrives on challenges and difficult situations, seeing them as opportunities, not obstacles.

"The customers here are used to big steaks. Slowly we have changed the habits of our customers away from their habit of eating big portions of overcooked, greasy foods to eating fresher foods with lots of vegetables. We use a short cooking time so that the vitality of the food that it takes from the ground can go directly into the body to nourish it and energize it."

We began with a salad of local lettuces—red oak leaf, arugula, and curly endive—in a nutty vinaigrette, the kind of salad you couldn't have had in Atlanta four or five years ago. But Gunter Seeger decided that he needed those kinds of salad greens.

"When I came to Atlanta three years ago it was like a desert. I didn't have the produce I wanted here, so I made it happen. I got in touch with growers around here and found a few good people who are proud of what they do—some people with intelligence and sensitivity. They experimented with growing things until they got it right, and now we're quite pleased. That's the great thing about the American market. It's flexible. If you ask for something, they'll make it for you."

Following the salad came a celery-scented golden consommé with chanterelle and hedgehog mushrooms and chives. The soup was pure essence of flavor, without any cream or butter to diffuse the delicate tastes of the vegetables.

Whereas the soup was a subtle evocation of flavor, Seeger's "dialogue" of vegetable purees left nothing to the imagination. Silky purees of beet, celery, parsley, and red bell pepper fanned out from the center of the plate like blades of color, each more vivid than its neighbor. The beet was an iridescent magenta, the celery an ivory-green, the parsley bright as an emerald, the pepper a harvest-moon orange—each one condensed to an intensity of flavor.

Seeger's food is high energy. He sees it as his mission to get the food to the table in such a way that you can still experience its taste with as little masking as possible. Thus, sauces often appear next to the main focus of a dish rather than poured on top. The *feuilletée* of cippoline onions is a perfect example: three small flat onions (not much bigger than a silver dollar) cooked until just sweet and tender, then placed side by side on a fragile leaf of caramel-coated pastry. Another leaf of pastry is topped by another row of onions, which is in turn topped by a final pastry sheet. The final touch is a burnt umber–colored crescent of curry sauce with overlapping penny-sized discs of fresh ginger root.

"Many chefs have no pride, so there is no personality to their cooking. There's no energy to their cooking. I'm serious about my work. I must be proud of everything I do. That's what gives my cooking its personality."

No better proof of this could be found than in Seeger's presentation of yam mousseline with black truffles. Seeger extracts everything he can from the black beauties: after placing a golden mousseline of yam back in its pale purple skin, he boils the truffles, reserving the water, and slices them thin, overlapping the black discs atop the yam. The yam is then placed in a clear sauce made from the reduced truffle stock.

Seeger realizes the importance of vegetables in our diet because he knows their potential for giving us health and energy. But he also knows he must move slowly in educating his tradition-bound audience to his way of cooking.

"You have to go in steps. And it takes time. I see the next step in cuisine going even more toward vegetables and less toward meat. Sometimes you think it's hopeless, but if out of one hundred diners I take ten to the next step, it's very nice. You see, the meat industry is very rich and very strong. It'll be a battle, because if people ever realize how unhealthful it is to eat all that heavy, greasy meat, the industry stands to lose billions of dollars. They don't want to lose that. So we try to do the best we can. I don't think we can change everybody overnight, but when you think about it, what is more important than food? If you don't have the right food in your body to give it energy, you can't do anything."

After the interview, my daughter showed me her sketch of the shaggy-maned, moustachioed chef. "Daddy, I *had* to draw him. He was so handsome when he talked. He was *so* into how he felt about his cooking."

"That's probably why his food is so good, don't you think?"

"Yes. I really could eat his food every day. And he was *so* handsome."

APPETIZERS AND FIRST COURSES

Dialogue of Vegetables

Cepe Carpaccio with Garlic Chives

PASTA AND RISOTTO DISHES

Yam Ravioli with White Truffles

PIERO SELVAGGIO

Valentino, Primi, Los Angeles, California

Still on Top

THERE I was at UCLA, after a two-year absence from college, on my first day of graduate school. I had been forced out of my original plan of going to graduate school right after college by fear of Lyndon Baines Johnson, General Hershey's draft board, and the whole wonderful Vietnam experience. But now, with hostilities nearing an end and a high draft-lottery number in my hip pocket, I was back, with a vengeance—ready to study obscure Russian poets of the sixteenth century. Yet at my moment of triumph, that first day of classes, I suddenly realized that that just wasn't what I wanted anymore. I had changed, and what they were teaching seemed totally irrelevant to my life.

It happens. Piero Selvaggio can tell you that. He was working on *his* graduate degree in literature at that same university when he decided he had to leave. "I realized that learning about the writings of Lope de Vega had nothing to do with my life."

When I came to my realization, I started studying yoga and meditation. Selvaggio started learning the restaurant business. After a few months managing somebody else's place, Piero and a friend bought a bar on a nondescript section of Pico Boulevard in Santa Monica and opened Valentino. Within three years, Selvaggio bought his partner out and, with his charm and personal attention, took Valentino straight to the top. That was sixteen restaurant years ago—a lifetime in people years. And Piero Selvaggio has been through what must seem like several lifetimes in his career as Los Angeles's (probably America's) foremost Italian restaurateur.

In the early seventies, diners in Los Angeles weren't too discriminating. Chefs stayed pretty much in the kitchen, and people were generally satisfied with the fare served up by Selvaggio and crew. It was a cut above the tomato-sauced, garlicky food found at the red-checkered tablecloth places with hanging Chianti bottles. The restaurant was packed most of the time. Then one night, the former chairman of Standard Brands Paint Company, who was a regular at Valentino (and who knew good food), told Selvaggio that what he was serving was "horrible."

It was 1977, and winds of change were blowing. Selvaggio knew he'd have to change his food or risk losing his edge to up-and-coming chefs like Michael McCarty and Wolfgang Puck. So he headed back to the old country. It had been nearly fifteen years since he had left his native Italy at the age of seventeen, knowing next to nothing about food. Now it was time to reintroduce himself to the cooking of his homeland.

He brought back with him the foods of Italy and the people to cook them. He introduced America to *mozzarella di bufala*, fresh porcini mushrooms, sun-dried tomatoes, radicchio, arugula, extra-virgin olive oil, and white truffles. Today, many of these foods are available in supermarkets, and some have become restaurant clichés. But the kind of food served at Valentino in the late seventies and early eighties was downright revolutionary in its time.

Valentino was now serving inventive and exciting Italian food that was both faithful to the traditions of Italy and up-to-date with current trends there. And toward the middle of the eighties, with Spago, Chinois, Trumps, and the West Beach Cafe drawing diners to eat in beautiful surroundings, Selvaggio decided the stodgy look of Valentino had to change as well.

In 1987, a totally redesigned Valentino emerged, including an outdoor patio complete with fountain, and a striking interior swathed in imported fabrics, Italian tiles, pastel-colored walls, and avant-garde lighting.

The menu doesn't begin to tell the story of the possibilities at Valentino. Yes, you can order from it and have a wonderful meal. For example, some recent starters were a fragrant vegetable and mushroom soup; a salad of warm radicchio, green beans, and mushrooms with *mozzarella di bufala*; and a savory baked eggplant parmigiana with ricotta and tomatoes. The pasta dishes included fettucine in a rich walnut-mascarpone sauce, as well as a chewy risotto with three kinds of mushrooms.

But to really experience Valentino, you must put yourself in the hands of the maestro himself, Piero Selvaggio. Suave, fit, good-looking, and eminently accessible, Selvaggio will see to it that your experience at Valentino is not just enjoyable, but memorable. If you display the slightest interest in whatever might be special that night, Piero will offer you foods you may not have heard of or tasted unless you've just come back from Italy (as he often just has).

"Cooking is a creation. As a creation, it is a personalized view of the way we like to express our feelings. It is how we share our sense of art, our knowledge, and our taste with other people. Food, after all, is not merely a product. It is necessary to our sustenance. Food is the support of life and is the center of the way we live when we take a moment to sit down and share life, share conversation, and share joy. That is the joy of cooking, which is a cliché, and yet it is the ultimate way we really fulfill ourselves and those around us. When people come to us and reach out to us, we must reach out with a very personalized and individual way of expressing our beliefs."

Piero may suggest cold appetizers such as a marvelous variation on the *tricolore* theme—a flag of basil, mozzarella, and tomato mousses—or *involtini di melanzane*, Italian goat cheese rolled in marinated, paper-thin slices of eggplant. Then he'll offer hot antipasti such as fresh porcini mushrooms, sliced thick, sautéed, and served smothered in *mozzarella di bufala*; or zucchini blossoms filled with Gorgonzola, breaded, and then deep-fried. The touch of a fork brings forth a creamy river of cheese from the golden puff.

After bringing you four carafes of olive oil, each from a different region of Italy, Piero will invite you to dip your bread in each to enjoy the subtle flavor variations.

He'll then recite a litany of pastas, risottos, gnocchis, and polentas that might include buttery *garganelli*, slender tubes of substantial pasta that seem to melt in your mouth; a creamy risotto simmered with corn and red bell peppers; or ricotta dumplings with basil and herbs.

Of all the countries in the world, Italy offers the most possibilities to a vegetable lover. Selvaggio agrees: "Vegetables are one of the key factors in Italian cuisine. We have a tremendous amount of appetizers, featuring things like eggplant, zucchini, and radicchio. Then there are the vegetable pastas, risottos, polentas. Vegetables are called on constantly to be the sole focus of the dish or to share the plate. All in all, they predominate in everything we do."

Valentino is expensive, especially if the tab reflects an endless string of unusual dishes Selvaggio has sent forth for your enjoyment. With the disappearance of tax-related expense accounts, the proliferation of many reasonably priced, casual restaurants, and the number of former singles who are now raising families, top-level restaurants are feeling the pinch, as Piero is well aware.

"The segment of the population that has money is getting bigger, but it's reticent to spend big money. Big money and a famous restaurant have become a difficult marriage. There aren't that many special occasions in everybody's life, so that other than Saturday nights, even with Mother's Day, birthdays, and anniversaries, you operate half-full. Then you throw in major sporting events on TV and religious holidays and it can get kind of depressing. In such a situation, a restaurant can only survive, like in my case, where there is the enthusiasm, the desire, and the professionalism that comes from the leader. It is enormously expensive to do a serious restaurant these days with qualified labor, great ingredients, and tremendous overhead.

"We have to face reality. And reality is that there are too many restaurants at the top. The restaurants at that level survive because of talent and dedication. But we must now appeal to the diner who is spending his own money, not the company's, and who has discovered the bistro, the cafe, the trattoria, the bar and grill."

Three years ago Selvaggio opened Primi, a few miles west of Valentino and also on Pico Boulevard. Primi was conceived as a restaurant where all the dishes were *primi piatti*, or first courses, so a guest could eat several small courses without ending up with a big bill. In practice, the tariff can mount rather steeply, but it is possible to order most soups, salads, antipasti, and pastas for $10 or under and still get the kind of superb food as is served at Valentino.

In the rear of the sleek pink, gray, and black dining room is an open kitchen out of which come such marvels as a soup of corn, barley, and Tuscan beans with a green-gold float of virgin olive oil; and an antipasto platter of assorted sweet peppers, beans, rich caponata, and cheese. The hot antipasti feature a cheese and eggplant timbale with olives and tomato, as well as a corn crepe with asparagus in a mushroom sauce. Salads feature generous chunks of the same buffalo mozzarella that Piero introduced at Valentino, and the pastas range from *rotolo*, pasta rolls with ricotta and spinach, to *margherite*, delicate daisy-shaped pasta filled with artichoke and served in a creamy asparagus sauce. Vegetable risottos, like the one with radicchio and Gorgonzola, or the *risotto verde* with an assortment of green vegetables, are only $11.

For Selvaggio, Primi is a stepping stone to what he sees as the future of American restaurants. As he puts it: "You keep looking around for the future, the reality of what's next, what people want. After all, we are in a people business." So he has two more restaurants on the drawing boards. One, Caffé Georgio (named for his toddler) will be, in the owner's words, "a restaurant of the nineties with more of a push toward real food, with a sense of art on the plate and the surroundings."

The other is Abbondanza, which, according to Piero, "will be a family restaurant in the true sense of the word. A restaurant where there will be a direction of eating from the center of the table. A tureen of soup, a bowl of salad, or pasta that can be shared by the family or party of four. It will be basic, old-fashioned, what I call 'revisited American-Italian-American food.' "

But don't look for spumone, spaghetti and meatballs, or wicker-covered Chianti bottles from the man with one of the largest wine cellars in America. Piero Selvaggio will uphold his standards as he explores new ways to reach the public, because he is first, last, and always a restaurateur, not merely a businessman.

"The music and sound of the show is the restaurant on a daily basis. If the music is exciting, made by happy people, grateful people applauding you and respecting you for what you do, this is the stimulus that makes you go to work every day, and makes you feel the sacrifice and difficult times are worthwhile."

SIDE DISHES

Involtini de Melanzane

Zucchini alla Menta

PASTA AND RISOTTO DISHES

Risotto con Asparagi

Corn and Peppers Risotto

JACKIE SHEN

Jackie's, Chicago, Illinois

"This is my baby; this is my joy."

JACKIE SHEN works with a steady but furious intensity for eighteen hours straight almost every day. She arrives at her restaurant about the time most people's REM sleep is ending, roughly 4:30 each morning. After making sure the kitchen gets properly cleaned, she orders the day's food. There's no time to drive downtown to shop for her daily provender, much as she'd like to, but Jackie has established a reputation with her purveyors of accepting only the finest.

And that's what they bring her. By 9 she has launched into lunch prep with only one assistant chef. As soon as the lunch rush is over Jackie starts getting ready for dinner, and by 10 that night (with only two other cooks) has turned out eighty remarkable dinners. Finally, around midnight, when all the necessary paperwork associated with the restaurant has been completed, Jackie calls it a day.

Of course, she could hire more staff, but Jackie learned early on that you can maximize profits when you do more things yourself. Besides, when you're the perfectionist Jackie is, you know that's also the way to be sure things are done right.

Jackie was born in Hong Kong and grew up in the British Crown Colony speaking both English and Chinese. At the suggestion of a high school teacher, she decided on a career in hotel management, since Hong Kong provided many opportunities to work in that profession. The teacher recommended that Jackie go to school in the United States to gain the necessary skills.

So Jackie enrolled in the University of Houston's hotel management school, and after graduation was hired by Chicago's Ritz-Carlton as an assistant manager. From there she went to the Park Hyatt as a restaurant manager. But when a position opened up to become a steward, Jackie leapt at the chance. Now, that position may sound glamorous, but here's the job description: Wash dirty pans in hot kitchen.

Shen stayed in the kitchen and managed the room service operation. By now the restaurant bug had bit her, so she bought a taco stand with seventeen counter seats and renamed it Uncle Pete's Snack Shop.

Uncle Pete's provided experience, but it could not offer Jackie Shen the training she needed to become a great chef. So she sold the coffee shop, and went to learn from Jean Banchet, who was then at La Mer. Before too long, her confidence and skills had grown, and Jackie's Restaurant was launched.

Jackie Shen's cooking is decidedly Western, though Asian ingredients do find their way into cooking. I began a meal at Jackie's recently with a svelte cream of watercress soup. I went on to one of her signature dishes, a delicate filo nest with exotic mushrooms.

But it's Jackie's pastas that best demonstrate her great versatility in creating light dishes. That evening she prepared a golden puff pastry cornucopia out of which flowed a swirl of orange pumpkin fettucine interspersed with four varieties of grilled mushrooms. A pioneer in the use of edible flowers, Jackie filled out the color spectrum with a crown of edible purple and yellow pansies and pink orchids. The flowers added not only color, but light herbal flavors to balance out the richness of the pasta and its lemon cream sauce. For still more color and flavor Jackie worked in bits of marjoram, sun-dried tomato, and fresh raw garlic, providing potent little flavor explosions to give verve and force to the dish.

Another pasta winner is the avocado, linguine, and tofu salad. The pasta is dressed in a hot peanut sauce and topped with avocado, Chinese marinated tofu, lotus root, and "flowers" made of pea pods and cherry tomatoes. Toasted sesame seeds provide a contrapuntal crunch.

Jackie is aware of how crucial the look of a dish is to its ultimate enjoyment, so she personally "paints" each plate with her arrangement of the food for maximum sensual appeal. The myriad colors and shapes of vegetables suit this purpose perfectly.

"Food is beautiful, and food is for one's inspiration. If one is inspired by the food we serve and he can find creativity and imagination, if he is uplifted by what he eats, then he will go from here and be more inspired, more creative."

Jackie keeps her food on the light side. "I'm aware now that the trend is that people are into lighter, healthier food. People come in here now and they eat well, but they do not eat to get fat. They eat to enjoy. And I like to see that what they eat they enjoy and that they enjoy their health. Since I'm doing it myself, I would like to help other people do it, too.

"I feel lean and healthy when I have just vegetables for a full meal. I can sleep better and work better. The food's vitamin value gives me the energy to carry on my work load. I still eat candies and chocolate sometimes, but I'm aware how much that affects my body. So it only happens once in a while," she laughed. "I try to remember what my mother said, as all mothers do: 'Eat your vegetables.' "

Recently, Jackie expanded her Lincoln Park storefront and added eight more tables. The hours are even longer now, and though Jackie is finally increasing her staff to help ease her load, she stays on the line to make sure her standards are kept high.

"I can hardly see myself missing an evening and not being able to cook. So in my own diet I'm really becoming cautious about what I eat so I can be here to give my best. What I eat affects my body, my health, and my mental state. When I'm thinking or doing something and I get tired, I wonder why I'm tired." (She wonders why she's tired?)

But Jackie Shen has worked too long to let fatigue get in the way of her success. She'll make whatever mid-course corrections are necessary to keep herself at her fighting weight. For Jackie's is not just a job for Shen. It's her life. As she puts it: "This is my baby; this is my joy."

APPETIZERS AND FIRST COURSES

Warm Asparagus with Pine Nuts and Orange Rind

SALADS

Avocado, Pasta, and Tofu Salad with Chinese Sesame Seeds

PASTA AND RISOTTO DISHES

Fettuccine with Sun-dried Tomatoes, Basil, Broccoli, and Goat Cheese

ANDRÉ SOLTNER

Lutèce, New York, New York

A Serious Chef

ANDRÉ SOLTNER has been cooking at Lutèce for an astounding twenty-eight years, creating wonderfully complex but uncomplicated dishes that rarely fail to satisfy. As Alice Waters told *USA Today* on the silver anniversary of Lutèce, "It is one thing to produce wonderful dishes for five years. For ten is another. But for twenty-five? That is rare."

Soltner doesn't stray far from his classical and regional French roots, and he doesn't make bizarre combinations of food in his search for new dishes.

"I hope that when people think of André they will remember me as a serious chef. You know, to be a chef you have to have a certain talent to cook. You are born with that, but the main thing is to be serious about it. A painter, for example, he has a talent, but he doesn't feel like painting every day, because he is an artist, you know. So it doesn't matter if his painting waits a week, a month; it doesn't matter. If you run a restaurant you cannot think this way because we are open every day and we have to deliver every day. So that needs real discipline. If I don't feel like getting up in the morning, I have to get up, because at noon, I have to deliver. So that is the difference between craftspeople and artists. I consider our work craft, which doesn't eliminate a certain talent or artistry, but we have to have the discipline."

Growing up in Alsace, André wanted to work with plane and chisel, like his father the cabinetmaker, rather than colander and whisk. But custom dating back to feudal times dictated that the eldest son take over the father's business. So André signed up for a three-year apprenticeship at the Hôtel du Parc in nearby Mulhouse. He still credits the chef de cuisine there with making him a professional, but his greatest teacher was the one who instilled the love of cooking in his heart. "It started with my mother. She was my first inspiration. She gave me the love of cooking."

And that is what André Soltner is all about. "There is an invisible string of love between me and my customers. The food I give pleases them. The love they give in return pleases me." The chef's regard for his customers is apparent the moment you enter Lutèce.

The skylit pink patio with a green flagstone floor, white trellises, and colorful wicker chairs is my favorite room at Lutèce. The two upstairs rooms with plush chairs, chandeliers, and carpets are formal without being stuffy. But wherever you dine at Lutèce, the solicitous staff and chef make you welcome.

The waitstaff performs well because they are treated well. Soltner pays his entire seventeen-person staff, right down to the dishwashers, more than just about any place in town. Some of the same people who were washing dishes here twenty years ago are still at Lutéce—cooking right next to Soltner. He has almost no turnover, and turns away about as many job applicants a day as he does would-be diners. Those fortunate enough to work for this caring man are provided with major medical insurance coverage, a pension plan, and even profit sharing.

When you go through the classical French chef's apprenticeship, as Soltner did, you learn how interrelated all the stations in the kitchen are in creating a memorable dining experience for the customer. What Soltner lacked when he left Hôtel du Parc at seventeen he picked up at restaurants in Switzerland and other parts of France until he landed his first chef de cuisine position at the Alsatian Chez Hansi in Paris. It was there that André Surmain met Soltner and in 1961 invited him to come to America.

In 1972, Surmain sold his interest in Lutèce to Soltner, who charted the restaurant's successful course to its current position as one of the preeminent French restaurants in America.

Those early days were difficult for the classically trained traditionalist, who sometimes felt his hands were tied by the lack of good raw materials. He wouldn't even attempt certain dishes because he couldn't get the quality ingredients necessary. Now he relishes the fact that he can get girolle mushrooms from the Northwest and real apples and peaches from upstate New York. Despite the burgeoning interest in growing organic greens, vegetables here still fall short of the chef's high expectations.

"We would like sometimes to be better served in vegetables. But if you compare with ten years ago, it's much better today. Back then they grew big vegetables here with no flavor. A potato had to be very big. It didn't matter if it had no taste and in the middle there was a big hole! There are local farmers now who are aware the demand is here and that they can make a living growing vegetables. But still the flavor and selection are not as good as in Europe. I think there should be a greater concern about growing good vegetables. They are absolutely integral in my cooking—as important as meat and fish."

When Soltner cooks he brings out the best natural flavors of his ingredients rather than making them vehicles for sauces. A case in point was a fresh vegetable plate I had recently at Lutèce.

The braised endive, the *haricots verts*, the pureed spinach, the artichokes—all tasted like primal vegetables, repositories of that vegetable's pure essential flavor. The tomato was bursting with late-summer ripeness and filled with a creamy ratatouille; the silver dollar-sized potato pancakes at the edge of the plate were light, yet hearty.

I had just eased back into my gently yielding wicker chair to reflect on the perfection of those greens, when the waiter presented me with an onion tartlet in a puff paste basket complete with handle. The filling was sweet and savory, the pastry almost translucent.

André is dedicated to maintaining this high level of food day in and day out. He has no intention of opening another restaurant. "I don't see why I should have two restaurants. I've got too many problems with this one," he laughed.

"But seriously, income-wise what I need I have from one. For pride or whatever, I can accomplish enough with this one. I don't see why I should have two. I couldn't be in both places at once, and if one slipped or if I enlarged and the food suffered, I might lose everything."

André's wife, Simone, is his hostess and cashier, and Lutèce is in the same building as their apartment, two floors above the dining room. So dining at

Lutèce is a little like dining at the elegant home of a friend who happens to be a magnificent cook.

André intends to keep bringing pleasure to others, as he has done for the past forty years. "I'm going to do it as long as I enjoy it. It will end, like it does for everybody, some day, but I'm happy, and I want to spend my fifteen or sixteen hours each day making those who come to Lutèce as happy as I am."

SIDE DISHES

Artichokes, Cauliflowerets, and Mushrooms a la Barigoule

. . . .

JOACHIM SPLICHAL

Patina, Los Angeles, California

The Great Unknown

HE'S THE greatest unknown chef in America. But that's going to change very quickly now that Joachim Splichal has opened Patina on the site of the venerable Le St.-Germain in Los Angeles.

For the last few years, you could only enjoy Splichal's food if you were a member of the exclusive Regency Club in Los Angeles. It's been a long time since anyone like you or me could just drop into a restaurant and taste the innovative cooking of this compact, German-born, French-trained chef. In the early eighties, Splichal, who worked with such Michelin three-star chefs as Louis Outhier and Jacques Maximin, cooked at Los Angeles's Seventh Street Bistro, where he introduced a vegetarian *menu dégustation* that he dubbed *les richesses du jardin*, "the riches of the garden."

You began with a salad of wild rice and asparagus with an herb vinaigrette, served in a radicchio cup. The second dish was an artichoke heart filled with ratatouille. Splichal then cooked up a colorful tomato and *haricots verts* sauté in a cream suace. Next was a plate of spinach with buttery sliced morels, followed by chanterelle and shiitake mushrooms poached in a consommé of chanterelles, then wrapped and briefly baked in an emerald green cabbage leaf that held in all the woodsy juices. Splichal saved the most stunning dish for last: a regal-looking vegetable napoleon of feathery puff pastry layered with asparagus, spinach, tomatoes, mushroom duxelles, and about fifteen other vegetables served on a pool of red pepper cream sauce.

Splichal is in a class by himself when it comes to vegetables. "When I came to the States and saw the overcooked vegetables you got when you asked for a vegetable plate, I decided to do something different. So I treat vegetables with the same respect and prestige you give a meat or fish dish. You put the different elements together and really present a beautifully thought-out dish. My philosophy about vegetables is that you can be very innovative and use things you normally don't use or use them in ways no one has thought of. And I just like the challenge of creating something new, I think. It's fun."

Splichal moved on to Max au Triangle, where he created such original vegetable dishes as a "sausage" of avocado in a broth made by steeping vegetables and herbs in water, as well as "weiner schnitzel" made from scalloped artichoke bottoms and served in a lemon sauce, not to mention "sushi" of wild rice and vegetables wrapped in spinach instead of seaweed.

Now, at Patina, Splichal has further refined and simplified his vegetable dishes. At a recent dinner, my companion opted for the vegetarian prix fixe "Garden Menu" while I happily jumped around the menu picking this and that. We were both well satisfied.

The "Garden Menu" begins with a "semi-carpaccio" of summer vegetables, including wafer-thin slices of broccoli, carrot, cauliflower, green beans, green onions, asparagus, artichoke hearts, and turnips, each cooked separately and served in warm olive oil with freshly ground black pepper. Next came what was advertised as a two-colored tomato tart, which was actually three-colored: alternating red, yellow, and green tomatoes, skinned, halved, and briefly baked cut-side-down over a sheet of puff pastry and served with niçoise olives. The final course was sweet white baby turnips, stuffed with a duxelles of shiitake and oyster mushrooms and served with a beurre blanc sauce. Sprinkled around the turnips were strips of the same mushrooms, cooked in wine.

My a la carte selections included a superior salad of Malibu field greens with crisp-on-the-outside, soft-on-the-inside polenta "croutons," a heap of French green beans with cream of summer tomatoes, and a remarkable soup billed as "metamorphosis of a potato, garlic, and thyme terrine in a soup." The waiter brought a slice of the colorful terrine to the table in a soup bowl and then poured one of the most delicious potato soups I've ever eaten over the terrine. The metamorphosis took place when the mashed potato portion of the terrine melted and merged with the soup. As a bonus dish, Splichal brought to us warm, gooey corn blini with shiitake mushrooms and green onions in celery-lemon sauce.

Splichal's restaurant is warm, white and woody, with touches of gray and copper. There is no art on the walls, but there is true artistry in the kitchen and on the meticulously presented plates. The years of planning that went into the design of Patina have paid off, and just in time, as far as Splichal is concerned.

Until plans were finalized for the restaurant, for two years Splichal was a globetrotting, eighteen-hour-a-day consultant for such concerns as Chicago's constantly packed Cafe 21. Traveling all over the world to vacation destinations, planning menus, and talking about gourmet food all day sounds glamorous. But that wasn't quite how it was.

"It's very tiring," said a relaxed Splichal as we talked about his consulting days at the Regency Club overlooking the Bel Air hills. "It's like each day you start another business. You go to Hong Kong. You get off the plane. Boom! They expect you to be on the ball when you get there. You leave, sleep on the plane, arrive in L.A., change your suitcase, and fly to Chicago. They're waiting there. The meter is ticking for them. So you have to come up with new ideas, new things, solve problems. It's very demanding because it's a one-man show. Although it was good from a financial standpoint, it was a tremendous stress. I decided I needed a base, so I'm ready to settle down with Patina."

Joachim and his wife, Christine, have thought carefully about their new venture. "We've created a restaurant of understated elegance, very anti-trend. No art on the walls, no huge dining room, no gigantic bouquets of exotic flowers, no noise. It's four little rooms in which we can serve up to a hundred people. What we're presenting here is innovative food based on what's fresh that day from the market, in an intimate environment of personal service. We won't be a restaurant that has to produce five hundred covers a day just to meet its expenses. So we don't have to order tons of this or that. We'll therefore be able to individualize a lot of things and come up with many interesting ideas."

Splichal wants to work with local growers to produce things specifically for him that might not be available anyplace else. "We'll have a contact with the product and be able to control it from its very creation."

The chef's varied experiences in the food world have taught him a lot about the restaurant business and about what he wants to achieve with his own place.

"I've changed over the last three years tremendously. I've really got a different viewpoint on what my life is all about, what I want to do in the long range, how I will structure my life. I've learned a lot about business in the last five years. You can have the best restaurant, but if you don't make the bottom line, you won't exist for long. I know how hard it is when you never make the bottom line, believe me. It was the best experience for me, though. It made me change. I am much more aware, tougher, in a sense. I know what I want. I have a direction."

That explains why Splichal gave his new restaurant the name Patina: it means the glow of something grown beautiful over time.

MAIN COURSES

Napoleon of Vegetables

JEREMIAH TOWER

Stars, 690, San Francisco, California, and Peak Cafe, Hong Kong

JEREMIAH TOWER likes people to have a good time. That's why he opened 690 Restaurant, with its wraparound mural of frolicking bathers. Tower calls it a place for those who've graduated from the Hard Rock Cafe but aren't ready for the more up-scale Stars.

Stars is a good time, too, with the baby grand driving a jazzy rhythm across the star-carpeted room, the warm yellow walls bouncing with the laughter of convivial company, and the great open kitchen pouring out delicious aromas. And, from the food to the decor, Jeremiah Tower designed it all.

Design figured in Tower's original career plans. He graduated from the

Harvard Graduate School of Design and was on his way to Hawaii to ply his trade when he ran out of money in Berkeley, California, in 1972. He heard there was a job opening at a restaurant called Chez Panisse and, while there to interview for the position, was pressed into service to save that day's soup. With a little white wine, cream, and seasonings, a star (and ultimately, Stars) was born.

Though Tower had no formal training in cooking, he had a great background for being a chef: an appreciation for fine food from his early years. The son of an executive, Tower traveled first class with his family more than twice around the world before he was sixteen.

Tower put his wide knowledge of food to use at Chez Panisse. In 1972, he originated a Northern California regional dinner there that, as Jeremiah puts it, "changed the face of American dining." Until then he had been cooking dinners based on the regions of France. This dinner, however, was based on local ingredients and updated American recipes. The new approach drew the attention of American food writers, and drove a wedge between Alice Waters and Tower, who eventually sold his shares and left for London to consult for Time-Life's *Good Cook* series, then returned to California to cook at the Ventana Inn in Big Sur and to teach at the California Culinary Academy.

He then signed on as chef and co-owner at the failing Santa Fe Bar and Grill in Berkeley, which he turned around in short order. He did the same at Balboa Cafe in San Francisco. It was obvious that Tower was ready for his own restaurant.

On July 4, 1984, twelve years after he arrived in Berkeley with five dollars to his name, Jeremiah Tower launched Stars. The three-tiered restaurant has the look of a Paris brasserie, which is exactly what Tower had in mind. Inspired by La Coupole, he created a restaurant where people of every station could feel comfortable dining several times a week.

Tower's cuisine relies heavily on the vegetables and herbs of California for flavor and color excitement. To assure that he has the finest ingredients, he has his own Napa Valley farm, a luxury that is reflected in the daily-changing menu.

One of my favorite salads is his crisp Northern California lettuces with pears, walnuts, and Roquefort cheese; another is his mango and avocado salad with tomato *concasse* and a chili-lime dressing. Then there's the warm okra, tomato, and basil salad, the okra coated in cornmeal and deep-fried.

Tower feels that vegetables can stand on their own in a wide variety of dishes. Often he'll make vegetable soups without the addition of chicken stock. His chilled sorrel and herb soup with tomato cream, for example, is water based, to "give a purer, truer taste of the greens themselves." His hot garlic soup with sage butter–filled profiteroles is another winner.

It is in his vegetable side dishes and main courses, however, that Tower really shines. His raclette with radicchio and cherry chutney is an unqualified triumph of California bistro fare. The pungent cheese is melted over boiled potatoes and raw radicchio tossed in a lemon vinaigrette. Then the tangy chutney is placed on top to bring out all the flavors. Another wonderful dish is squash blossoms stuffed with goat cheese and served with both a tomato vinaigrette and an herb mayonnaise. And for sheer heartiness, the grilled polenta with tomato *concasse* is hard to beat.

As a main course, nothing can take the chill out of a foggy San Francisco night better than Tower's warm vegetable stew. Tower also makes this dish with water rather than chicken stock, because "the resulting sauce has a much fresher and purer vegetable taste." The stew is prepared by sauteing blanched vegetables like carrots, cauliflower, broccoli, baby corn, green beans, and assorted squashes in a savory mixture of onions, thyme, and tarragon. The touch that lifts this dish above the ordinary is the addition of minced garlic and a mixture of fresh herbs and olive oil at the end.

But the dish that Tower fears will form part of his epitaph ("He invented black bean cake") is a favorite of mine and thousands of other Stars-gazers: black beans cooked with a spicy mixture of ancho chili powder, cumin, fresh hot green chilies, and cilantro leaves, then pressed into cakes and deep-fried. Served with sour cream and a salsa of tomato, onion, chilies, and lime juice, it's no wonder this dish helped make Tower famous.

Tower also uses beans in soups and side dishes. I still remember his spicy lentil soup served with red bell pepper cream. The chef will also curry lentils and serve them with sage aïoli and fried mint alongside couscous. One of his best dishes is an artichoke hearts stuffed with fava bean puree. The beans are cooked with savory and pureed, then placed in the artichoke, which sits in concentric pools of yellow and red pepper sauce.

Stars appeals to a wide variety of diners. Whether you nip in for mid-afternoon pizza with barbecue sauce and asiago cheese, come on a one-hour lunch break, or settle in for a three-hour dinner, you can be a star at Stars.

For committed budget gourmets Tower recently opened the teensy Star Cafe next door. You can drop in here anytime and, at very reasonable prices (nothing over $8.75, and much less for vegetable fare), eat some resoundingly good food without a lot of fuss. Fresh vegetable risotto with asparagus and Parmesan, pasta with wild mushrooms, and grilled cheese sandwiches with French fries are some of the choices.

690 vibrates with a tropical beat, featuring foods from around the world such as Rio potatoes (mashed with banana and jícama); spicy lentils with couscous and curry aïoli; and salad with black bean puree, saffron, and toasted cashew cream. My favorite dish is the garlic bread box, a hollowed brioche filled with smoky creamed lentil puree, topped with preserved fruit relish, and surrounded by mint vinaigrette and minced vegetables.

When the Peak Cafe opens in Hong Kong this year, diners halfway around the world will get a chance to experience what San Franciscans have enjoyed for the last decade: great food and a good time.

SALADS

Pasta Salad with Baked Eggplant and Tomatoes

Herb Oil

Tomato Concasse

MAIN COURSES

Warm Vegetable Stew

BARBARA TROPP

China Moon Cafe, San Francisco, California

"You are the servant of the food as much as the food is the expression of your particular personality and perspective."

"THE BEST dish I had in two years of traveling over China was a dish of bean sprouts. It was exquisite, made of sprouts grown in rainwater, sautéed very simply. Delicious, fresh, and lovely." That remark tells a lot about Barbara Tropp, and why her China Moon Cafe is so different from all other Chinese restaurants in San Francisco.

Like most Chinese restaurants, China Moon serves water chestnuts. But they're not the pallid, canned variety—she serves fresh ones, sweet and crunchy, alone as a dinner appetizer. There's not a drop of red food dye or MSG on the premises. Instead of the well-worn, hundred-item menu with everything from hundred-year-old eggs to egg foo yung (the same menu, it seems, in almost every Chinese restaurant in town), she changes her fifteen-to-twenty-item menu, save for a few favorites, in its entirety every three weeks. And the decor is neither plastic nor pretentious, but as spare and elegant as a Chinese silk painting. One of the most beautiful things I've ever seen in a restaurant was a Chinese porcelain bowl filled with green apples on the counter at China Moon Cafe.

If the Chinese restaurants in this country were like the ones in Taiwan, there would have been no need for this dark and lovely woman to open her own place. At the end of the seventies, Tropp was the author of a successful cookbook on Chinese culinary technique, *The Modern Art of Chinese Cooking*, and had a loyal following of students for her cooking classes. But the more she lived in San Francisco, with its great number of Chinese restaurants, the more their cooking style gnawed away at her sensibilities and her knowledge of what Chinese cooking could be.

"The Chinese food in this city was so far from the Chinese food I had eaten in Taiwan! And to my mind, vastly inferior! The fundamental tenets of the cuisine, like freshness, were not being upheld at all. Meanwhile, Chinatown was full of extraordinary ingredients like Chinese—and I don't mean Japanese—eggplant and fresh water chestnuts: ingredients that were not only *not* being used in Chinese restaurants, but also *not* being used by all the so-called "new American chefs," who really had the same old fears about entering Chinatown. I felt there was something to be said. So when it came to whatever year-in-a-row it was that the Mandarin was voted best Chinese restaurant in town, I threw down my towel and said, 'All right, I'm going to jump in and open a noodle stand.' "

But Barbara was savvy enough to know you don't just "jump in" to something as complex as the restaurant business. So from 1980 to 1982, she worked in the kitchen at the elegant vegetarian restaurant Greens, which is owned by the Zen Center of San Francisco.

"At that time, it was a totally Zen kitchen. Myself excluded, everyone there was a Zen student. It ran in a very particular way. We worked in silence. There were prayers twice a day. It was very, very special. It had a lot to do with the food itself taking priority even over the cook involved with it. It taught me a very healthy respect for ingredients, for the limits of one's knowledge, and for the ideal of simplicity. It was very formative for me in many ways."

She started at Greens as the bell pepper chopper, a post she held for three months, and would gladly have held another three. She saw what she describes as the "glory" in bell peppers and the "beauty of repetition." So, in the three years that China Moon has been open, there is one dish she has made most every week: strange flavor eggplant with croutons. She does it because she learns from the dish every time she makes it. Strangely wonderful it is, too: meltingly soft, all at once sweet and hot like chutney, and tasting of ginger, garlic, vinegar, and brown sugar.

"I feel like I begin to understand the seasons of eggplant, the seasons of garlic and ginger, through this one very simple dish and the repetition of it. That's the part of the kitchen I love. If it falls to one of the others in the kitchen to do it, for them it's a whiz-bang job. For me, it's kind of looking into one little astral corner of the universe. There's a real beauty, particularly in vegetables, about that, because you are so attached to the seasons of the year."

If Barbara Tropp does not speak the same language as other chefs, it is because, unlike many, she did not set out to be a chef; she was a scholar who fell in love with the food of China while studying Chinese poetry in Taiwan. When she returned to continue her graduate studies at Princeton, she taught herself Chinese cooking as a way of overcoming her homesickness. She got so good at it that she began teaching and catering, until finally she gave up writing her thesis to do a cookbook on Chinese technique.

Tropp has left her academic studies behind, but she has not abandoned her thoughtful, philosophical approach to life; it informs every aspect of how she approaches cooking. "It's a very interesting business. It dangles in front of you great issues of ego and humility. Not in any kind of Eastern religious sense—I mean it in a very simple, feet-on-the-ground way of how you go about your life. If you are tangling with the issues of ego, as all of us do, there is almost no better format than the restaurant business. The beauty is that it's the food that tells everything, and that's very humbling."

The lunch menu at China Moon, titled "Dim Sum and Then Some," offers cold items like Peking pickled cabbage, lemon-pickled fresh lotus root, caramelized tofu triangles, spicy pickled red onion, dry-fried Szechwan green beans, and the aforementioned eggplant dish. Each cold dish is served on small, lovely teal blue porcelain plates—the kind of detail that makes dining at China Moon special. Other starters have featured caramelized pecans and spicy cucumber fans.

The items from the cold tray are prepared in the downstairs kitchen, the entrees (or as Barbara calls them, "kitchen selections"), in the small but efficient wok kitchen by the front window,.

The menu suggests two or three kitchen selections per person, so I began with chili-orange noodles with fresh bean sprouts and coriander, followed by Tropp's famous Buddha buns, filled with curried vegetables and cellophane noodles and topped with black sesame seeds. The buns are baked, not fried, and are rich but not heavy.

One of the items that remains on the menu at all times is Tropp's pot-browned noodle pillow. Somewhat reminiscent of soft shredded wheat in texture, the chewy pillows absorb plenty of sauce before disintegrating. Mine came in a coconut curry sauce with stir-fried onions, three kinds of peppers, button and wild mushrooms, scallions, and spinach. There's a spicy orange sauce that is equally delightful. Other recent menus have featured Tropp's *pizzetta*, sort of a Szechwan pizza topped with colorful peppers and shiitake mushrooms, and a sand-pot casserole bubbling with fresh-shucked peas, baby corn, tomatoes, and mushrooms in a rich, long-simmering stock.

Like many of the better restaurants in China, China Moon Cafe uses only fresh vegetables. There is nothing canned or frozen—not even peas. But Tropp does not claim to be a traditional Chinese cook. "I don't have the taste for the extremes of oil, sweet, and salt that the Chinese do. My own tolerances for those things are far less. So my cooking is lighter than the traditional form. I am also a fairly simple cook, this partly by virtue of the background in Taiwan, so my food tends to be far more home style than restaurant style. Because I'm living in California, I cook like a Chinese home cook who is living in California, meaning that I will not use a canned bamboo shoot, even though it's Chinese, because it's canned. It's not fresh; it tastes and smells awful. I will go out to the market and use a red bell pepper, or a Fresno chili, which Chinese don't have, because it's the fresh thing. So the food winds up looking not Chinese in its ingredients. But I operate from an absolutely classic Chinese base in terms of flavor balances and standards."

Tropp laughs when people call her cooking French- or Japanese-influenced, since there is almost nothing from those cooking styles going on in her kitchen. It's just that people are so unfamiliar with what she's doing and with the true cooking of China that, in this label-conscious society, they try to give it some kind of name. Barbara jokingly calls what she does "home-style Chinese cooking of north and central China that's done by a somewhat eccentric person who lives in San Francisco."

Has China Moon helped change Chinese restaurant food in San Francisco? Barbara is realistic. "The heyday of glop and goo Chinese cooking is still very much with us. The only places I think that have integrity are ones opened by recently arrived immigrants who have not yet learned to conform to Chinese-American cooking standards. You find much more liveliness in Thai, Vietnamese, and Cambodian restaurants now. I think Chinese cooking, not only in this country, but also in large areas of Chinese Asia, has been slowly killed by excessive amounts of chemicals, cornstarch, and food that is not fresh."

Barbara Tropp is a reformer, but not an evangelist. "China Moon is very small. I am not an entrepreneurial chef. I want only the one restaurant. Its purpose is the food. Not the dollars. It only has to make enough dollars to support ourselves. That's the limit. It puts me in a different circle from many chefs. The beauty is that you can share standards and share a palate with many different kinds of people."

PASTA AND RISOTTO DISHES

China Moon Cafe Pot-browned Noodle Pillow Topped with Curried Vegetables

MAURO VINCENTI

Rex Il Ristorante, Pazzia, Fennel, Los Angeles, California

"Am I Colorful Enough?"

REMEMBER WHEN you went into an Italian restaurant to get chow instead of *ciao*? It was probably called Tony's or Mario's, not Scuzzi or Sfuzzi. In those days you ate lasagna, manicotti, and spaghetti. Nobody'd ever heard of pasta, and Italian food was red.

Then in 1977 Mauro Vincenti left Italy for America, and though he'd never run a restaurant before, opened Mauro's. "I needed something to do, and since I couldn't find any good Italian restaurants in Los Angeles, I decided to open one." It wasn't long before people took to the freeways to dine at Mauro's in out-of-the-way Glendale. They lined up to experience Mauro's fresh, house-made pasta; crusty, just-baked bread; and creamy gelato.

In 1981 Mauro sold Mauro's and opened Rex Il Ristorante in moribund downtown Los Angeles, spending over a million and a half dollars to painstakingly restore the room to its Art Deco thirties radiance. Rex became wildly successful. Tutto Los Angeles came to the restored Oviatt building to dine in the magnificent, high-ceilinged, wood-paneled former haberdashery. They sat at tables spaced a decorous distance apart (a Vincenti trademark he maintains today) and dined in what the owner describes as "the most elegant restaurant with the smallest portions, the highest prices, and the best chefs in Los Angeles." The formality of the room, and service best described as "empathetic," brought an elegance to Los Angeles dining not seen since the heyday of places like Perino's and Chasen's. The food, though preposterously expensive and doled out in admittedly Scrooge-like portions, was exquisite. After supper people repaired to the jewel-box dance floor loge overlooking the diners at their Lalique tables below and whirled into the single digits.

But with expense account lunches going the way of casual sex, Mauro doubts Rex can long survive. "I won't give an inch," says Mauro, tilting his head to draw a perfect bead on my eyes while he draws on his third or fourth Marlboro.

"Ingredients are expensive. Quality has to be maintained. I want to be proud of what I serve, and if I can't do that, Rex will have to disappear."

But the ever-upbeat Vincenti has already opened two other restaurants: Pazzia—which means "craziness" in Italian—is his sleek new entry into the red-hot West Hollywood dining scene. All concrete, marble, and glass, with an open kitchen and striking Aldo Rossi chairs, Pazzia features the best of refined southern Italian cuisine. Almost simultaneously Mauro opened Fennel, overlooking the Pacific on what has become Santa Monica's restaurant row: Ocean Avenue. Fennel is a collaboration between Vincenti and four of France's hottest young chefs, each of whom will commute between the continent and L.A. polishing and freshening the menu monthly. The Provençal-based menu works nicely on *our* Riviera, with its similarity of climate and the availability and variety of fine produce in Southern California.

Mauro describes the cuisine at Pazzia as a new version of southern Italian fare. "In the south we have no room to raise a lot of cattle. So there's very little dairy products or meat in the diet. Of course, southern Italians don't recognize the food here because we made it modern." (Southerners also won't recognize some of the dishes on the menu—such as *pappa al pomodoro*—unless they're from south Tuscany, where the food is decidedly northern Italian.) "At Pazzia you won't find more than two or three items on our menu with any butter or cream. And it's hard to cook with olive oil. Cream and butter expand when you cook. You can make mistakes and hide them. Not with olive oil. It's a lot of trouble."

Pazzia features a long, gently cascading fountain and an open courtyard framed by a gelato-espresso bar on the right and the glass-walled restaurant on the left. It's packed with fashionable art and entertainment folk slurping up such dishes as the aforementioned *pappa*, made with day-old Tuscan bread simmered in a rustic tomato ragú, then dressed in olive oil and a little Parmesan. The salad of caponata and mozzarella is another favorite, the caponata composed of tiny cubes of tender eggplant and grilled peppers, balanced by raisins and set off with mozzarella. California's newest designer fruit stars in the pear-apple, *caciotta*, and walnut salad: a juicy, crunchy starter tangy with pungent Italian cheese. The yellow-pepper soup is a delightful puree made without cream or butter, its depth provided by the smoky peppers.

The pasta dishes are no less imaginative. *Pennette*, or "little quills," with ricotta, eggplant, tomato, and basil, vies with the ravioli of lentils and pine nuts for my favorite. The chewy breads Mauro serves are perfect for ensuring that you don't miss a drop of any of the perfectly sauced dishes.

At Fennel the collective genius of Michel Rostang of Michel Rostang, Yan Jacquot of Toit de Passy, Michel Chabran of Pont de L'Isère, and André Genin of Chez Pauline combines with Mauro's stewardship to bring to our shores what must certainly be among the most contemporary of French culinary ideas. You may be fortunate and catch Fennel on a night when one of the fab four is there.

If not, don't worry in the least. You'll be in the hands of one of the finest young chefs in America: resident chef Jean-Pierre Bosc. Bosc has the culinary equivalent of perfect pitch. Choose his savory tomato and artichoke tart with a Lenotre-like crust. Another clear winner is his zucchini flowers stuffed with eggplant, mushrooms, olives, and tomatoes in a creamy basil puree. On a warm California night, nothing refreshes like the chef's colorful carrot and zucchini vichyssoise. Bosc's rendition of *pistou*, a basil and fennel-accented Mediterranean vegetable soup, is as cooling as a Pacific Ocean breeze. And whatever you do, don't miss his fried mashed potatoes. They're as unusual and homey as they sound.

With over seven hundred recipes available on the seasonally changing menu, it is doubtful Fennel will ever fall into a rut. And Bosc's specials freshen the carte daily. At Fennel you'll "share the experience" (as Californians are wont to say) with diners at only fifteen tables, spaced a considerate distance from each other, in a room designed to draw your attention only to the ocean or the food.

Mauro's menus are designed with vegetables in mind. He told me: "About 10 percent of all of my customers won't eat meat at all. So I serve more vegetables. It's 40 percent of my menus. And it's not that vegetables cost less. [Here he made some gesture possibly known only to Italians to signify just how costly greens have grown.] But I'm doing it because that's what people want. And I've go to do the best. Don't forget, you're talking to someone for whom ego comes before money. Pride is important to me. I know I have the best ingredients and the best chefs, so I'd be embarrassed to serve something bad. I'd want to hide, but I can't hide. I'm very proud of what I serve. I'm not an old restaurant chef who says 'my customer is always right.' The dish may not be to the customer's liking or to his taste, but the dish itself will be perfect. The customer is sometimes wrong."

Mauro Vicenti has his own colorful way of doing things; when you talk with him he entreats you to look into his eyes to see the sincerity there. He touches you frequently to emphasize his points and establish human contact. He is all about revolutionizing restaurants and the way people eat. Vincenti is bringing the freshest, newest food to Los Angeles without being trendy. As he says: "I don't want to do what's in fashion; I want to create fashion."

APPETIZERS AND FIRST COURSES

Caponata

SOUPS

Yellow Bell Pepper Soup

Pappa al Pomodoro

Carrot and Zucchini Vichyssoise

JEAN-GEORGES VONGERICHTEN

Lafayette, New York, New York

Cuisine des Cinq Sens

WHEN THE Marquis de Lafayette came to Philadelphia for a visit in 1777, little did the colonists imagine that a short three years later he would return with an expeditionary force from France to defeat the British commander Cornwallis at Yorktown, thus securing America's independence. Two centuries later, French super-chef Louis Outhier came to the United States, and, within a few years, the venerable name of Lafayette became involved in another American revolution.

This was a revolution in the way we eat, a subject no less dear to the French than *liberté, egalité, et fraternité.* The revolution took place in Boston and New York with the opening of two restaurants, both bearing the name Lafayette. It was a revolution that would change our way of thinking about French food, for the restaurants Lafayette introduced us to a newer, lighter way of saucing than we had experienced.

Swisshotels hired Outhier of the Michelin three-star rated l'Oasis to come to the United States and do for hotel dining here what he had done for the Mandarin International Chain in Bangkok, Singapore, Hong Kong, Geneva, and London.

Outhier was a consultant for Mandarin, and in every city he had turned their restaurant operation around for them. One of the chief factors in this remarkable success was a young chef from Alsace named Jean-Georges Vongerichten, who was at Outhier's side in most of the cities where the *mâitre-cuisinier* consulted. So Outhier know he had the right man to open the Marquis de Lafayette in Boston.

After building the Lafayette into what *Boston Magazine* called the best hotel restaurant in the city, Vongerichten was asked by Outhier to come to New York in July 1986 to run the kitchen at the Drake Hotel's Lafayette. One year later, Vongerichten had earned for the restaurant a four-star rating from the *New York Times*.

What is so revolutionary about Vongerichten's cooking? "When I make a vegetable sauce, I simply call it vegetable juice. I don't use any water; I just bring a freshly squeezed juice like zucchini, beet, carrot, or celery to the first boil with a few spices and a little butter or olive oil. I might add a little lemon juice if the vegetable is sweet, like carrots. Then some salt and pepper, and that's it. The flavor is so intense that it stands up to anything I put on the menu, like yellow pike in celery juice or scallops in leek juice. These are strong flavors, but my vegetable sauces are strong enough to provide a good contrast."

It might shock some of Vongerichten's early mentors, like Paul Haeberlin of the Michelin three-star L'Auberge de l'Ill, to hear the radical ideas of his young protégé. But then again, Haeberlin, Bocuse, and Outhier, all of whom Vongerichten trained with, didn't get their stars by doing slavish reproductions of Escoffier's dishes. It was Haeberlin (who also trained Jean Joho of Chicago's Everest Room and Hubert Keller of San Francisco's Fleur de Lys) who took a sixteen-year-old Vongerichten under his wing for 3½ years and then saw to it that he would get further training with Bocuse and Outhier and in Germany's only three star restaurant, L'Aubergine in Munich. It was during Vongerichten's stint there that Outhier called with the offer to go to Bangkok.

Of all the places Jean-Georges Vongerichten has cooked, New York is his favorite. "This is the best. The customers are good, because they are very cosmopolitan and they understand good food. The produce here is good too. The only thing I have to import is truffles. It may not be the same produce as I get in France, but it's just as good. It would have to be, because vegetables mean a lot to me."

Jean-Georges had to have top-notch produce, not just for his morning tonic of fresh carrot juice, but for his innovative cuisine. Though at first his customers didn't know what to make of his vegetable preparations and sauces, within six months they were asking for them.

The chef explained the importance of vegetables in his cooking: "Vegetables are important to balance the dish. You must have them for texture and balance. Otherwise, everything on the plate is soft. I abhor that same-textured food. For me, all the senses have to be involved. That's why I call my cooking *cuisine des cinq sens*, cooking for the five senses. If the food has texture and crunch, you can hear it, you can feel it. I encourage people to pick up the food and eat it with their fingers. And they do here, because Americans are sensual. So there's something crispy they can feel and hear on every plate. It looks appealing, and, naturally, if I've done my job right, it smells and tastes good too."

Such modesty is refreshing, but Vongerichten needn't worry. The *menu dégustation* he prepared for me was ground-breaking in terms of flavors, textures, and designs.

My first course at Lafayette was a *börek:* a flaky, three-cornered-hat-shaped filo dough pastry filled with a mushroom duxelles. Surrounding the pastry at perfect intervals were three dark pools of truffle vinaigrette. Truffles being as precious as they are, I didn't want to miss a drop of the vinaigrette even though I had finished the *börek*. The crusty whole-wheat rolls provided the solution. The dish involved all the senses, looking, smelling, feeling, tasting, and sounding altogether appealing.

A salad of avocado, warm asparagus, and enoki-dake mushrooms on a bed of assorted greens with endive spears fired on all cylinders, too. An unusual soy vinaigrette reinforced the delicate mushrooms, putting an Asian spin on the dish.

My medallions of artichoke with a julienne of vegetables was a thing of beauty: tender hearts of artichoke were filled with al dente strips of turnips, leeks, zucchini, and carrot, then another artichoke heart was placed on top, and the resulting "sandwich" was breaded and then sautéed in olive oil until golden. The chokes were then cut in half vertically, so that when they fell open, they revealed the colorful vegetable mélange in the center. The fried bread crumbs gave the dish such a nutty flavor that I thought the chef had used almond meal. I was also fooled by the chef's sauce, which I mistook for beurre blanc. But Jean-Georges had merely reduced the vegetable cooking liquid, whisked in some olive oil and nutmeg, and created a rich-tasting sauce, which he poured around the medallions.

Next, a gâteau constructed of brilliant green Swiss chard was served in a sauce of sweet paprika oil, garnished with fresh tomatoes and a sprig of thyme. The paprika oil is characteristic of Vongerichten's approach to innovative saucing. He also combines herb-infused olive oil with pastas and other dishes.

Asian chefs are fond of constructing delicate baskets and nests, usually out of noodles, to hold a delicious something or other. Vongerichten one-upped them by fashioning a wispy basket of leeks in which he placed a few juicy sautéed mushrooms. The whole sat in a potent leek sauce made with nothing more than the barely boiled vegetable juice, a little olive oil, herbs, salt, and pepper.

The main-course platter featured small portions of such exotica as fried lotus root chips done to a golden, crunchy turn; fresh chestnut puree with pomegranate seeds and huckleberries; mushroom caps on faux stems made of deep-fried mashed potatoes; and diced kohlrabi with spaghetti squash and parsley.

A few nights later Vongerichten came up with an entirely new tasting menu: skewered polenta with *confit* of shallot and ginger; a cold *chartreuse* of fall vegetables; lentils with a whisper of curry; a fluffy potato and Parmesan cake; and a bit of eggplant cannelloni with goat cheese in a sauce of brilliant green zucchini juice.

If this seems like largesse, consider the vegetarian who has Vongerichten prepare him a twenty-two-course cavalcade of dishes once a month: the ultimate meatless feast. He's chosen the perfect spot. Vongerichten realizes that everyone wants food that appeals to the five senses. He delivers.

PASTA AND RISOTTO DISHES

Cannelloni of Eggplant Caviar Glazed with Goat Cheese and Zucchini Juice

ALICE WATERS

Chez Panisse, Café Fanny, Berkeley, California

"I hope there will be a time when we can just take it for granted that what we're eating is really good for you."

AFTER A trip to France at the age of nineteen, Alice Waters had assimilated enough French cooking to dazzle her friends at dinner. "Why don't you open a restaurant?" they urged. (How many times have you heard that?) When she came to California from Chatham, New Jersey, the former Montessori teacher with no restaurant experience actually did it, and Chez Panisse was born.

It began as a neighborhood breakfast-lunch hangout in a two-story frame house behind a honeysuckle hedge. Dinner was added to help the restaurant make a go of it when inflation started to take off in the seventies. By 1974, higher prices were necessary to ensure quality ingredients, making it difficult for Alice's original clientele to afford Chez Panisse, so she opened an a la carte cafe upstairs.

Why the auburn-haired chef with the clear gray eyes and the sweet, lyrical voice has been so influential can be attributed directly to her interest in vegetables. In fact, salads and vegetables are what originally brought fame to Chez Panisse.

"Salads and various vegetables were the first focus here. We were always trying to get the ones that tasted good and tender from the very day we opened this place. In fact, we used to spend nearly eight hours sorting and washing the salad ingredients in those days. It was absurd. There just wasn't anything available. And you wound up sorting through the inferior stuff and making it look like it was small, tasty lettuce. We would have to put together little pieces of this and that to get the variety we wanted in a dish. We couldn't always get the basics—carrots, onions, celery—in the quality we needed here. We ended up compromising somewhere."

If Alice Waters was compromising, nobody noticed it. Instead people got excited about how good and fresh the food was at Chez Panisse. Customers noticed Alice's attention to quality and her commitment to serving the best produce available.

Even today, Waters attributes her reputation in large part to the high-quality vegetables she serves.

"It's been a long process from a very narrow range of vegetables that were available when we started to the horn of plenty we have now. I guess the biggest revelation to me is what the possibilities are. I think this is what has provided the real interest for the cooks and really is probably what the restaurant is best known for."

And that's the key: providing the interest for the cooks. From Alice Waters, the original chef, through Jeremiah Tower and Joyce Goldstein to Paul Bertolli today, Chez Panisse's chefs have based their menus on what's great in the market that day. The restaurant has a different prix fixe menu every night. Although they plan the menu a week in advance, there can be last-minute changes when ingredients don't measure up to expectations. Tower was known for occasionally changing the menu one hour before opening if he didn't get what he needed to make a dish memorable.

The chances of that happening now are rarer and rarer as a burgeoning market for fresh produce has made the level of vegetables so consistently high and reliable that you can bank on it and draw interest.

"Things we didn't know what to do with five years ago, we're using now. We have three produce sources. We get a box a week from Chino's in San Diego County that's primarily used in the downstairs dining room. Salad Gardens in Berkeley provides all the lettuces for downstairs. And Bob Cannard's garden in Sonoma is incredible. We get 80 percent of his produce. I hate to use the word *organic* with Bob. He's way beyond that. He's an extraordinary gardener. The cooks themselves go out there and pick lettuces three times a week. It's grown from getting a few things from him to the point that almost everything we get is from him. And I must say, it has changed the cooking here tremendously."

People are fond of calling Alice Waters's cooking "California cuisine," but she'll have none of it.

"Everyone wants to call this California cuisine. I don't think we have a cuisine yet; it's just the beginnings of cooking in California. I think there are a lot of things that give character to the cooking here. At Chez Panisse we look to the Mediterranean influences: the French first, Italian second. But we also have a lot of Asian influence in California, and that is shaping our cooking here, too. As ingredients like lemon grass become more available to us here it changes our cooking."

The concept of ingredients defining a cuisine was revolutionary when

Alice started. For years, restaurants in America had cooked whatever was available from purveyors. Now it's a buyer's market, with demand creating an increase in the quantity and quality of what growers supply.

Notes Waters: "I'm very interested in the marketplace. I think it's encouraging to see all the farmer's markets that have popped up. I am very interested in central urban markets that have the possibility of giving the farmer the volume he needs to make a living. We at Chez Panisse want the possibility of buying the things we want there and getting a good price."

I believe the reason Alice Waters has survived and prospered over the past score of years is that she has never put profits above conscience. In the early days, she charged as little as possible so that as many people as possible could afford to eat at Chez Panisse, and at the same time she insisted on the finest ingredients. That insistence on giving her customers quality built a loyal following and a reputation that stays with her even today. And now that she can get the best produce, out of her sincere desire for her customers' good, she wants to get the purest.

"Ninety-nine percent of everything we serve here is organically grown. That's made a big difference to me. I don't know if it's evident to people that come here, but they do remark on the flavor of things. I'm certainly aware of how good it is for you. I hope there will be a time when we can just take it for granted that what we're eating is really good for you."

Not only does Alice Waters want Chez Panisse to be able to buy the best organic produce, she wants us all to be able to.

"I'm hoping there will be a whole cottage industry cropping up all over the country. Little farmers need the right outlets, and restaurateurs and homemakers alike need competitive sources for produce, not just supermarkets where you get one choice for everything. This kind of competition brings out the best in the vegetables, because growers are encouraged to grow the best ones. Right now we're alienated from the source of our food and the people who grow it. We need to get back in touch with the way things look: whole and just picked from the ground."

To this end, every summer Alice strongly supports the Tasting of Summer Produce, where over three hundred farmers bring their produce in for people to taste. White eggplant, blue tomatoes, and the best little wild strawberries you've tasted outside of Europe are some of the wonderful things that have premiered at the show. There's a comparative tasting, where people can compare who has the best melons, tomatoes, or peaches. After the tasting, the public shares their reactions with the growers so they can see what they need to do to please the customer. This kind of marketplace approach to shopping disappeared during the middle of this century, and Waters sees its return as absolutely necessary to the health of the nation.

"I think there's a reason there are marketplaces. It's a great way to shop, to integrate all sorts of people. I believe the marketplace is a way to make some really basic social change in our alienated society where you don't even know your neighbors. I don't know how we're going to continue if we don't relate in a way like this that brings all ages, all segments of society together in a healthy competition that's the best environment for the shopper and farmer alike. I'm completely excited by the marketplace."

That excitement for obtaining the best is shared by all those who eat at Chez Panisse Restaurant and Café, as well as at Café Fanny, Waters's tiny stand-up breakfast/lunch place named after her daughter.

We benefit indirectly from Alice Waters's pioneering work in developing better produce each time we go to the grocery store. Supermarkets have to carry better fruits and vegetables now, because the people who shop there have had dishes made with great produce in restaurants like Chez Panisse and just won't settle for inferior stuff when they cook at home.

But for those who'd rather enjoy those benefits and have someone else do the dishes, it is happy news indeed that after nearly two decades the cooking at Chez Panisse has never been better and continues to evolve. In the cafe you can always get the legendary salads, like the one with the finest Sonoma goat cheese obtainable, breaded and warmed to runny perfection and served with a dressing of California olive oil, vinegars, and fresh herbs. The individual pizzas and calzones, which Wolfgang Puck, California Pizza Kitchen, and a host of others have Alice to thank for, are still as great as ever. And daily specials, like a recent hearty Tuscan bean soup with rosemary oil, or parsley *pappardelle* noodles with wild mushrooms, leeks, and white truffle oil, make the cafe a reasonable and no less satisfying alternative to the more expensive prix fixe menu downstairs.

In the restaurant, where the dinners are now (like everything else in America) five times what they cost in 1971, the cooking is as fine as at any restaurant in America. One recent Chez Panisse dinner featured as a starter a salad of fennel, boletus mushrooms, Parmesan cheese, and white truffles. It was a perfect marriage of texture, taste, and color. Nights can be chill most of the year in Berkeley, so the warm bowl of Perfection squash and celery root soups was most welcome. The sweet golden squash soup was well balanced by the deeper, earthy celery root flavor, and the visual treat of two soups in one bowl made the dish all the more appetizing. A savory ragout of Chino Ranch turnips was a perfect main course.

What's next for Alice Waters? Well, there won't be any more restaurants. She's taking sabbaticals these days, traveling, enjoying her family, and turning her attention to social issues that will improve the quality of life for the farmer and the general public. For Alice is first and foremost a teacher, whose love for her subject is her primary motivation.

SOUPS

Carrot and Red Pepper Soup

SIDE DISHES

Fava Beans with Olive Oil, Garlic, and Rosemary

Potatoes and Onions Roasted with Vinegar and Thyme

PIZZA AND BREADS

Pizza

JONATHAN WAXMAN

New York, New York

"I feel people have given vegetables a bum rap. They were always the misunderstood, misbegotten, and ill-prepared portion of our diets. Today will bear out that we decided to change this course . . . and have a lot of fun doing it."

UNTIL JONATHAN Waxman shuttered his namesake restaurant in New York and disengaged himself from the city's Hulot's and Bud's as well as London's Jams, he was one of the city's foremost celebrity chefs.

The reasons for his success are manifold. The chef trained in France at La Varenne Cooking School and with Ferdinand Chambrette. In this country there were stints with Alice Waters and Michael McCarty. One of his cooking hallmarks was his special way with vegetables.

"It's funny," said Waxman recently. "When we ran Jonathan's people would ask me where the vegetables came from. I didn't think they'd notice. They'd say: 'Why do these taste different?', and I'm totally flabbergasted by that. We brought in freshly picked produce, and only enough to last two days. People would ask, for example, where a radish came from because it was so fresh and peppery. They loved it."

The tall, teddy-bear chef with the modified sixties beard and mane loves vegetables, too. "I've been exercising so much. I run several miles every day. And as I get in better and better shape my body is telling me what it needs to be really healthy. I almost think that to be in great shape people should eat only vegetables. I really like meat, but the more I exercise, the more I run, the less I feel the need to have meat. It's real strange, I don't know what it is, but I feel I could be very happy just having vegetables and starch. I've actually started to go vegetable crazy! I think my body is sort of looking for it."

His decision to get out of the restaurant business for a while has prompted a physical and mental stock-taking.

"I want to find out who I am. And I don't want to do it when I'm eighty. That's ridiculous. So I'm just going to take a break. I'm retiring, gonna do some vacationing and traveling and just stay on the loose."

After leaving Michael's, Waxman took California to New York and opened Jam's on East 79th. He parlayed his success in the sparely decorated place into a four-restaurant partnership with wine maven Melvyn Master. The partnership dissolved, leaving Waxman with ownership of Jam's and a financial stake in Hulot's and London's Jams.

Waxman's influence is still felt on the East Coast in the kitchens of the chefs he trained and in the palates of the customers he pleased.

Local farms were important to Waxman's success. "People are growing all sorts of different things. We don't have the Chino* thing on the East Coast, but what we do have is all these old-time farmers growing normal lettuce and spinach done in such a wonderful way. They grow organic because they believe the old-fashioned way gives you the same yield but with better flavor. It's difficult to keep it organic because neighboring farms have helicopters spray insecticide on their crops, and

**Tom Chino's organic north San Diego County ranch supplies Wolfgang Puck, Alice Waters, and other California restaurateurs with much of their produce.*

it drifts. The growers I used to buy from have to fight incredible traffic coming in from Long Island, and then somebody has to stay in the truck at all times or the truck'll get stolen. And the farmers are so willing to work with you. I got produce from a little place in Cutchogue. They sent me incredible vegetables that I couldn't get even a year before. They would send me wonderful asparagus that was picked that morning! I was shocked. They asked me if this was the *size* I needed. Can you believe it?"

Though Waxman gives the farmers the credit, somebody has to *do* something with all those vegetables. One of my favorite Waxman creations was a sauté of garlic, baby peppers, and new potatoes with lots of fresh herbs. Then there was his eggplant with marinated celery root, and what must have been the best mashed potatoes in the galaxy.

"Vegetables don't have to be a main dish. They can be a first dish, small and a lot of fun. I don't mean too small either, like just two spears of something. That just wouldn't make sense. But if you get a whole roasted onion with the top taken off, scooped out, and stuffed with vegetables, people really like that. And I listened to the people. They're more interested in vegetables now. In the old days it was just meat. Meat and potatoes. French fries was the only vegetable they'd eat. People wouldn't eat the vegetables on the plate. Instead they'd have a huge piece of dessert. Now they might order only fruit for dessert. And they *eat* the vegetables. I used to do eggplant with celery root sauce, and it was just eggplant with celery root that is marinated. I thought it would sell. And boy, did it sell! You know what I think is funny? People go crazy for vegetables. I know I do. The look of a plate of tomatoes is really exciting to me now. People know they're supposed to eat vegetables, and the way we made them, our customers really liked them. We didn't see plates coming back with the same vegetables we sent out. Our customers realized that vegetables aren't just for color or to complement a dish, but are to be enjoyed in their own right. And that's a real change."

Waxman describes his cooking as "earthbound": food that isn't fussy, time-consuming, or precious. "I decided to bring food down to its basics and say okay: let's make a dish with four things rather than twenty things or fifteen things or eight things. So instead of having a dish that would have foie gras and truffles and everything but the kitchen sink, why not make a dish with duck breasts served with mashed potatoes, lima beans, and beets, and that's it? And it seems to work. I've always been what I consider a middle-of-the-road cook. I don't go too far in being too plain or too outlandish. Then I can do something frivolous or basic and people won't get upset. When you simplify things people seem to be real happy with it; it makes for a cleaner sort of thing. It feels better, and people really respond to it."

Part of the reason that people responded to Waxman's streamlined approach is that his presentations only *seemed* simple. The underpinnings of each dish were the years the chef had spent mastering technique and working with ingredients to develop flavors, textures, and colors. "I used to think that good food, the way I like food, was something that was just from your heart, but I realized that was not true. You have an intellectual basis, you read cookbooks, you see what other people do, you steal as much as you can, and you're influenced by a lot of other people."

Several of Waxman's dishes stand out in my memory. His cream of broccoli soup should go into the Culinary Institute curriculum. A salad of avocado with papaya relish soothed, sweetened, and gently bit the lips and tongue . . . like a lover. And his tomato salad showed up on menus all over Manhattan the summer after it premiered at Waxman's: vine-ripened Cutchogue tomatoes in all their sassy glory, punched up by a vinaigrette that tasted like liquid chutney. And I'll miss his Chez Panisse-inspired honest, crusty whole-grain bread.

Let's hope he's not gone for too long.

JASPER WHITE

Jasper's, Boston, Massachusetts

Point-Blank Cooking

"I GREW up on a farm near the south Jersey shore," says Jasper White. "It was a rural area covered by lots of little farms. To control insects there they sprayed all the time. Now I understand all these people there are dying of cancer of the organs: esophagus, stomach, kidneys. There's a suspicion that all those chemicals sprayed on the crops—well, it's common sense. When I think of how many tons of chemicals they dumped on us . . ." His voice trails off.

So White insists on getting as much local, organically grown, unsprayed produce as possible. "I buy most of my produce for about six months of the year from local farmers, mostly organic farmers. There's an intensity to the fruits and vegetables grown by these people that you can taste. You can taste the work that went into them, the love and the *integrity* in the true sense of the word, meaning 'complete, whole.' And when you compare the flavors of my strawberries that come from little organic farms in New Hampshire, Massachusetts, and Rhode Island to the Z-25's from California, I mean you're talking about two completely different entities. There's just no comparison. And the lettuces that are grown for us—sure, every now and then you'll see a few little holes in it from bugs or what have you, and we have to wash it thoroughly to get out a few ladybugs, but at least it's not covered with spray. Maybe the average customer doesn't realize it—I mean, I don't wave a flag saying we serve organic produce—but they know there's something special about the way it tastes."

"Organically grown," contrary to popular belief, is not some quirky new way of cultivating ugly, smallish vegetables, fruits, grains, and nuts. In fact for all of recorded history, except for the last twenty to thirty years, people have farmed organically, quite successfully, thank you, without chemical fertilizers or poisonous insecticides. But multi-billion-dollar chemical companies are quite good at marketing, and it wasn't long before the nation's farmlands were awash in compounds toxic to both insects and humans alike.

Some farmers, especially in traditional New England, never bought the con job. They'd figured out long ago how to control pests—and how to avoid creating new, pesticide-resistant strains by never spraying in the first place. It is these flinty New Englanders, plus a few younger, forward-thinking, backward-looking, return-to-the-soil types, that White relies on to bring better-tasting, chemical-free produce to his table.

Jasper White has always been a traditional guy. His first teacher was his grandmother from Rome. Remembers White: "She was so fussy, actually kind of neurotic about her cooking. Everything had to be just right. She was from the old country, Italy. It wasn't enough to just grow the food in her garden. There were certain times of day to pick it for maximum flavor. There was a whole ritual about it."

And from the time the big, bearded chef began training professionally at the Culinary Institute of America, and as he developed in restaurants around Boston with co-chef Lydia Shire, White remained a traditionalist, working to do time-honored things as well as he possibly could. He found that the basic dishes, cooked simply with the best ingredients, were far more successful than dishes that strained and overreached.

"I call it 'point-blank cooking.' I've been cooking for twenty years. It would be easy for me to fool people. But I'm trying to be direct with people. So I like to fall back on cuisine that's already been done and done well. If you look at some of these classic French or Moroccan dishes that took a thousand years to develop, you're not going to make them any better than they are. The more I cook, watching all the transitions that American food has gone through, thinking it was quasi-French and so on, the more I come back to classic dishes, classic combinations that people still love, like potato pancakes with applesauce."

Because Jasper's restaurant is in Boston, most of his cooking is done with seafood. But that's all the more reason his vegetables have to be right. For Jasper believes in the traditional concept of the protein-starch-vegetable meal. And if the vegetables aren't right, the whole dish fails.

"I'll give you an example of what I'm doing now: grilled tuna, and pepper stuffed with eggplant and some bread crumbs in a puree of fresh tomatoes. The puree is what makes this dish work. It's what holds it together. You couldn't really do the dish with just the tuna and the pepper. They would have nothing to do with each other. The additional vegetable (the tomato puree) ties the starch (the eggplant with bread crumbs) to the protein (the tuna). That's another important role that vegetables play. Obviously, it's not the only role. The most important are nourishment and good cooking. To you they're everything because you're a vegetarian. But in a sense, to me they're everything too, because they're equal to every other component in the dish. You can't serve a beautiful fresh fish and old mushrooms.

"In fact, vegetables are such an integral part of all the food, that without them there is no cooking. Outside of the vegetable dishes themselves, most every sauce and stock we make here has onion, carrot, celery, and garlic. You can't really talk about food and not talk about vegetables. There's no getting away from them."

And Jasper White is a genius with vegetables. From inventive salads, like spring vegetables and grilled leeks stuffed with goat cheese, or wild-mushroom tempura with spinach and radishes, to soups like corn and pumpkin bisque, White brings out the best in vegetables.

"My whole philosophy about cooking is strictly to go for flavor. And vegetables provide so much of those flavors. I try to cook with my eyes closed, taste the food, and when I have the flavors at the intensity I'm comfortable with, then I'll go back and look at colors and everything else."

On a recent visit to Jasper's I started with johnnycakes, an old New England favorite that may predate the 1802 molasses warehouse in which the restaurant comfortably resides. Only I doubt our founding fathers would have eaten them with sautéed chanterelles, leeks, and carrots in an herb-laced vinaigrette. The nutty corncakes offset the tangy vegetable composition perfectly.

My salad of mixed greens, red onions, green beans, blue cheese, and olives was colorful and alive with flavor. What was it about those vegetables that made them so potent?

The short, intense growing season is part of the reason, thinks White: "If you can grow vegetables here, you can grow them anywhere. In fact, we have some specialties that you don't get anywhere else. The first thing that comes to mind are the vegetables that are affected by frost, and root vegetables. You might have good root vegetables in other parts, but people up here truly believe that you shouldn't even pick rutabagas, sweet potatoes, turnips, and parsnips until there's a frost. There's a sweetness that comes to them that, if you're not in a cold region, you can't even compare to. The same way with our Jerusalem artichokes that originated in Cape Cod. Sometimes I do a gratin of Jerusalem artichokes, potatoes, and rutabagas like the classical French *dauphinois* with cream and a little garlic that's wonderful."

When the weather starts to warm, White serves one of his favorite dishes, spring-dug parsnips. There is one parsnip farmer, with over two hundred acres under cultivation, who leaves about twenty acres in the ground right through winter, letting them freeze solid. In spring, they are dug up, and when White gets his hands on them, he knows just what to do.

"When I get them they're very intense, very sugary. I'll make a puree with water and a little butter, and I don't put another thing in them. I get the same questions all the time: 'What spices did you put in that puree? It has these really exotic overtones of ginger and nutmeg. Why do they taste so peppery and spicy?' I don't know if things in the soil penetrate those parsnips or what, but they must take on something extra from the earth during the winter. They're certainly unique to this cold climate."

Although Jasper feels the region he comes from defines his cooking, allowing him the security of falling back on well-loved favorites like chowders and steamed brown breads, he couldn't have risen to national prominence if he cooked only humble Boston fare. His gastronomic sense allows him to go forward from his regional base to explore new territory.

So I was delighted, but not surprised, to be served a main course of lemon fettuccine with corn, pumpkin, pumpkin seeds, chanterelles, spinach, and fennel. The tender autumn vegetables were tossed with the smoky, almost bacony, crunchy seeds, and the pasta was laced with shreds of lemon—a dish to compare with the best new Italian cuisine.

Jasper White's updated traditional New England fare can make us proud of what's cooking in America.

BARRY WINE

The Quilted Giraffe, New York, New York

An Ethical Chef

WE GO to restaurants to eat, to have fun, to relax. It is of little interest to us whether waiters pay taxes or not. And no matter how extravagant the restaurant or how special the occasion, we'd like to get out of the evening as inexpensively as possible. Thus some people object to Barry Wine's mandatory service charge (even though it amounts to only a paltry 2 to 3 percent over the normal tip), since it's already added on when you get the bill, inducing a temporary "check shock." But after you hear Barry Wine on the subject you understand why he does it.

"Tipping is responsible for training young kids, behaviorally, to become law breakers, by setting up a system in which it's commonly accepted practice that they don't pay taxes. You take all these kids age eighteen to twenty-five and run them through this system, and at age twenty-five how do you expect them to have any respect for the law? It's semi-condoned by the federal government. Nobody really does anything about it. Being condoned that way, it's going to really screw up these kids. 'What's going on here?' they'll think. 'Why are we getting away with this?' "

So when people complain about paying $75 or $100 for prix fixe dinners at the Quilted Giraffe they might take into account what they are paying for in addition to some of the highest-quality ingredients, preparation, dishware, decor, and real estate in Manhattan. As Wine puts it: "Our restaurant is as expensive as it is because it operates without tipping and because we don't give payoffs, kickbacks, or free dinners. We're probably 20 percent more expensive than the restaurant down the street, and the reason is because the people here are working on the books. We're paying them more, and we're paying taxes on it. Besides, quality ingredients are very expensive here. What we charge reflects the real cost of preparing and serving it."

Wine's service-included policy is customary in Europe and Japan, but the Quilted Giraffe has more than this in common with Japanese restaurants. The Wines have slowly gravitated towards a greater and greater use of Japanese ingredients, cooking methods, and tableware over the past five years. "I've been to Japan three times. I've started to move a little away from the use of butter and more into the Japanese style. We'll use a dashi seaweed base to which we'll add soy sauce, ginger, mirin, or sake and cook some dishes with no fat, butter, or cream." The restaurant serves its food on beautiful Japanese lacquerware and offers an eight-course *kaiseki* meal of small, exquisite portions that is priced according to the ingredients used.

But don't go into the Quilted Giraffe expecting to remove your shoes. There are little touches of Japan in the decor, but the restaurant is as much an ode to Manhattan as Woody Allen's movie by the same name.

The interior is reminiscent of the 1939 New York World's Fair, with Art Deco motifs everywhere, from the skyscraper-like steel light fixtures right down to the salt shakers. Gray granite and terrazzo abound, and the battleship-gray leather banquettes pick up the steel leitmotif, which is carried right up the walls to the Art Deco ceiling. This steely-gray atmosphere shows off huge sprays of fresh flowers and the pale green service plates. The plates also introduce you to the broken pediment design that crowns the AT&T building, in whose shadow the restaurant is nestled.

In terms of food, however, the Wines stand in no one's shadow. From the time they started with a Saturday-nights-only restaurant in their Victorian mansion in suburban New Paltz, through their French phase on Second Avenue, to today when every dish is cooked to order, *à la japonaise*, the Wines have gone their own way.

"Susan and I are recognizing more than ever how special the Quilted Giraffe is. As the food business gets taken over more and more by big companies and chain restaurants, it's more clear to us what business *we're* in. We're in a certain level of quality that those people aren't in. They're in new ideas, marketing, advertising. They need to operate in a certain way, and we're not in that business. A certain segment of the public and a certain segment of the food press doesn't always know the difference. It may be only a 5 percent difference. Those restaurants are operating at 94 percent of perfection, and we may be operating at 99 percent. That little 5 percent difference probably requires 100 percent more work and attention to detail and much greater expense on our part. That's why I closed the Casual Quilted Giraffe. That was my foray into the 94 percent kind of restaurant. I didn't like it. It wasn't good enough for me to be only that good. I want to be one of those 99 percent restaurants. You can only be that if you're always here, stay essentially this size, and have basically no outside ventures."

I ventured into the Quilted Giraffe for lunch recently (a marvelous way to have the same great food at about a 40 percent discount from dinner) and composed, with the help of the kitchen, a series of courses to my liking.

I began with a warm tomato consommé: a clear golden, slightly sweet broth with chives and golden chanterelles. The soup contained no cream, butter, or oil, but was as satisfying, and a good deal lighter, than any cream-based soup. I dipped the wonderful whole-grain sourdough-molasses roll into the soup so I wouldn't miss even a drop.

My main course was a series of little offerings served in Japanese lacquer bowls and plates neatly arranged on a lacquer tray. I wasted no time in tying into the crispy, light corn and zucchini fritters and a chunky ratatouille accompanied with boiled red potatoes. Each of the simply cooked vegetables kept its own discrete flavor, yet worked in perfect harmony with the others.

On another plate were plump won ton filled with wild mushrooms, Swiss chard, and sun-dried sour cherry tomatoes. A red bowl was filled with sautéed Brussels sprouts, onions, daikon, and red peppers with hen of the woods and matsutake mushrooms. Sitting in their own juices, the vegetables had a simple, natural sweetness that would have been missed in a creamy sauce.

All the dishes were served with sauces, and you could choose to dip or not to dip in bracing wasabi mayonnaise, tangy barbecue with pepper, hot mustard, or soy sauce with chives.

The food at the Quilted Giraffe is clearly a departure from most restaurants. "We've played a part in changing the restaurant business, but that change has not come easily," said Barry. "Starting with American cooking in 1975 in New Paltz, we moved to doing French food in Manhattan. Someone who was not French running an expensive, fancy restaurant was unheard of at that time. We got a lot of flak then. But today, because of our pioneering, it's much more accepted. At every stage we've gone through right up to 1982 or '83, people have said: 'Who do these guys think they are? They're not French.' But we did things the way we did because we felt it was right."

And now that the restaurant has moved away from French food to a more Japanese style, no one bats an eye. People who before criticized them as being iconoclasts now call them trendsetters. Barry describes their new cooking style:

"Everything in this restaurant is cooked to order. There's always a risk that we'll undercook it or overcook it, but that's the fun of it, too. Every table is an adventure. Some restaurants are in the 'dish-out' business. There's no adventure there. Traditional French cooking is very much like this, in the sense that you cook this food in a copper casserole, give it to the waiter, and he 'dishes it out.' That's a whole different approach than we have. For them, recipes are very important. In this kind of cooking, we have a way of doing things, but nothing's done ahead. In fact, your next course is not cooked until you've eaten your prior course. We go from raw to served in from 3 to 10 minutes. It's a whole new approach to cooking."

Obviously, vegetables, with their short cooking time, distinct flavors, and vivid colors, lend themselves well to this kind of cooking, and Wine keeps from ten to twelve kinds on hand at all times. There are also intense purees of beet, celery, and cauliflower that are to vegetables what fine sorbets (for which Wine is also known) are to fruit.

Wine prides himself on his produce. During the hot summer months he brings in from his land in New Paltz, where he employs a full-time farmer to till the soil, 100 percent of the restaurant's requirements. The farm provides asparagus, peas, lettuces, early radishes, herbs, tomatoes, peppers, and squash in summer, and during colder months provides such vegetables as Brussels sprouts, turnips, leeks, and fennel. And the fresh-cut flowers come from the same plot.

It helps having plenty of good, fresh produce on hand. Not only do many people order vegetarian fare, Wine has a vested interest in having lots of vegetables in the kitchen.

"I think when you're running this kind of operation, there's a tension in the kitchen. Every time you send out a plate, you're worried it won't be as perfect as you want it to be and the customer expects it to be. To keep myself from becoming unbelievably fat from eating due to nervous tension, I eat a lot of vegetables."

To talk to the relaxed, silver-thatched former lawyer with the infectious grin, you'd never think he'd get nervous. In the main, he's not. He and Susan just keep rolling along, and if they create waves, they leave those behind and move ahead. The press is mostly on their side these days, and now that they're in their "dream kitchen" in "a wonderful facility" they've become a welcome fixture on the New York restaurant scene.

Says Barry: "What's interesting about the Quilted Giraffe is that for the amount of publicity that you see, it has this image of being an institution. We do a lot of interesting things and always have over the years. We don't have a press agent. The press has just sort of found us. But the Quilted Giraffe is really just this little restaurant with me and Susan."

PASTA AND RISOTTO DISHES
Shiitake and Sour Cherry Ravioli

ROY YAMAGUCHI

Roy's, Honolulu, Hawaii

You Can Go Home Again

ROY YAMAGUCHI'S career as a chef began in a high-school home economics class in Honolulu. Later that year, his high school counselor exhorted him to apply to the Culinary Institute of America. Roy was accepted and came to the mainland. What he didn't know was that success in the restaurant business was still fourteen long, hard years away, and that, ironically, his ultimate fulfillment would come in a restaurant not far from his high school.

Yamaguchi learned the basics at the Culinary Institute. His development continued in Los Angeles under Jean Bertranou and Michel Blanchet at L'Ermitage and Michael McCarty of Michael's. By the time Yamaguchi opened Le Gourmet for

Sheraton Hotels, he was a fully realized chef, and he was only twenty-two. Until then, not many people went out of their way to have dinner at a Los Angeles airport hotel. Certainly not Sheraton's La Reina, one of the uglier hotels in the world. (It looks like it was constructed in about fifteen minutes out of giant aluminum plates.) But in the early eighties, knowledgeable Los Angeles diners flocked to a small green restaurant in the corner of the hotel, where Yamaguchi ran his seventy-seat dining room. There, for the first time, in his own kitchen, Yamaguchi could express his own culinary ideas. The classical and nouvelle French training he had received, combined with his flair for Japanese ingredients and presentation, created some of the first Franco-Japonaise food in America.

A frequent habitué at Le Gourmet was Dr. Stan Kandel, one of the men who put together the Spago partnership. He loved Yamaguchi's cooking, so he proposed to Roy that they do a restaurant together. Yamaguchi was ecstatic. At age twenty-six he was about to have his own place. No doubt he was the right man. Unfortunately, it turned out to be the wrong place.

The site selected was a former poly-nothing barn of a place on North La Cienega Boulevard in Los Angeles. Though it's called Restaurant Row, the street might more aptly be named Restaurant Graveyard. All up and down the boulevard are huge hulks of discarded restaurants, paint peeling, signs bleached and broken, the buildings themselves listing a few degrees to port or starboard, sitting in masses of uncut weeds. And yet restaurateurs and investors continue to pour millions into recycling these buildings, trying to draw diners in with some new idea.

Roy Yamaguchi certainly offered something new. No one cooked quite like him, with his vegetable *gyozas:* succulent pot stickers stuffed with minced grilled vegetables and served in a red bell pepper sauce. Lemon grass and ginger sparked his salads and sautés. Corn and feta cheese joined sesame oil vinaigrettes in salads so original I often saw patrons asking their waiters for the recipe.

But the place was just too big. 385 North, as it was called, could seat over 250 people. And despite a kitchen and crew designed for maximum output, there was no way Yamaguchi could attain the culinary heights at 385 North he had achieved at Le Gourmet.

So, after a good run of a few years, during which I enjoyed many a plate of delicately flavored slivered vegetables strewn with wild mushrooms sautéed in butter and soy, and platters of bright green long beans, mâche, Belgian endive, and radicchio with papaya in a honey-sesame oil vinaigrette, 385 was shuttered.

Yamaguchi returned to Honolulu. When I finally caught up with him in the fall of 1988 he was hard at work on a new place, which he had decided to call Roy's, and which was located in a residential-commercial suburb east of Honolulu.

"There's hardly any wall space," beamed Roy. He was right. Its 150 linear feet of glass allows magnificent views of Maunalua Bay, Diamond Head, and Koko Head from the second-story restaurant. Inside, the color scheme is turquoise, mauve, and peach. But nothing distracts from those gorgeous vistas.

"The chefs don't mind cooking here, because from the open kitchen in the center of the restaurant, they can always look out those windows and see the Hawaiian sunsets," observed Roy, pointing westward to the bay dotted with windsurfers' bright sails.

The views are nice, but Honolulu is full of good views. What the city has lacked for years is a fine restaurant. Now it finally has one, with moderately priced selections that draw families in to dine twice a week or more.

A pizza covered with Big Island tomatoes and Maui onions, cooked in Roy's wood-fired oven and served with a salad of wilted greens, Japanese mushrooms, macadamia nuts, and feta cheese, is probably the best light meal you can get on the islands. And it'll set you back about ten bucks.

Those same sweet onions and tomatoes appear in a salad with balsamic vinegar and olive oil. It's about as basic a salad as you can get. One of the most popular entrees, though it's not on the menu, is a vegetable plate featuring marinated and mesquite-broiled onions, peppers, mushrooms, and whatever's fresh from Roy's growing cadre of dedicated small farmers.

"Vegetables play a big part in my cooking because I use them in so many ways," says the strikingly good-looking chef. "From appetizers to entrees, I use vegetables with different vinegars and oils, different herbs and spices to achieve the taste I'm looking for. Then I cut the vegetables in julienne, dice, or in some Oriental fashion to add an extra dimension."

Roy calls his style "Euro-Asian." His customers just call it "good." Long hungry for some decent, reasonably priced local food, Honoluluans flock to Roy's. With two hundred dinners a night (250 on weekends), Roy has, since his December 1988 opening, turned away about as many as he's served. And the people from the mainland haven't even discovered Roy's yet.

Roy has geared his restaurant toward the people he lives and works with: Hawaiians.

"Restaurants here used to take diners for granted because tourists were the main patrons. They didn't have to do much because they figured the people were here one day, gone the next. They knew they'd get the business anyway. And people who lived here were stuck with inferior food. Now, hotels are becoming very competitive in the quality of the food they serve, and the locals are becoming discriminating. They'll pick one over another based on the level of food they can get, so all the restaurants have to improve or they won't survive. That brings the level of all our restaurants up."

Roy's will set the standard for Honolulu's restaurants for some time to come. A "vermicelli" of zucchini threads in twin sauces of red and yellow bell peppers, and grilled Mongolian-style eggplant napoleons with a caper-enhanced basil butter sauce are examples of his revolutionary cooking: an intelligent, vegetable-centered cuisine.

"Without vegetables," emphasizes Yamaguchi, "you can't make any dish complete. You can't make any cuisine complete. It just won't work."

Roy and his partners are considering other ventures now that they've seen the demand for good food outstrip their restaurant's ability to supply it. Their eyes are on a Waikiki location.

But Yamaguchi plans to invade the mainland soon, and he's considering Newport Beach as a possible site. But something tells me that Los Angeles, where Roy Yamaguchi toiled for so many years, will one day be home to a Roy Yamaguchi venture. It was here he got his real training and achieved his first success. I wouldn't be surprised to see this talented chef return to Los Angeles to triumph.

APPETIZERS AND FIRST COURSES

Spring Vegetable Gyoza with Roasted Red Bell Pepper Butter Sauce

SIDE DISHES

Zucchini "Vermicelli" with Two Sauces

MAIN COURSES

Grilled Mongolian-style Eggplant Napoleon with Caper, Garlic, Basil, and Butter Sauce

PART TWO

APPETIZERS

AND FIRST COURSES

Provençal Artichoke Tarts

Leslee Reis

Makes six 4-inch tarts

Savory little vegetable pies for a first course, lunch, or picnics.

1 recipe Pastry Dough, page 327

FILLING

- 6 to 8 *artichokes*
- 5 *tablespoons butter*
- 2 *medium yellow onions, coarsely chopped*
- 15 *garlic cloves*
- 2 *heads Bibb or Boston lettuce, cut into shreds*
- 1/4 *teaspoon dried thyme, or leaves from 1 fresh thyme branch*
- 2 *bay leaves*
- *Salt and freshly ground white pepper to taste*
- 2 *red bell peppers, seeded and cored*
- 1/4 *cup sugar*
- 1/4 *cup red wine vinegar*
- 1/4 *cup dry white wine*
- 1 *tablespoon minced fresh basil*
- 3 *tablespoons minced fresh Italian parsley*

Preheat the oven to 400°. Prepare the pastry dough and roll out on a floured surface. Cut 6 circles 1 inch larger than the 4-inch tart pans, fit them into the pans, and crimp the edges. Prick the bottom of the shells with a fork, line with aluminum foil, fill with dried beans, and bake the shell until golden brown, about 15 to 20 minutes.

Remove all the leaves and stems from the artichokes and scoop the chokes from the hearts. Quarter the artichoke hearts.

Heat 3 tablespoons of the butter in a heavy saucepan and add the artichokes, onions, garlic, lettuce, and seasonings. Cook over low heat for 30 to 45 minutes, or until tender.

Meanwhile, cut the peppers into 1/2-inch julienne. Melt the remaining 2 tablespoons butter in a sauté pan or skillet and sauté the peppers for 2 or 3 minutes. Add the sugar, vinegar, and wine and sauté a few minutes longer, or until tender.

Mix the peppers with the artichoke mixture and half of the fresh herbs. Adjust the seasoning. Divide evenly among the prebaked tart shells and sprinkle the remaining fresh herb mixture on top. Serve hot.

Warm Asparagus with Pine Nuts and Orange Rind

Jackie Shen

Serves 2

A beautiful opener, with a touch of Grand Marnier and a garnish of edible flowers.

- *12 pencil-thin asparagus stalks*
- *¼ cup olive oil*
- *Grated zest of ½ orange*
- *Salt and pepper to taste*
- *4 oak leaf lettuce leaves*
- *2 teaspoons pine nuts*
- *3 edible (unsprayed) pansies**
- *Orange Sauce, following*

Peel the tough part of the asparagus stalks with a potato peeler. Then blanch the asparagus in boiling water to cover for about 2 minutes; make sure they keep their bright green color. Cool in ice water. Cut off any tough stems. Heat the olive oil in a sauté pan or skillet and sauté the asparagus, orange zest, salt, and pepper until the asparagus is heated through.

Arrange the lettuce leaves on a serving platter at 12, 3, 6, and 9 o'clock. Arrange the asparagus in the center of the serving platter. Sprinkle with pine nuts and garnish with edible pansies. Pour the orange sauce over the asparagus and serve at once.

ORANGE SAUCE

- *1 teaspoon Grand Marnier*
- *2 tablespoons fresh orange juice, or to taste*
- *Grated zest of ½ orange*
- *¼ cup hazelnut oil*

Place the Grand Marnier in a small saucepan and boil until reduced by half. Allow to cool. Whisk in the remaining ingredients.

**Edible pansies are available at specialty food stores.*

Asparagus with Hazelnut Vinaigrette

Joyce Goldstein

Serves 4

1 pound asparagus, trimmed

VINAIGRETTE

¼ cup hazelnuts
2 tablespoons hazelnut oil
2 tablespoons light olive oil
2 tablespoons balsamic vinegar
Salt and pepper to taste

Steam the asparagus over boiling water till al dente. Transfer to a bowl of ice water, then drain and dry on paper towels.

Toast the hazelnuts carefully in a dry sauté pan or skillet until they are lightly browned, then chop them.

In a bowl, mix all the remaining ingredients for the vinaigrette and pour over the asparagus. Sprinkle with the toasted nuts.

Cèpe Carpaccio with Garlic Chives

Gunter Seeger

Serves 6

A vegetarian version of the usual Carpaccio.

- *1 pound large fresh cèpes*, cleaned with paper towel*
- *Fine sea salt and fresh black pepper to taste*
- *2 tablespoons fresh lemon juice*
- *6 tablespoons extra-virgin olive oil*
- *1 bunch garlic chives, minced*

With a mandoline or a sharp knife, slice the cèpes very thin and arrange on 6 plates. Add salt and pepper, lemon juice, olive oil, and a sprinkle of chives.

**If cèpes (fresh porcini mushrooms) are not available, use fresh oyster, chanterelle, or matsutake mushrooms.*

. . . .

Sautéed Wild Mushrooms with Sage Sauce

Thierry Rautureau

Serves 4

1 *tablespoon butter*
1 *tablespoon vegetable oil*
1 *pound mixed wild mushrooms*
3 *shallots, sliced*
½ *bunch purple sage*
½ *bunch sage*
1 *cup dry white wine*
2 *cups Vegetable Stock, following*
Salt and pepper to taste
¼ *cup olive oil*

In a sauté pan or skillet, heat the oil and butter; sauté the mushrooms until tender; set aside.

In a saucepan, heat the butter and sauté the shallots for 2 minutes. Add half the purple sage and sage, and cook for 1½ minutes. Add the white wine and cook to reduce by two thirds. Add the stock and again reduce by two thirds. Strain and adjust the seasoning. Place the mixture in a blender or food processor and gradually add the olive oil while the motor is running. Pour into a saucepan. Julienne the remaining sage leaves, add to the sauce, and heat for 1 minute. Serve warm under sauteéd wild mushrooms.

VEGETABLE STOCK

1 *bunch celery chopped*
1 *carrot, peeled*
1 *fennel bulb*
¼ *cup extra-virgin olive oil*
½ *cup coarsely chopped white onion*
2 *thyme branches*
3 *bay leaves*
1 *tablespoon fennel seed*
1 *tarragon branch*

Coarsely chop the celery, carrot, and fennel. In a pot, heat the oil and add all the ingredients except the tarragon; sauté for 5 minutes. Add water to a level 1 inch above the vegetables. Bring to a boil and simmer for 25 minutes. Add the tarragon and simmer 5 minutes more. Strain and cool. This stock will keep covered, in the refrigerator for 2 to 3 days. It may also be frozen and kept for up to 1 month.

Onion and Black Olive Tarts

Michel Richard

Makes four 4-inch tarts

These are easy to make if you use frozen puff pastry.

- 6 *tablespoons olive oil*
- 8 *cups very thinly sliced Spanish onions (about 2 pounds)*
- 1 *garlic clove, minced*
- 8 *ounces fresh or defrosted frozen puff pastry*
- 8 *black niçoise olives, pitted and cut into quarters*

Heat the olive oil in a nonstick sauté pan or skillet over medium-high heat. Add the onions and cook to a deep golden brown, stirring often to prevent burning (this will take about 30 to 40 minutes); set aside.

On a floured surface, roll out the puff pastry to a thickness of ⅛ inch. Cut out four 4½-inch circles and place them on a parchment-lined or greased baking sheet. Place in the freezer until ready to use.

Preheat the oven to 400°. Remove the baking sheet from the freezer. Place a 4-inch tartlet tin in the center of each circle of dough. Place another baking sheet on top of the tins. Then place a heavy pan on top of the second baking sheet. (This is done so that as the puff pastry bakes, it creates a large "cup," which you will later stuff with the cooked onions.) Bake the puff pastry in the preheated oven for 15 minutes, or until golden brown.

Divide the onions between the 4 tarts, spreading to cover the bottom. Top with the quartered olives and serve immediately.

Stuffed Zucchini Flowers with Tomato Marinade

Thierry Rautureau

Serves 4

TOMATO MARINADE

- 2 *medium tomatoes, diced small*
- 4 *shallots, chopped*
- 2 *garlic cloves, chopped*
- ⅓ *cup julienne-cut basil leaves*
- ½ *cup balsamic vinegar*
- ¼ *cup extra-virgin olive oil*
- ½ *teaspoon ground black pepper*

ZUCCHINI STUFFING

- 2 *zucchini, coarsely chopped*
- 2½ *tablespoons olive oil*
- 2 *tablespoons chopped shallots*
- ½ *tablespoon chopped garlic*
- 6 *tablespoons chopped fresh thyme*
- ½ *bunch basil, chopped*
- ¼ *cup bread crumbs*
- *Salt to taste*
- ¼ *teaspoon black pepper*
- 8 *baby zucchini with flowers attached*
- 1 *tablespoon olive oil*

To make the tomato marinade, combine all the ingredients in a bowl and refrigerate covered for at least 1 hour.

To make the zucchini stuffing: in a saucepan, sauté the zucchini in the olive oil until tender. Add the shallots and sauté for 3 minutes. Add the herbs, then puree the mixture in a blender or a food processor. While blending add the bread crumbs, salt, and pepper. Cool in the refrigerator at least 15 minutes. Preheat the oven to 350°. Blanch the zucchini flowers in boiling water for 2 minutes. Plunge into ice water, drain, and dry. Fill a pastry bag with the zucchini stuffing. Carefully open the zucchini flowers and pipe in the zucchini stuffing. Twist the flower gently to close. Place olive oil in a baking pan, add the zucchini flowers, and bake in the preheated oven for about 2 minutes.

Pool the tomato marinade on each of 4 plates. Place 2 stuffed zucchini flowers over the tomato marinade. The zucchini may be sliced thin, beginning at the tip and slicing toward the flower, while leaving the flower intact, and fanned, if desired.

Dialogue of Vegetables

Gunter Seeger

Serves 6

Five different vegetable purees on each plate: green, white, yellow, red, and purple.

- *4 fresh parsley sprigs*
- *1 cup vegetable stock*
- *2 tablespoons clarified butter**
- *4 whole salsify roots,** peeled and chopped*
- *2 shallots, chopped*
- *Salt and pepper to taste*
- *3 yellow bell peppers, quartered, cored, and seeded*
- *1 tablespoon extra-virgin olive oil*
- *1 teaspoon honey*
- *Dash of sherry vinegar*
- *4 tomatoes*
- *1 fresh rosemary sprig*
- *1 fresh thyme branch*
- *1 garlic clove, chopped*
- *4 beets, peeled and diced*
- *Dash of balsamic vinegar*

Blanch the parsley for about 1 minute in salted boiling water to cover and cool in ice water. Place the parsley in a blender or food processor with ½ cup of the vegetable stock and puree until very smooth; strain. In a saucepan, heat 1 tablespoon of the clarified butter and sauté the salsify and shallots until the shallots are translucent. Add ¼ cup of the vegetable stock and cook, covered, until tender, about 10 minutes. Season with salt and pepper. Puree in a blender or food processor until absolutely smooth. Strain through a sieve and set aside.

Cut the peppers into pieces and cut the skin from the peppers with a sharp knife. In a sauté pan or skillet, heat 2 teaspoons of the olive oil and cook the peppers until tender, about 10 minutes. Puree in a blender or food processor until absolutely smooth. Season with honey, vinegar, and salt. Strain through a sieve and set aside.

Blanch the tomatoes in boiling water for about 10 seconds and cool in ice water. Peel, cut in quarters, and squeeze the seeds out. Heat the remaining 1 teaspoon of the olive oil in a sauté pan or skillet and cook with the rosemary, thyme, and garlic. Season with salt and pepper to taste. Remove the herbs and puree until absolutely smooth. Strain through a sieve and set aside.

In a sauté pan or skillet, heat the remaining 1 tablespoon of the butter and sauté the beets for about 15 minutes. Add a dash of balsamic vinegar and reduce heat. Add ¼ cup of the vegetable stock and cook until tender. Season with salt and pepper to taste. In a blender or food processor, puree until absolutely smooth. Strain through a sieve and set aside.

Make sure all the purees have the same consistency: using the thickest one as your ideal, thin any of the others by reducing over medium heat until they reach the same consistency as the ideal. Place each of the purees, one spoonful at a time, on each of 6 large white serving plates. After each puree has been completely served, rotate the plate one fifth of a turn and serve the second puree, and so on, until all five have been served. Each puree should be contiguous with the two neighboring purees, and all should touch in the center. Serve warm or cold.

**Melt 4 tablespoons unsalted butter over low heat. Skim off the foam and pour off the clear yellow clarified butter, leaving the milky residue.*

***Salsify, or oyster plant, is available in winter months.*

Filo Pastry Stuffed with Ratatouille and Tofu

Thierry Rautureau

Serves 4

RATATOUILLE STUFFING

- ¼ *red onion*
- ½ *red bell pepper*
- ½ *zucchini*
- 1 *Japanese eggplant*
- 1 *tablespoon olive oil*
- ½ *tablespoon chopped garlic*
- ½ *thyme branch, chopped*
- ½ *cup ¼-inch-diced tofu*
- 2 *filo sheets*
- 3 *tablespoons olive oil*
- *Tarragon and Chervil Cream Sauce, following*
- *Chervil and tarragon branches for garnish (optional)*

Cut the onion, pepper, zucchini, and eggplant into ¼-inch dice. In a saucepan, heat the olive oil and sauté the onion for 1 minute, then repeat with the pepper, zucchini, and eggplant, sautéeing each in turn for 1 minute. Add the garlic and thyme and cook 1 minute more. Set aside to cool.

Preheat the oven to 375°. Place one eighth of the ratatouille and 1 tablespoon of the tofu on a filo strip 1 inch from one end. Fold into triangles like a flag, beginning at the filling end. Repeat with the other 2 strips. Place the triangles on a lightly greased baking sheet and bake in the preheated oven until golden brown, about 10 to 15 minutes.

Serve over pools of tarragon and chervil cream sauce. Garnish with branches of chervil or tarragon, if desired.

TARRAGON AND CHERVIL CREAM SAUCE

- *3 tablespoons butter*
- *4 shallots, chopped*
- *1 carrot, peeled and minced*
- *1 cup dry vermouth*
- *3 cups vegetable stock*
- *1 cup heavy cream*
- *½ bunch tarragon*
- *½ bunch chervil*
- *Salt and pepper to taste*

In a saucepan, heat the butter and cook the shallots and carrot over medium heat for 3 minutes. Add the vermouth and cook to reduce by two thirds. Add the stock and cook to reduce again by two thirds. Add the cream and the herbs, and once again reduce by two thirds. Strain. Season to taste and serve hot.

Crispy Tarts with Vegetables

Leslee Reis

Makes six 4-inch tarts

Vegetable tarts with a touch of fresh goat cheese.

TART SHELLS

- *1 package frozen filo dough, defrosted*
- *2 tablespoons melted butter*

FILLING

- *3 zucchini*
- *3 yellow squash*
- *3 Japanese eggplants*
- *Olive oil*
- *2 garlic cloves, minced*
- *Salt and pepper to taste*
- *6 ounces fresh chèvre cheese*

Preheat the oven to 400°. To make the tart shells, cut circles out of the filo sheets just a little bigger than the 4-inch tart pans. Cut 4 circles of dough per tart. (While working with the filo, keep the unused portion covered with a damp cloth.) Layer the dough into the tart shells, pressing it gently into the shell and brushing lightly with melted butter between each layer until all 4 layers are used for each shell. Bake in the preheated oven for 10 to 12 minutes, or until golden; be careful they do not burn. Remove from the oven and cool on racks. Wrap any unused pastry sheets tightly in plastic wrap and store in the refrigerator or freezer for another use.

Reduce the oven heat to 375°. To make the filling, slice the vegetables ¼ inch thick. Heat a film of olive oil in a sauté pan or a skillet. Add the garlic. Season with salt and pepper. Taking care not to crowd the vegetables, sauté them one layer at a time. Set aside.

Divide the chèvre into 12 pieces. Dot 1 piece of the chèvre onto the bottom of each baked tart shell. Arrange the sautéed vegetables in the tarts, alternating colors. Dot the top of each tart with another piece of chèvre. Bake in the 375° oven for 3 to 5 minutes, or until the cheese has melted; serve warm.

New Mexican Sushi

John Sedlar

Serves 6

John Sedlar, who virtually started the whole Southwest regional cooking movement, recognizes no geographic boundaries in his culinary sojourn, as this recipe demonstrates.

- ½ *cup short-grain glutinous rice,* rinsed*
- 1 *tomato, peeled, seeded, and cut into ½-inch cubes*
- 1 *jalapeño chili, seeded and minced*
- 1 *garlic clove, minced*
- ½ *green bell pepper, cored, seeded, and cut into ½-inch dice*
- ½ *small onion, minced*
- ¼ *cup corn kernels, cooked*
- 1 *teaspoon salt*
- ½ *teaspoon ground black pepper*
- *Three 7-by-8-inch sheets* nori* *(sushi sheets)*

Place all the ingredients except the *nori* sheets in a small saucepan. Add cold water to cover by about ½ inch. Bring the mixture to a boil over moderate heat. Stir once and cover. Reduce the heat and simmer until all the liquid has been absorbed, about 15 to 20 minutes. Remove from heat and let the rice sit, covered, for about 10 minutes. Stir the rice briefly with a two-pronged fork or chopsticks. Cover it again, and let it sit 10 more minutes. Then uncover the rice mixture and let it cool to room temperature.

Place one of the rectangular sheets of *nori* on the work table, shiny surface up, with the long side of the *nori* facing you. Brush the sheet very lightly with water. Spread about 6 tablespoons of the rice mixture in a rectangle about 3 inches wide by 8 inches long and approximately 1 inch from the edge closest to you. Roll up the *nori* sheet, compacting it and the ingredients inside with your hands as you roll. Then use a very sharp knife to cut the roll into 8 small portions. (Some *nori* sheets are scored to show you where to cut.) Repeat the process with the other two sheets.

Serve 3 pieces of sushi to each person, with assorted condiments and dipping sauces such as soy sauce or a marinade of vinegar and shredded ginger.

**Available in most supermarkets or Japanese markets as well as natural foods stores.*

Vegetable Timbales with Peanut Sauce

Seppi Renggli

Serves 6

Brilliant-colored timbales in an Asian-inspired peanut sauce.

- *Two 6-ounce Japanese eggplants*
- 1 *large (6-ounce) carrot, peeled*
- 6 *Chinese cabbage leaves*
- 1 *garlic clove*
- 1 *large jalapeño chili, halved and seeded*
- 1 *small (4-ounce) onion*
- 1 *medium (4-ounce) zucchini*
- *One* each *4-ounce red and green bell pepper, halved, cored, and seeded*
- ¼ *cup water*
- 6 *ounces firm tofu, diced*
- *Peanut Sauce, following*

Slice 1 unpeeled eggplant and the carrot each into 6 long slices and blanch in boiling water to cover with the Chinese cabbage for about 2 minutes. Cool off in cold water, drain, and pat dry.

Mince the garlic and jalapeño, and dice the rest of the vegetables, including the remaining eggplant, into ½-inch cubes. In a small sauté pan or skillet, bring the water to a boil, add the vegetables, cover the pan, and cook for 3 to 5 minutes. Remove the lid and boil to reduce the liquid completely. Add the diced tofu; let cool.

Line each of 6 small custard cups with 1 Chinese cabbage leaf, 1 carrot slice, and 1 eggplant slice; fill the center with the steamed vegetables. Place in a large steamer over boiling water, cover, and steam for 5 to 7 minutes.

Pour 2 tablespoons peanut sauce on each of 6 preheated plates and turn a hot timbale out of the cup and onto the sauced plates. Serve at once.

PEANUT SAUCE

- 1½ *teaspoons Asian (toasted) sesame oil*
- ⅓ *cup minced shallots*
- 1 *garlic clove, minced*
- 1 *teaspoon chopped fresh lemon grass, or ¼ teaspoon* sereh *(dried lemon grass)**
- 1 *teaspoon* oelek sambal *(Indonesian spicy chili paste)**
- 1½ *tablespoons* ketjap manis *(Indonesian soy)**
- 1½ *teaspoons grated fresh ginger*
- ½ *teaspoon* each *ground cumin and ground coriander*
- 5 *tablespoons crunchy peanut butter*
- 1 *cup plain low-fat yogurt*

Heat the oil in a saucepan. Add the shallots, garlic, lemon grass, and *sambal.* Cook over medium heat until lightly brown. Stir in the *ketjap*, ginger, cumin, and coriander until smooth. Add the peanut butter and whisk until smooth. Remove the pan from the heat and whisk in the yogurt until the sauce is smooth.

**Available in Indonesian or Asian food stores.*

. . . .

Spring Vegetable Gyoza with Roasted Red Bell Pepper Butter Sauce

Roy Yamaguchi

Serves 4

Pot stickers with a California-style beurre blanc.

- 2 *teaspoons peanut oil*
- 1/3 *cup chopped zucchini*
- 1/3 *cup chopped yellow squash*
- 1/4 *cup chopped Japanese eggplant*
- 2 *tablespoons chopped Bermuda onion*
- 1/3 *cup chopped peeled and seeded tomato*
- 4 *fresh basil leaves cut into fine julienne*
- 1/3 *cup crumbled French feta cheese*
- 1/4 *teaspoon salt*
- 1/8 *teaspoon ground white pepper*
- 16 gyoza *wrappers (available at Japanese food stores)*
- 1/2 *tablespoon cornstarch mixed with 3 tablespoons water*
- 1/4 *cup peanut oil*
- 1/2 *cup water*
- *Roasted Red Bell Pepper Butter Sauce, following*
- *Watercress sprigs*

Heat the peanut oil in a very hot sauté pan or skillet and sauté the zucchini, yellow squash, Japanese eggplant, and onion for 10 seconds. Add the tomato and basil and continue to sauté for another 20 seconds. Remove the vegetables from the pan and place in a ceramic bowl. Add the crumbled feta cheese, stir, and season with salt and pepper.

Lay the 16 *gyoza* wrappers flat on the table and place 1 tablespoon of the vegetable mixture in the center of each wrapper. Place your finger into the cornstarch paste and rub it against the sides of a *gyoza* wrapper. Then fold the skin over to form a half moon. Repeat to seal each *gyoza*. Heat the 1/4 cup of peanut oil in a large sauté pan or skillet. Place all the *gyozas* in the pan and lightly brown them on one side. Then quickly add water and cover. Continue to cook until all the water is evaporated and the side that is facing down is crispy.*

Place 4 tablespoons of the red bell pepper sauce on each of 4 plates. Arrange 4 *gyozas* one each plate and garnish with watercress.

**The objective is to cook the* gyozas *all the way through without turning them over. The process starts by browning the skin in oil. The water cooks them thoroughly, and when no water is left in the pan, the oil will make the soggy skins crisp again.*

ROASTED RED BELL PEPPER BUTTER SAUCE

- *1 medium red bell pepper*
- *1 shallot, minced*
- *1 cup dry white wine*
- *2 tablespoons white wine vinegar*
- *¼ cup heavy cream*
- *2 cups (4 sticks) unsalted butter, softened*

Cut the pepper in quarters; remove seeds and core. Char the bell pepper evenly, skin side up, under a broiler. Place in a plastic bag, close tightly, and let sit for 15 minutes. Rub the blackened skin from the pepper and chop the pepper.

In a small saucepan, boil the shallot, white wine, and white wine vinegar until the liquid is reduced to 2 tablespoons. Then quickly add the heavy cream and cook to reduce again by half. Over very low heat, whisk in the butter a few tablespoons at a time until the sauce is thick. Place the pepper in a blender or a food processor with the white butter sauce and puree. Strain through a sieve; keep warm in a double boiler.

Vegetable Fritters with Chick-Pea Batter

Susan Feniger and Mary Sue Milliken

Serves 4 to 6

Spicy vegetable fritters with two dipping sauces.

- ¾ *cup chick-pea flour,* sifted*
- ½ *teaspoon salt*
- ½ *teaspoon ground turmeric*
- ½ *tablespoon ground cumin*
- ½ *tablespoon ground coriander*
- 1 *tablespoon black mustard seeds*
- ½ *tablespoon* garam masala*
- 1 *teaspoon dried red pepper flakes*
- ½ *teaspoon baking powder*
- ½ *teaspoon cornstarch*
- ½ *teaspoon vegetable oil*
- *About ½ cup water*
- 36 *bite-sized pieces of assorted vegetables: bell peppers, eggplant, zucchini, broccoli, cauliflower, green onions, mushrooms, green cabbage*
- *Peanut oil for frying*
- *Salt to taste*
- *Yogurt Sauce, following*
- *Fresh Green Chutney, following*

Place the first 10 ingredients in a large bowl and blend thoroughly. Add the oil and water and mix slightly, just until a pancakelike batter forms. Do not overmix. More water may be necessary if this batter appears too thick. Cut the assorted vegetables into bite-sized pieces.

Fill a large, heavy pot with at least 3 inches of peanut oil and heat. Drop a little batter into the hot oil to test the temperature: the batter should sizzle, rise to the top, and brown quickly. Using tongs, dip each piece of vegetable in the batter and drop one by one into the hot oil; cook in batches so that the pot is not too crowded. Cook until golden brown. Remove the vegetables with a slotted spoon, drain on paper towels, and salt lightly. Serve immediately, with yogurt sauce and green chutney as dipping sauces.

**Available at Indian markets or natural foods stores.*

YOGURT SAUCE

- *3 tablespoons peanut oil*
- *1 tablespoon black or yellow mustard seeds*
- *2 tablespoons grated fresh ginger*
- *1 tablespoon pureed garlic*
- *½ teaspoon ground turmeric*
- *¼ teaspoon salt*
- *1 cup plain yogurt*

Heat the oil in a sauté pan or skillet. Add the mustard seeds, cover the pan, and, when they begin to pop, add the ginger, garlic, turmeric, and salt. Fry and stir for a few moments, until the aromas are released. Remove from the heat, add the yogurt, and chill for about 1 hour.

FRESH GREEN CHUTNEY

- *½ cup packed fresh cilantro leaves*
- *½ cup packed fresh mint leaves*
- *1 teaspoon pureed garlic*
- *2 teaspoons grated fresh ginger*
- *½ teaspoon salt*
- *2 to 3 serrano chilies, seeded and minced*
- *1 tablespoon fresh lime juice*
- *1 teaspoon olive oil*

Chop the mint and cilantro leaves very finely and mix with the remaining ingredients.

. . . .

Grilled Mozzarella in Romaine with Sun-dried Vinaigrette

Mark Militello

Serves 8

The perfect first course for an all-grill menu.

SUN-DRIED VINAIGRETTE

- 1 *cup extra-virgin olive oil*
- ¼ *cup balsamic vinegar*
- ¼ *teaspoon chopped garlic*
- ¼ *teaspoon chopped shallot*
- ½ *teaspoon sugar*
- 2 *tablespoons sun-dried tomatoes, chopped*
- 1 *tablespoon chopped fresh basil*
- *Salt and freshly grated black pepper to taste*
- 8 *large romaine leaves*
- 1 *pound whole-milk mozzarella, cut into 8 pieces*
- 8 *fresh basil leaves*
- 1 *cup olive oil*
- *Mixed lettuce leaves*

Light a charcoal fire in an open grill. Mix all the ingredients for the vinaigrette and store in the refrigerator until ready to use.

Steam the romaine leaves over boiling water until soft and pliable, about 1 minute. Place a basil leaf on top of each piece of cheese. Place each piece of cheese in the center of a romaine leaf and carefully wrap each into a bundle. Place in a shallow dish, pour the olive oil over, and chill. Over a medium-hot fire, grill the romaine bundles until the mozzarella is soft to the touch. Place on a bed of greens and pour vinaigrette over the top.

Pan-fried Goat Cheese with Roasted Peppers

Thierry Rautureau

Serves 4

GOAT CHEESE SAUCE

- 1½ *cups dry white wine*
- 3 *shallots*
- ¾ *cup heavy cream*
- 3 *ounces Montrachet goat cheese*
- 1 *tablespoon butter*
- *Dash white pepper*
- 2 *tablespoons minced fresh chives*
- 2 *tablespoons olive oil*
- 4 *bell peppers (2 red, 1 yellow, and 1 green), cored, seeded, and diced*
- 1 *tablespoon chopped shallots*
- ½ *tablespoon chopped garlic*
- 1 *tablespoon chopped fresh tarragon*
- *One 5½-ounce log Montrachet goat cheese, chilled*
- ¼ *cup milk*
- 5 *tablespoons fresh bread crumbs*
- 1 *tablespoon olive oil*

To make the sauce: In a saucepan, boil the white wine and shallots until the liquid is reduced by half. Add the cream and again reduce by half. In a blender or a food processor, puree the wine-cream mixture with the goat cheese and butter until smooth. Strain and season. Add the chives and set aside.

In a sauté pan or skillet, heat the olive oil and sauté the peppers until tender, about 5 minutes. Add the shallots, garlic, and tarragon and remove the pan from the heat.

Slice the goat cheese into 8 circles. Dip the circles in the milk and dredge in the bread crumbs. In a sauté pan or skillet, heat the olive oil and sauté the goat cheese over medium heat for 1 minute on each side.

Place the peppers in a 3-inch circle on one side of each of 4 plates. Place 2 goat cheese circles on the other side of each plate, so they just touch the peppers, but not each other. Pour the sauce in the open spaces around the peppers and goat cheese. Serve immediately.

Caponata

Mauro Vincenti

Serves 4

Serve this eggplant appetizer with thin oven-toasted baguette slices.

- *1 tablespoon raisins*
- *2 medium eggplants*
- *2* each *red and yellow bell peppers, halved, cored, and seeded*
- *2 medium zucchini*
- *3 tablespoons extra-virgin olive oil*
- *1 medium red onion, minced*
- *2 medium celery stalks, minced*
- *1 teaspoon honey*
- *1 teaspoon balsamic vinegar*
- *1 tablespoon pine nuts*
- *Salt and pepper to taste*

Soak the raisins in warm water to cover until plump; drain and set aside. Chop the eggplant, peppers, and zucchini into large cubes approximately 1 inch in size. In a sauté pan or skillet, heat 2 tablespoons of the oil and sauté the vegetables until slightly browned and tender; set aside.

Heat the remaining 1 tablespoon of olive oil in another sauté pan or skillet and sauté the onion and celery until golden. Add the honey, balsamic vinegar, pine nuts, and plumped raisins to this mixture. Add this mixture to the cubed vegetables and cook over low heat for 1 minute. Season to taste and serve at room temperature.

SOUPS

Vegetable Stock

Vincent Guerithault

Makes about 1 quart

This is an all-purpose salt-free vegetable stock that can be used in many of the recipes in this book. It's also good as a clear broth to serve at the beginning of a meal.

- 2 *quarts water*
- 1 *whole onion, chopped*
- 2 *whole cloves*
- 1 *carrot, peeled and sliced*
- 1 *celery stalk, sliced*
- 1 *leek, white and pale green part, chopped*
- 2 *whole tomatoes, chopped*
- 1 *small bunch parsley*
- 2 *bay leaves*
- 1 *small bunch rosemary*
- 1 *small bunch thyme*
- 1 *teaspoon black peppercorns*

In a large pot, bring all the ingredients to a boil, then reduce heat and simmer for approximately 1 hour. Strain through a sieve. Store, covered, in the refrigerator for up to 1 week, or freeze.

Artichoke and Asparagus Soup

Roland Liccioni

Serves 4 to 6

A pale green, creamy soup for a special dinner.

- *1 pound asparagus*
- *8 artichokes*
- *1½ tablespoons olive oil*
- *1 onion, chopped*
- *2 quarts water*
- *2 to 3 cups milk, half and half, or cream*
- *Salt and freshly ground pepper to taste*

Remove the tips from the asparagus and reserve them for garnish. Trim and chop the stems (peel them first if they are very large). Remove the leaves and stems from the artichokes. Scoop out the chokes with a spoon and coarsely chop the hearts. In a large, heavy 4-quart pot, heat the olive oil, then add the onion and sauté until it is translucent. Add the artichokes, cook 5 minutes, then add the water. Bring to a boil, reduce heat, add salt, and simmer slowly for 25 minutes. Add the asparagus stems and simmer another 10 minutes. Drain off the water and puree the vegetables in a blender or a food processor, then pass through a sieve.

Steam the asparagus tips over boiling water until just tender, 3 to 5 minutes, depending on the size of the asparagus. In a saucepan, heat the puree and add milk, half and half, or cream to attain the consistency desired. Season, ladle into soup bowls, and garnish with the asparagus tips.

Café Provençal's Avocado-Cucumber Soup with Fresh Oregano

Leslee Reis

Serves 4 to 6

A refreshing soup to serve hot or cold. Increase the seasoning if you plan to serve it cold.

- 3 *tablespoons butter*
- 1½ *large yellow onions, coarsely chopped*
- 2 or 3 *cucumbers, peeled, seeded, and diced*
- 2 *garlic cloves, mashed*
- 1 *tablespoon chopped fresh oregano, or ½ teaspoon dried oregano*
- *Salt, fresh-ground white pepper, and cayenne to taste*
- 3 to 5 *cups rich vegetable stock, preferably homemade*
- 3 *ripe avocados, peeled and seeded*
- *Chopped green onions, blanched diced cucumber, and fresh oregano leaves for garnish*

Melt the butter in a sauté pan or skillet and sauté the onions, covered, until translucent, about 10 minutes. Add the cucumbers, garlic, oregano, and seasonings; cook another 10 minutes. Add 1 cup of the stock and cook 5 minutes.

Puree in a blender or a food processor; strain if desired. Add the avocados and puree again. Add more vegetable stock until a light cream consistency is achieved. Taste for seasoning. Garnish with green onions, cucumber, and oregano.

Chef Vincent's Cream of Avocado Soup

Vincent Guerithault

Serves 4

This luxurious soup is also good with a little cayenne pepper.

- *2 cups Vegetable Stock, following*
- *1 tablespoon olive oil*
- *1 tablespoon chopped shallots*
- *2 large ripe avocados, peeled, pitted, and pureed*
- *1 cup heavy cream*
- *Salt, pepper, and freshly grated nutmeg to taste*
- *1 tomato, peeled, seeded, and diced*

Chill the vegetable stock in the refrigerator. In a small sauté pan or skillet, heat the olive oil and sauté the shallots until translucent. Set aside and let cool. In a bowl, combine the avocado puree, chilled vegetable stock, cream, and shallots. Season with salt, pepper, and fresh nutmeg. Pour into chilled bowls or cups and sprinkle diced tomato over each serving.

VEGETABLE STOCK

Makes 1 quart

- *1 medium leek, sliced*
- *2 carrots, peeled and sliced*
- *1 onion, sliced*
- *½ celery stalk, sliced*
- *1 parsley sprig*
- *1 bay leaf*
- *1 fresh thyme sprig*
- *1 fresh rosemary sprig*
- *1 clove*

Place the water in a large pot. Add the leek, carrots, onion, and celery. Tie the herbs into a bundle with cotton string and add to the pot. Bring to a boil, then simmer for 1 hour. Strain the stock through a sieve.

Cold Beet and Buttermilk Soup

Cindy Pawlcyn

Serves 6 to 8

- 1½ *pounds beets*
- 1 *quart buttermilk*
- 1 *cup sour cream*
- 1 *tablespoon sugar*
- *Pinch of Dijon mustard*
- 1½ *tablespoons red wine vinegar*
- 1 *English cucumber, peeled, seeded, and minced*
- 5 *green onions, minced*
- *Salt and white pepper to taste*
- *Chopped fresh dill for garnish*

In a large covered saucepan, boil the beets in water to cover until tender, about 40 to 45 minutes. Let cool, then peel and coarsely chop. Place in a blender and blend in batches with a little of the buttermilk to make a coarse puree. Pour into a bowl and add the remaining buttermilk, the sour cream, sugar, mustard, and vinegar. Add the cucumbers (reserving ½ cup for garnish) and onions to the beet mixture. Add salt and pepper. Chill before serving. Place a spoonful of sour cream on top of each serving, and sprinkle with chopped cucumber and dill.

Black Bean Soup with Hot Green Chutney

John Downey

Serves 8

Start this Mexican-style bean soup the day before you plan to serve it.

- *8 ounces dried black beans*
- *1 medium onion*
- *2 medium carrots*
- *2 celery stalks*
- *1 small leek*
- *2 whole garlic heads, broken into cloves and peeled*
- *1 tablespoon corn oil*
- *2 bay leaves*
- *1 tablespoon chopped fresh thyme*
- *2 teaspoons anise seed*
- *2 tablespoons chopped fresh rosemary*
- *½ tablespoon whole black peppercorns*
- *2 quarts water*
- *Salt to taste*
- *Juice of ½ lemon*
- *Hot Green Chutney, following*

Wash the beans and pick out any small stones. Soak the beans overnight in a large container of cold water. Be prepared for expansion.

Coarsely chop the onion, carrots, celery, leek, and garlic. Heat the oil in a large, heavy pot and cook the vegetables and herbs slowly in the oil until tender, about 15 to 20 minutes.

Drain the beans and add to the pot of vegetables along with the fresh water. Bring to a boil, reduce heat, and simmer slowly, covered, for 2 hours, stirring occasionally. Remove from heat and puree in small batches in a blender or a food processor. Pass the soup through a sieve and reheat. Add salt and lemon juice to taste. Serve hot in soup bowls with a spoonful of chutney in the center.

HOT GREEN CHUTNEY

Makes about ½ cup

2 Anaheim (long green) chilies
¼ to ½ jalapeño chili
Juice of 1 lime
¼ teaspoon Garlic Salt, following, or commercial garlic salt
Leaves from ½ bunch cilantro

Remove the cores and seeds from the chilies. Chop roughly, then puree with all the other ingredients in a blender or a food processor.

This chutney should be very spicy. It will keep for about 2 days refrigerated.

GARLIC SALT

3 garlic cloves, minced
⅓ cup fine sea salt

In a large bowl, mix the garlic and salt until well combined. Store tightly covered in a spice jar in the refrigerator.

Yellow Bell Pepper Soup

Mauro Vincenti

Serves 4

A silky golden soup with no butter or cream.

- 2 *tablespoons extra-virgin olive oil*
- 1 *medium carrot, peeled and minced*
- 1 *medium celery stalk, minced*
- 1 *small red onion, minced*
- 3 *medium white potatoes, peeled and chopped*
- 4 *yellow bell peppers, cored, seeded, and chopped*
- 3 *cups water*
- *Salt and pepper to taste*
- *Olive oil, freshly grated Parmesan cheese, and chopped fresh Italian parsley for garnish*

Heat the olive oil in a saucepan and sauté the carrot, celery, and onion until tender. Add the potatoes and peppers and sauté over medium heat for about 5 minutes. Add the water. Bring to a boil, reduce heat, and simmer, covered, until tender, about 30 minutes.

Pass the vegetable mixture through a food mill or a coarse sieve to remove the skin of the bell pepper. Place in a blender or a food processor and blend until very creamy.

Reheat over low heat for about 1 minute. Add salt and pepper, pour into soup bowls, and sprinkle with olive oil, cheese, and parsley.

Carrot and Red Pepper Soup

Alice Waters

Serves 8

"Our first double soup: The carrot soup acts as a backdrop for the more pungent red pepper flavor. Pour the carrot soup first into the bowl, then add a smaller portion of the red pepper soup; the red pepper taste will remain distinct and provide a spicy accent to the deeper carrot flavor." —*Alice Waters*

CARROT SOUP

- 4 *tablespoons unsalted butter*
- 6 *cups water*
- 6 *large carrots (1 pound, 4 ounces), peeled and cut into ½-inch dice*
- ½ *yellow onion (4 ounces), cut into ¼-inch dice*
- 1 *teaspoon salt*
- ⅛ *teaspoon freshly ground black pepper*
- 1½ *teaspoons fresh lemon juice*

RED PEPPER SOUP

- 2 *tablespoons unsalted butter*
- 3 *medium red bell peppers (14 ounces), cored, seeded, and diced*
- ⅔ *cup water*
- ¼ *teaspoon salt*
- *Red wine vinegar to taste*
- ⅛ *teaspoon freshly ground pepper*
- *Chopped fresh chervil leaves and crème fraîche thinned with warm water for garnish (optional)*

To make the carrot soup, melt the butter in a 6-quart stainless steel soup pot. Add 1 cup of the water, the carrots, and onion. Bring to a low simmer, cover, and stew for 30 minutes.

Remove the cover from the pot. The vegetables should be very tender and the water almost entirely evaporated. If not, continue cooking them until they are. Add the remaining 5 cups water and bring to a boil. In a blender or food processor, puree the soup in batches for 3 minutes each. Season with salt, pepper, and lemon juice. The soup should have a velvety consistency and be slightly thicker than heavy cream.

To make the red pepper soup, in a 3-quart saucepan melt the butter. Add the peppers and water, bring to a simmer, and cook, uncovered, for 20 minutes, or until the peppers are very tender. Most of the water will have evaporated during this time.

In a blender or food processor, puree the peppers with ½ cup of the water and pass the puree through a medium-fine sieve to catch any bits of skin. Return the soup to the pan and add salt. If the pepper soup lacks depth, correct it with a few drops of red wine vinegar. If necessary, thin the red pepper soup with a little of the remaining water so that its consistency is similar to that of the carrot soup.

Pour ⅔ cup of the carrot soup into each of 8 warm bowls. Stir 2 tablespoons of the red pepper soup into the center. Garnish with chervil and crème fraîche if desired. Draw the cream over the surface with the tines of a fork.

Carrot and Zucchini Vichyssoise

Mauro Vincenti

Serves 8

A beautiful three-colored vichyssoise containing no potatoes.

- *½ cup olive oil*
- *2 large onions, coarsely chopped*
- *5 to 6 cups coarsely chopped carrots (about 1½ pounds)*
- *4 to 5 cups minced zucchini (about 1½ pounds)*
- *1 large leek, white part only, chopped*

In a saucepan, heat half of the olive oil and sauté the onions and carrots until the onions are translucent. Cover the vegetables with water and cook slowly for 35 minutes. Drain and puree in a blender or food processor, in batches if necessary, until smooth, adding a little water if necessary. Chill 1 hour.

Trim and peel the zucchini, reserving the skins. Mince the zucchini. In another saucepan, heat the remaining olive oil and cook the leek until it is translucent; add the zucchini. Cover with water and cook 15 minutes. Drain and puree in a blender or a food processor, in batches if necessary, until smooth adding a little water if necessary. Chill 1 hour.

Blanch the zucchini skins in boiling water for 3 minutes. Drain the skins and place them in cold water to stop the cooking process. Puree in a blender or a food processor with a little water, if necessary. Chill 1 hour.

Just before serving, add water if necessary to make all three purees the same consistency. Slowly pour the purees, one at a time, into shallow bowls. As you add the second and third purees, play with the colors to create any design you like.

Recipe created by Jean-Pierre Bosc

Mexican Cauliflower Soup

Joyce Goldstein

Serves 4 to 6

- 6 *tablespoons unsalted butter*
- 6 *cups sliced yellow onions (about 1¾ pounds)*
- 1 *tablespoon chili powder*
- 2 *tablespoons ground cumin*
- 2 *medium heads cauliflower, cored and cut into small chunks (about 8 cups)*
- 1 *large russet potato, peeled and diced (about 1 cup)*
- 6 *cups or more vegetable stock*
- 1 *teaspoon salt or to taste*
- ½ *teaspoon ground black pepper*
- *Diced tomatoes, and/or chopped cilantro, diced avocado, grated cheese, or oregano leaves for garnish*

In a large saucepan, melt the butter. Add the onions and cook until they are translucent. Add the chili powder and cumin and cook for a minute or two, stirring well. Add the cauliflower, potato, and vegetable stock and bring the mixture to a boil. Reduce heat and simmer until the cauliflower and potato are soft, about 25 minutes.

Puree the soup in batches in a blender or food processor. Thin with more stock if necessary. Season with salt and pepper. Garnish with one or more of the suggested garnishes and serve.

Puree of Celery Soup

Seppi Renggli

Serves 6

An unusual celery soup to serve hot or cold.

- 1 *bunch celery*
- 3 *tablespoons corn oil*
- 1½ *cups coarsely chopped onions*
- 3 *garlic cloves, coarsely chopped*
- 1 *large jalapeño chili, or 1 small dried chili pepper, coarsely chopped*
- 1 *cup water*
- 1¼ *cups chopped parsnips*
- 1 *quart strong vegetable stock*
- 2 *tablespoons oatmeal*
- *Salt and freshly grated nutmeg to taste*

Cut half the celery into ¼-inch dice and half into 1-inch pieces; set aside. In a large saucepan, heat the oil and sauté the onions, garlic, and jalapeño for 5 minutes.

Meanwhile, in a saucepan, bring the water to a boil, add the diced celery, and cook for 15 minutes; set the celery aside with its cooking water.

Add the 1-inch pieces of raw celery, parsnips, vegetable stock, and oatmeal to the onion mixture. Bring to a boil, reduce heat, and simmer for 30 minutes.

Puree the onion-celery mixture in batches in a blender or food processor. Strain through a sieve into a saucepan. Add the diced celery with its liquid and bring to a boil. Reduce heat and simmer until slightly thickened. Season with salt and nutmeg. Serve warm.

To serve cold: Chill very well and blend again with ½ cup soda water. Serve in chilled bowls and garnish with pale green-yellow celery leaves.

Cream of Celery Root Soup

Georges Perrier

Serves 8

This brightly flavored soup is also good without the cream.

- ½ *onion*
- 1½ *leeks, white part only*
- 2 *garlic cloves*
- 1½ *ripe tomatoes*
- 1 *carrot*
- 1 *shallot*
- 2 *knobs celery root*
- 1½ *teaspoons butter*
- 1 *bay leaf*
- 1 *fresh thyme branch*
- 2 *quarts (8 cups) vegetable stock*
- 1 *cup heavy cream*
- *Salt and pepper to taste*
- *Garlic Croutons, following*

Coarsely chop all vegetables. In a large pot, melt the butter over low heat and sauté the vegetables, bay leaf, and thyme until the onions are translucent, about 15 minutes. Add the vegetable stock and simmer for 1 hour over medium heat. Puree in batches in a blender or a food processor and strain through a sieve. Whisk in the heavy cream. Season with salt and pepper and sprinkle with the garlic croutons. Serve at once.

GARLIC CROUTONS

- 2 *tablespoons olive oil*
- 2 *garlic cloves*
- *Salt and pepper to taste*
- 1 *tablespoon chopped fresh parsley*
- 4 *slices bread, crusts removed*

Preheat the oven to 350°. In a blender or a food processor, puree the oil, garlic, salt and pepper, and parsley until smooth. Spread the mixture on both sides of the bread. Cut the bread into ¼-inch cubes. Place the cubes on a baking sheet and bake in the preheated oven for 10 to 15 minutes, or until golden. Stir a few times during baking.

Grilled Corn Soup with Ancho Chili and Cilantro Cream

Stephan Pyles

Serves 4 to 6

A smoky, spicy soup with colorful cream accompaniments.

- 2 *garlic cloves*
- 1 *teaspoon olive oil*
- 4 *ears fresh corn, husked*
- ½ *cup chopped carrots*
- ¼ *cup chopped celery*
- ½ *cup chopped onion*
- 1 *jalapeño or serrano chili, seeded and chopped*
- 1½ *cups water or vegetable stock*
- 1 *cup heavy cream*
- *Salt to taste*
- *Cilantro Cream, following*
- *Ancho Chili Cream, following*

Preheat the oven to 300°. Place the olive oil and garlic cloves in a small pan and roast for 15 minutes.

Prepare a charcoal fire in an open grill. When the coals are gray, grill the corn for 5 minutes on each side. Remove and let cool.

In a saucepan, place the carrots, celery, onion, garlic, and chili. Cover with the water or stock and bring to a boil. Let simmer for 5 minutes. Cut the kernels from the corn. Add to the pan and simmer 10 minutes longer.

Pour the mixture into a blender or food processor and puree for 2 minutes, in batches, if necessary. Strain the liquid through a sieve and return it to the saucepan. Add the cream, place over low heat, and simmer for 5 minutes. Add salt, pour into soup bowls, and float a spoonful of each cream on each serving. Serve at once.

CILANTRO CREAM

- 2 *cups water*
- 5 to 8 *spinach leaves, stemmed*
- 1 *cup loosely packed fresh cilantro leaves*
- 3 *tablespoons milk or half and half*
- 2 *tablespoons sour cream or crème fraîche*

In a saucepan, bring the water to a boil. Add the spinach leaves and cook for 1 minute. Drain off the liquid and place the leaves in ice water for 1 minute. Place the cilantro, milk, and spinach leaves in a blender or food processor. Blend until smooth. Pass the mixture through a fine sieve over a bowl. Whisk in the sour cream.

ANCHO CHILI CREAM

- 1 *small ancho chili, halved, cored, and seeded*
- 3 *tablespoons milk or half and half*
- 2 *tablespoons sour cream or crème fraîche*

Preheat the oven to 400°. Place the ancho chili in the preheated oven for 45 seconds. Remove from the oven, place in a bowl, and add warm water to cover. Let stand for 10 minutes, or until softened. Remove the chili from the water and place it in a blender or a food processor with the milk. Blend until smooth. Pass the mixture through a fine sieve over a bowl. Whisk in the sour cream.

Shoe Peg Corn and Peanut Soup

Marcel Desaulniers

Serves 8

"The peanuts used in this recipe are purchased at the Peanut Shop in Williamsburg. They are prepared by an old country recipe almost impossible to duplicate. They are ready to eat as purchased. We split the whole peanuts in half and toast them, which intensifies the peanut flavor. Any quality unsalted peanut can be used in this recipe if a trip to Williamsburg is not convenient.

"Many varieties of corn are available on a year-round basis. If the white shoe peg corn suggested in this recipe is not available, do not hesitate to use other sweet corn (white, yellow, or bi-colored). If you must use canned corn, substitute a 16-ounce can of white shoe peg corn.

"Crunchy-style peanut butter may be used rather than the suggested creamy style. However, it should be an adult-style product; that is, one that is low in sugar and salt."

—*Marcel Desaulniers*

- 1 *cup unsalted shelled Virginia peanuts*
- 7 *tablespoons safflower oil*
- 1 *tablespoon water*
- 4 *celery stalks, diced*
- 1 *medium onion, diced*
- *Salt and pepper to taste*
- 2 *cups white corn kernels, shoe peg variety if possible (about 6 ears)*
- 6 *cups vegetable stock*
- ½ *cup unbleached all-purpose flour*
- ¼ *cup creamy peanut butter*

Preheat the oven to 300°. On a baking sheet, toast the peanuts in the preheated oven for 25 to 30 minutes, or until golden brown; set aside. In a 5-quart saucepan over medium heat, heat 1 tablespoon of the safflower oil and the water. Add the diced celery and onion. Season with salt and pepper and sauté for 5 minutes. Add the corn kernels and sauté for an additional 5 minutes. Add the vegetable stock. Bring to a boil, then lower the heat and simmer for 10 minutes.

In a 2½-quart saucepan over low heat, heat the remaining 6 tablespoons of the safflower oil. Add the flour to make a roux. Stir and cook for 1 minute. Add the peanut butter, and continue to cook for 6 to 8 minutes, or until the mixture bubbles; stir constantly to prevent scorching. Strain 4 cups of the simmering stock into the roux. Whisk vigorously until smooth. In a 5-quart saucepan, place this mixture along with the remaining stock and vegetables. Whisk until well combined. Simmer over low heat for 10 minutes. Adjust the seasoning.

Place equal portions of soup into each of 8 warm serving dishes. Garnish with toasted peanuts and serve immediately.

Note: If the soup is not to be served within 2 hours, cool it properly and refrigerate for up to 3 days. Reheat the soup slowly to a simmer before adding the toasted peanuts.

Cream of Eggplant Soup

Emeril Lagasse

Serves 8

- *4 tablespoons butter*
- *1½ cups minced onions*
- *1½ cups minced celery*
- *1½ cups finely diced peeled potatoes*
- *2 large eggplants, peeled and finely diced*
- *1 teaspoon curry*
- *½ teaspoon chopped fresh thyme*
- *½ teaspoon chopped fresh basil*
- *1 quart vegetable stock*
- *2 cups heavy cream*
- *Salt to taste*

In a 6- to 8-quart saucepan, melt the butter and sauté the onions, celery, potatoes, and eggplants until tender, about 25 minutes. Add the curry, thyme, and basil. Cook until the ingredients begin to stick to the bottom of the pan.

Whisk in the vegetable stock and continue cooking until the mixture begins to thicken, about 30 to 45 minutes. Remove from heat, add the cream, and season with salt if needed. Serve immediately.

Mexican Corn Chowder

Susan Feniger and Mary Sue Milliken

Serves 6 to 8

Although this chowder has no milk or cream, it's rich and hearty, with a festive orange hue and a warming bite.

- *Corn oil for frying*
- 1 *flour tortilla, cut into thin strips*
- 4 *corn tortillas*
- 1 *large red bell pepper*
- ½ *tablespoon oil*
- 1½ *cups minced onions*
- 1 *teaspoon minced garlic*
- 5 *cups fresh corn cut from the cob (about 10 ears), or frozen corn kernels*
- 1 *small tomato, peeled and seeded*
- ½ *jalapeño chili, seeded and chopped*
- 6 *cups vegetable stock or water*
- 6 to 8 *lime wedges for garnish*

In a large, heavy skillet heat ½ inch of oil and fry the tortilla strips until golden brown; remove to paper towels with a slotted spoon. Repeat with 1 whole corn tortilla at a time.

Cut the bell pepper into quarters; remove the seeds and core. Char the skin evenly under a broiler. Place the pepper in a closed plastic bag for 15 minutes. Rub off the charred skin and coarsely chop the pepper; set aside. Let the whole tortillas cool, then crush them into pieces with your hands or a heavy object.

In a saucepan, heat the oil and slowly cook the onions until they are golden brown. Add the minced garlic and sauté briefly to release its flavor. Add the corn, red pepper, crushed corn tortillas, tomato, jalapeño, and vegetable stock or water. Bring to a boil, reduce heat, and simmer for 45 minutes, stirring frequently to avoid scorching. Remove from heat, cool, puree in a blender and strain through a sieve.

If a lighter soup is preferable, thin with additional stock or water. Garnish each serving with the fried flour tortilla strips and a lime wedge. Serve at once.

. . . .

Curried Green Bean Potage

Madeleine Kamman

Serves 6 to 8

"This is a nice company dinner potage, which should be served without bread, since it constitutes the beginning of a meal rather than a whole meal. . . . Preferably use the fat round green beans that have almost no strings."

—*Madeleine Kamman*

- *1 red bell pepper*
- *1 yellow bell pepper*
- *4 tablespoons butter*
- *8 garlic cloves, minced*
- *2 leeks, white and pale green parts only, chopped*
- *2 large onions, chopped*
- *1 tablespoon very fresh curry (available at Indian markets)*
- *2 pale green leaf lettuce hearts, chopped*
- *1 pound green beans, chopped*
- *6 cups water*
- *1 bay leaf and several fresh parsley and thyme sprigs*
- *1 cup sour cream*
- *Salt and pepper to taste*
- *2 tablespoons chopped fresh parsley*

Cut the peppers into quarters, core, and seed. Char the peppers evenly under a broiler. Place them in a sealed plastic bag for 15 minutes, rub off the blackened skin, and set the peppers aside.

In a large soup pot, melt 3 tablespoons of the butter. Add three fourths of the garlic and sauté until golden. Add the leeks and onions and sauté until translucent. Add the curry and cook for a few minutes. Add the lettuce and green beans.

Add the water, bring to a boil, and reduce heat to a simmer. Tie the bay leaf, parsley, and thyme into a bundle with a cotton string and add to the pot. Cook until the green beans are tender but still green, 5 to 7 minutes, depending on the size of the beans. Remove the herb bundle and puree the soup in batches in a blender or a food processor, adding ½ cup of the sour cream. Return the soup to the pot.

Dice the peeled peppers into ⅓-inch squares. In a small sauté pan or skillet, heat the remaining 1 tablespoon of butter and sauté the peppers for 2 minutes. Add them to the soup. Season the soup with salt and pepper and heat thoroughly.

Mix the remaining sour cream with the remaining garlic and the parsley. Pour the hot soup into bowls and spoon a generous dollop of herbed cream in the center of each serving. Serve piping hot.

Mushroom Won Ton

Seppi Renggli

Serves 4 to 6

Light, savory won ton in a vegetable stock.

FILLING

- 1 *pound mushrooms (about 6 cups)*
- 2 *tablespoons almond oil*
- ½ *cup thinly sliced shallots*
- 2 *garlic cloves, chopped*
- 2 *medium jalapeño chilies, seeded and chopped*
- ½ *cup chopped mixed fresh herbs (parsley, tarragon, sage, and chervil)*
- 6 *ounces firm tofu, cut into ¼-inch dice*
- *Freshly ground black pepper to taste*
- 32 to 40 *round won ton wrappers*

BROTH

- 2 *cups salt-free vegetable stock*
- ¾ *cup snow peas*
- 20 *thin asparagus stalks*
- 1 *medium tomato, peeled, seeded, and diced*
- 6 *fresh cilantro sprigs*
- *Freshly ground black pepper to taste*
- 1 *teaspoon shredded fresh ginger*

Stem and set aside 2¼ cups of the best mushroom caps and slice the remaining mushrooms.

In a sauté pan or skillet, heat the oil and sauté the shallots, garlic, and chilies until the shallots are golden. Add the herbs, sliced mushrooms, and tofu. Cook over high heat until all the liquid has evaporated. Season with pepper. Allow the mixture to cool.

Place a spoonful of filling in the center of each won ton wrapper. Moisten the edges of the wrapper with a little water and press one edge of the wrapper against the other edge, forming a half-moon shape. Press tightly to seal the entire edge. Steam over boiling water in a covered pot for 4 minutes; set aside.

In a large pot, boil the vegetable stock to reduce it to ¾ cup. Add the reserved 2¼ cups mushrooms, snow peas, asparagus, and tomato. Simmer for 2 minutes; add the won ton and cook until heated through. Pour into preheated plates. Garnish with cilantro sprigs and season with freshly ground black pepper and ginger.

Butternut Squash Soup

Leslee Reis

Serves 6

About 2 pounds butternut squash
1 *tablespoon butter*
2 *tablespoons olive oil*
1 *large Spanish onion, chopped*
2 *medium leeks, white part only, sliced*
4 to 5 *cups vegetable stock*
Salt and pepper to taste
About 1 cup half and half

GARNISH

3 *tablespoons* each *crème fraîche and heavy cream, mixed with 2 tablespoons minced fresh chives, or 1 Granny Smith apple, peeled, cored, and coarsely chopped*

Preheat the oven to 350°. Cut the squash in half and remove the seeds. Dot each half with butter and bake skin side down on a baking sheet in the preheated oven until tender, about 30 minutes to 1 hour, depending on size (pierce with a knife to test). Scoop the squash from the skins.

In a large saucepan, heat the olive oil and sauté the onion and leek over medium heat until tender. Add the squash and cook 2 to 3 minutes more. Add the stock, bring to a boil, and simmer for about 45 minutes. Add salt and pepper. Puree the soup, in batches, if necessary, in a blender or a food processor. Bring the half and half to the scalding point. Reheat the soup, adding scalded half and half by spoonfuls until the desired consistency is achieved. Adjust the seasoning. Serve the soup, garnishing each bowl with 1 tablespoon of the cream, crème fraîche and chive mixture, or the chopped apple.

Roasted Tomato and Chili Soup

Mark Miller

Serves 4 to 6

Roasted tomatoes give a deep, smoky taste to this Southwestern soup.

- 8 *large vine-ripened tomatoes, cored*
- 6 *large fresh poblano chilies*
- ¼ *cup virgin olive oil*
- 1 *large white onion, cut into thin slices*
- 2 *garlic cloves, minced*
- 1 *bunch fresh cilantro, tied with a string*
- 6 *cups vegetable stock*
- *Peanut oil for frying*
- 6 *corn tortillas, cut into ⅜-inch-wide strips*
- ¼ *cup fresh lime juice, plus 1 lime cut in slices*
- *Salt and pepper to taste*
- *Grated* queso fresca*, *crumbled natural cream cheese, or grated Monterrey jack*
- *Fresh cilantro sprigs for garnish*

Char the tomatoes over an open gas flame, on a charcoal grill, or under a broiler. Do not discard the blackened parts. Chop the tomatoes finely or puree them coarsely in a blender or a food processor.

Cut the chilies into quarters, core, and seed. Char the chilies evenly skin side up under a broiler or on a charcoal grill. Place in a tightly closed plastic bag for 15 minutes. Rub off the blackened skin. Cut the chilies into strips.

In a saucepan, heat the olive oil and sauté the onion, garlic, and cilantro over low heat. Cover the pan and let the onions steam for about 20 minutes. Add the chili strips, tomatoes, and stock. Cook for 10 minutes.

Meanwhile, heat ¼ inch of peanut oil in a sauté pan or skillet and fry the tortilla strips until crisp. Remove from the pan with a slotted spoon and drain on paper towels. Finally, season the soup with lime juice, salt, and pepper. Remove the cilantro just before serving. Garnish the soup with the strips of fried corn tortillas, a slice of lime, some grated *queso fresca*, and sprigs of cilantro. Serve at once.

**Available at Mexican or Central American food stores.*

Pappa al Pomodoro

Mauro Vincenti

Serves 8

The best soup you ever chewed: Tuscan bread soup full of tomatoey goodness.

- 1 *medium carrot*
- 1 *medium red onion*
- 2 *medium celery stalks*
- 3 *tablespoons extra-virgin olive oil*
- 10 *fresh tomatoes, peeled, seeded, and pureed, or 7 to 8 cups canned pureed tomatoes*
- ½ *pound dried peasant bread, cubed (about 8 cups)*
- 2 *garlic cloves, minced*
- 10 *whole fresh basil leaves*
- *Freshly grated Parmesan cheese (optional)*

Chop the carrot, onion, and celery into small cubes. Heat 2 tablespoons of the oil in a large pot and cook the vegetables until they are crisp-tender. Add the tomato puree to the vegetable mixture, bring to a boil, and add the bread. Simmer this mixture for 15 minutes, then add the remaining 1 tablespoon of the oil, the garlic, and basil. Serve warm in soup bowls and add Parmesan cheese, if desired.

Herbed Tomato and Onion Soup

Marcel Desaulniers

Serves 4

A garden of fresh herbs and vegetables for a light lunch or a first course.

- 3 *tablespoons safflower oil*
- 1 *medium onion, thinly sliced*
- *Salt and pepper to taste*
- ¼ *cup dry white wine*
- 2¾ *cups vegetable stock*
- ½ *tablespoon water*
- 1 *small onion, chopped*
- 1 *celery stalk, chopped*
- 1 *small carrot, chopped*
- 3 *medium tomatoes*
- 1 *tablespoon chopped fresh basil*
- 1 *tablespoon chopped fresh tarragon*
- ½ *tablespoon chopped fresh dill*
- ½ *tablespoon chopped fresh thyme*
- 3 *tablespoons unbleached all-purpose flour*

In a nonstick sauté pan or skillet over medium heat, heat ½ tablespoon of the safflower oil. Add the sliced onion, season with salt and pepper, and sauté for 15 minutes, stirring frequently, until the onion becomes translucent. Add the wine and 1¼ cups of the vegetable stock. Bring to a boil, lower heat, and simmer 15 minutes. Remove from heat and set aside at kitchen temperature.

In a 2½-quart saucepan over medium heat, heat ½ tablespoon of the safflower oil and the water. Add the chopped onion, celery, and carrot. Season with salt and pepper and sauté for 5 minutes. Core 2 of the tomatoes and cut them into quarters. Peel, seed, and chop the remaining tomato. Cut into ¼-inch pieces. Cover with plastic wrap and refrigerate until needed. Add the quartered tomatoes to the saucepan with the chopped vegetables. Add the remaining 1½ cups vegetable stock.

Thoroughly combine the chopped herbs. Tie 1½ tablespoons of the herbs in a small piece of cheesecloth with a cotton string and add to the pan. (Cover and refrigerate, until needed, the remaining chopped herbs.) Bring to a boil, then lower heat and simmer 30 minutes.

In a separate 1½-quart saucepan over low heat, heat the remaining 2 tablespoons of the safflower oil. Add the flour to make a roux. Cook and stir till the roux bubbles, being careful not to let it brown. Strain 1 cup of the simmering chopped vegetable-stock mixture into the roux. Whisk vigorously till smooth. Add this mixture to the large saucepan with the remaining stock and chopped vegetables. Whisk until well combined. Simmer for an additional 10 minutes.

Remove the soup from heat. Remove and discard tied herbs. Puree the soup, in batches, if necessary, in a blender or a food processor. Strain through a sieve and return to low heat in the same saucepan. Bring to a simmer. Add the cooked onion mixture, remaining chopped tomato, and remaining chopped herbs. Adjust the seasoning with salt and pepper. Serve immediately.

Tomato Stew

Michael Roberts

Serves 4 to 6

Another version of Tuscan bread soup.

- 3 *pounds ripe tomatoes*
- ½ *cup olive oil*
- 1 *small onion, finely diced*
- 4 *garlic cloves, finely sliced*
- 2 *fresh marjoram sprigs*
- *Salt and pepper to taste*
- ½ *stale French baguette*
- *Freshly grated Parmesan or Romano cheese*

Cut the cores out of the tops of the tomatoes. Fill a 3-quart pot with water. Bring to a boil over high heat and plunge in the tomatoes, about 3 or 4 at a time. Let cook 1 minute after the water returns to the boil. Remove and let cool until comfortable to handle. Repeat until all the tomatoes are done. Using a small knife, remove the skins from the tomatoes, then halve them crosswise and squeeze out the seeds. Chop the pulp coarsely and set aside.

In a medium skillet over low heat, combine the olive oil, onion, and garlic and cook, stirring, for 5 minutes, or until the onion is translucent.

Add the tomatoes, marjoram, salt, and pepper and cook, stirring, for 15 minutes. Crumble the bread into the pan, decrease the heat to low, cover, and cook another 30 minutes.

Remove the rosemary sprigs and serve the stew in soup bowls. Accompany with grated Parmesan or Romano cheese.

SALADS

Avocado, Pasta, and Tofu Salad with Chinese Sesame Seeds

Jackie Shen

Serves 2

Linguine and marinated tofu with a spicy peanut butter dressing.

TOFU MARINADE

- *1 teaspoon Chinese hot mustard*
- *½ cup light soy sauce*
- *1 bunch green onions, chopped*
- *1 thin slice fresh ginger, cut into julienne*
- *1 tablespoon Asian (toasted) sesame oil*
- *1 pound firm tofu, drained and cut into 1-inch cubes*

LINGUINE DRESSING

- *2 tablespoons peanut butter*
- *1 tablespoon rice wine vinegar or distilled white vinegar*
- *2 tablespoons soy sauce*
- *1 tablespoon Chinese chili oil**
- *Salt and pepper to taste*
- *8 ounces linguine*
- *8 to 10 radicchio leaves*
- *1 avocado, sliced*
- *8 watercress sprigs*
- *6 snow peas*
- *2 cherry tomatoes*
- *8 ¼-inch diagonal slices peeled lotus root, soaked in water and lemon juice (optional)*

Mix together in a large bowl all the ingredients for the tofu marinade; add the tofu and marinate at room temperature for 1 hour. Mix together in a bowl the linguine dressing. Cook the linguine in a large amount of boiling water until al dente; drain and toss the linguine with the linguine dressing.

Arrange the radicchio leaves in the center of 2 plates. Place the dressed linguine in the center of the radicchio. Fan out the sliced avocado over the noodles. Arrange the watercress around the radicchio, then place a few tofu squares in alternation with flowers made of 3 snow pea "petals" and a tomato center. Intersperse lotus root slices, if using, with the tofu and snow pea flowers. Serve at once.

**Available at Asian markets.*

Braised Belgian Endive and Baby Bok Choy

Michael Roberts

Serves 4

The *miso* vinaigrette deepens and sweetens this baked vegetable dish.

- ¼ *cup virgin olive oil*
- 4 *heads Belgian endive*
- *Salt and pepper to taste*
- 4 *baby bok choy*
- 2 *cups* mesclun *(mixed baby lettuces)*
- *Miso Vinaigrette, following*

Preheat the oven to 375°. Add the olive oil to a small roasting pan and place over medium heat on top of the stove. Add the endive and cook until brown on all sides, about 7 minutes. Sprinkle with salt and pepper. Add water to a ½-inch depth. Bring to a boil and place, uncovered, in the preheated oven. After 5 minutes, turn the endive and add the bok choy. Replace in the oven for another 7 to 10 minutes, or until vegetables are just tender.

Remove from the oven and place the endive and bok choy on a plate. Let them cool to room temperature. Pour the vegetable cooking liquid into a saucepan or skillet and cook over medium high heat until the liquid is reduced by about half and becomes thick. Scrape into a mixing bowl to cool; reserve.

On each of 4 salad plates, arrange a bok choy and a head of endive on a bed of *mesclun*; drizzle with vinaigrette and serve immediately.

MISO VINAIGRETTE

Reserved braising juices from endive and boy choy, above (about 1/3 cup)

2 *tablespoons dark* miso*

1/4 *cup rice vinegar*

2 *tablespoons Asian (toasted) sesame oil*

1/3 *cup vegetable oil*

1 *small bunch fresh chives, minced*

Combine the braising juices, *miso*, and vinegar in a blender or a food processor and blend until smooth. Slowly add both oils in a stream. Pour the vinaigrette into a bowl and add the chives.

**Soybean paste. Available at natural foods stores and Asian markets.*

Grilled Eggplant and Chili Salad

John Downey

Serves 4

- 1 *red bell pepper*
- 1 *cup Chili Vinaigrette, following*
- 1 *red onion, peeled and cut into ⅜-inch slices*
- ½ *cup mild-flavored olive oil*
- *Salt and pepper to taste*
- 1 *firm large eggplant*
- 1 *Anaheim chili*
- 1 *small bunch cilantro, stemmed, washed, and dried*
- *Whole lettuce leaves (optional)*

Prepare a wood or charcoal fire in an open grill. While the coals are still red, place the red pepper on the grill and char, turning it until it blisters and becomes evenly blackened on all sides. Place the pepper in a plastic bag, close tightly, and let sit for 15 minutes. Rub off the blackened skin and remove the seeds and core. Cut the pepper into julienne strips and marinate in the Chili Vinaigrette.

Keep the rings of the onion together by inserting a bamboo skewer or 2 toothpicks from the outer ring to the center, or use a grill basket. Brush lightly with some of the olive oil and season with salt and pepper. Cook on the grill until both sides are a very dark brown. (A little charred black will enhance the flavor of the salad.) Set the onion aside.

Cut the unpeeled eggplant into ⅜-inch-thick slices. Season with salt and pepper, brush with olive oil, and grill until dark brown, using a grill basket if possible. Caution: Eggplant goes from medium brown to black in a second on the grill—be careful not to blacken the slices.

Split the Anaheim chili and remove the seeds and core. Cut in julienne strips the same size as the red pepper. Add to the Chili Vinaigrette.

Remove the skewers or toothpicks from the grilled onion slices, and cut each slice into 6 segments. Add to the vinaigrette.

Cut the grilled eggplant into large dice and toss carefully with the rest of the ingredients. Be careful not to mash the eggplant. Check the seasoning of the salad mixture. Garnish with cilantro leaves. Serve as is or on a bed of lettuce leaves.

CHILI VINAIGRETTE

Makes 2 cups

- 1 *cup plus ½ tablespoon mild olive oil*
- ½ *jalapeño chili, cored and minced, with a few of its seeds*
- ½ *small red bell pepper, cored, seeded, and chopped*
- ½ *cup cider vinegar*
- 1 *shallot, minced*
- 1 *teaspoon Dijon mustard*
- 1 *teaspoon Worcestershire sauce*
- *Freshly ground salt and black pepper to taste*
- *Leaves from 1 bunch fresh cilantro, chopped*

In a sauté pan or skillet, heat ½ tablespoon of the olive oil and sauté the chopped peppers gently until tender; set aside.

In a bowl, combine the vinegar, shallot, mustard, Worcestershire, salt, and pepper. Whisk in the remaining 1 cup of olive oil slowly. Add the parsley, cilantro, and sautéed peppers. Allow to stand for at least 1 hour to marry the flavors. Stir again and correct the seasoning if necessary.

According to personal preference, you may add extra jalapeño pepper, but be careful. Store in a covered glass jar in the refrigerator. This will keep for 1 week.

Strawberries, Belgian Endive, Macadamia Nuts, and Balsamic Vinegar Dressing

Marcel Desaulniers

Serves 4

A surprising combination of ingredients that works beautifully.

- ½ *cup unsalted macadamia nuts*
- ½ *pint large strawberries, hulled and sliced ¼ inch thick*
- 3 *tablespoons balsamic vinegar*
- ½ *cup vegetable oil*
- *Salt and pepper to taste*
- 1 *medium head Bibb lettuce*
- 2 *large Belgian endive*

Preheat the oven to 325°. Toast the macadamia nuts on a baking sheet in the preheated oven for 12 to 15 minutes. Remove from the oven and cool to room temperature. Split the macadamia nuts in half; set aside.

Puree ⅓ cup of the sliced strawberries with the balsamic vinegar in a blender or a food processor. Add the vegetable oil and process for 30 to 40 seconds until the dressing is creamy and well combined. Adjust the seasoning with salt and pepper.

Place 2 leaves of Bibb lettuce on each of 4 chilled salad plates. Arrange 7 leaves of endive, like spokes, stem ends in the center, evenly spaced on the Bibb lettuce. Carefully layer 8 slices of strawberries in the center of the endive stems. Spoon 3 to 4 tablespoons of dressing directly over the strawberries and top with the toasted macadamia nuts.

Wild-Mushroom Salad

Michael McCarty

Serves 4 to 6

3 *bunches mâche*

2 *bunches arugula*

2 *heads baby limestone lettuce*

1 *head baby radicchio*

¼ *cup walnut oil*

8 *ounces* each *chanterelles, shiitakes, and oyster mushrooms, cut into ½-inch pieces*

¼ *cup pine nuts*

2 *large shallots, minced*

2 *medium garlic cloves, minced*

¼ *cup sherry wine vinegar*

2 *tablespoons* each *minced fresh basil, tarragon, thyme, and chives*

Salt and freshly ground pepper to taste

Break the greens into leaves and toss them together in a salad bowl. In a large sauté pan or skillet, heat 3 tablespoons of the walnut oil over high heat until it just begins to smoke. Add the mushrooms and let them sear without stirring for about 30 seconds; then sauté them, stirring constantly, for about 2½ minutes more, until nicely browned. Add the mushrooms to the salad bowl.

In the same pan, still over high heat, add the remaining 1 tablespoon oil and the pine nuts and sauté until golden, about 1 minute. Add the shallots, stir them quickly, then add the garlic and sauté about 30 seconds more. Add the vinegar and stir and scrape to deglaze the pan, then stir in the herbs, salt, and pepper.

Pour the hot dressing into the salad bowl and toss immediately to coat all the greens. Mound the mixture on salad plates and serve immediately.

Pasta Salad with Baked Eggplant and Tomatoes

Jeremiah Tower

Serves 4

"There are as many pasta salads as there are ingredients. But the method of cooking pasta to serve cold is always the same, and the secret is to prevent the pasta from sticking together by stirring the boiling water when you add the pasta. When draining the pasta, you can rinse it under cold water until it is cold and then toss it in oil. While this is the easiest method, it robs the pasta of maximum flavor. More trouble but more delicious is to toss the pasta in olive or herb oil while it is hot so that it absorbs the oil's flavors. You must toss the pasta until it is well coated but not swimming in oil, put it in the refrigerator, and take it out every minute or so and toss it again, until the pasta is cold and all the pieces are separate."

—*Jeremiah Tower*

- *8 ounces dried spaghetti or linguine*
- *1 cup Herb Oil, following*
- *Salt and freshly ground pepper to taste*
- *10 Japanese eggplants*
- *1 teaspoon chopped garlic*
- *1 teaspoon balsasmic vinegar or red wine*
- *½ cup coarsely chopped fresh cilantro leaves*
- *2 tablespoons fresh lemon juice*
- *⅓ cup olive oil*
- *½ cup Tomato Concasse, following (2 medium tomatoes)*
- *12 small yellow tomatoes*

Heat the oven to 350°. Bring a large pot of salted water to the boil, add the pasta, and stir for 1 minute. Cover the pot so that it comes back to the boil as soon as possible and then remove the cover. Cook until the pasta is al dente, about 10 minutes. Drain the pasta and immediately toss with ½ cup of the herb oil. Cool as described in the headnote. Season with salt and pepper.

Cut the eggplants lengthwise into ¼-inch-thick slices and put the pieces on a baking sheet. Brush with the remaining herb oil and season. Cover the pan with aluminum foil. Bake in the preheated oven until tender, about 20 minutes. Set aside 24 of the best-looking middle slices of the eggplants. Chop the remaining eggplant and then mix with the garlic, vinegar, cilantro, and salt and pepper to taste.

In a small bowl, mix the lemon juice with a little salt and pepper and whisk in the olive oil. Put the eggplant slices on 4 plates. Divide the pasta into 4 portions, place in the center of each plate, and twist the pasta into nests. Spoon the chopped eggplant over the pasta. Toss the tomato *concasse* with some of the lemon dressing and put on the plates. Cut the yellow tomatoes in half, toss with the lemon dressing, and put them on the plates. Pour any remaining dressing over the eggplant and serve at once.

HERB OIL

If you have herbs from which you have already taken the leaves, bruise the stems of thyme, tarragon, rosemary, and savory, or a combination of them, cover with olive oil in a jar or bottle, and soak for 3 days. Otherwise, chop herb stems and leaves coarsely and cover them with oil for 2 days or more.

TOMATO CONCASSE

The French are economical in culinary terminology: often one word requires a whole phrase in English. So I have used tomato *concasse*, an anglicized form of the French term *tomatoes concassées*, to mean tomatoes that have been peeled, seeded, and chopped.

Put ripe tomatoes in boiling water for about 5 seconds. Do not overcook or the tomatoes will turn to mush. Plunge them immediately into a bath of ice and water for 30 seconds. Peel off the skin and discard. Cut across the equator of the tomato only, for if you cut through the stem and down you will seal off some of the seed chambers. Hold the tomato halves cut side down and squeeze out the seeds. Core and dice the tomatoes and put them in a sieve over a bowl to drain.

Curly Endive with Warm Grilled Vegetables and Mustard Dressing

Marcel Desaulniers

Serves 8

"The coarse-grained mustard used at the Trellis is widely available Pommery Meaux. There are now dozens of mustards available; many are too sweet or too vinegary for my taste. Choose whichever mustard you prefer, but taste it before making the dressing.

"Many other vegetables are adaptable to grilling and could be added to this salad. Marinated raw asparagus, eggplant, tomatoes, yellow squash, fennel, and blanched potatoes are only a few of the vegetables that we have grilled for salads or served with entrees at the Trellis.

"The flavor of vegetables grilled over a charcoal or wood fire is unique; unfortunately, similar results cannot be achieved under a gas or electric broiler."

—*Marcel Desaulniers*

MUSTARD DRESSING

- 6 *tablespoons cider vinegar*
- 3 *tablespoons fresh lemon juice*
- 2 *tablespoons coarse-grained mustard*
- 4 *teaspoons Dijon mustard*
- ½ *teaspoon salt*
- ¼ *teaspoon fresh-ground black pepper*
- 1½ *cups vegetable oil*
- 8 *small leeks, white part only*
- 2 *medium red onions, peeled and cut into ¼-inch-thick slices*
- 2 *medium red bell peppers, cored, seeded, and cut into 16 strips 2½ inches long and 1 inch wide*
- 2 *medium (12 ounces total) zucchini, cut into quarters lengthwise*
- 16 *mushrooms*
- *Salt and pepper to taste*
- 1 *medium head (¾ pound) red leaf lettuce*
- 1 *large head (1 pound) curly endive, cut into ¾-inch pieces*

To make the dressing, whisk together in a ceramic bowl the vinegar, lemon juice, both mustards, salt, and pepper. Continue to whisk the mixture while pouring in a steady and slow stream of the vegetable oil. When all the oil has been added, adjust the seasoning with salt and pepper and combine thoroughly. Cover with plastic wrap and refrigerate until needed.

Prepare a charcoal or wood fire in an open grill. Blanch the leeks in boiling salted water for 5 minutes. Transfer to ice water; when cool, remove and drain thoroughly. Split the leeks in half lengthwise.

Place the leeks, onions, peppers, zucchini, and mushrooms in a ceramic bowl and season with salt and pepper. Add 1 cup of the mustard dressing and marinate for at least 30 minutes in the refrigerator.

Arrange 2 to 3 whole leaves of red leaf lettuce on each of 8 chilled 10-inch plates. Portion the curly endive on the red leaf lettuce. Grill the marinated vegetables over a medium-hot fire for 4 to 5 minutes.

Arrange the grilled vegetables on the curly endive on each plate. Dress each salad with 2 tablespoons of dressing and serve immediately.

Note: For crisper greens, refrigerate the red leaf lettuce and curly endive for 1 to 2 hours before serving.

. . . .

Arugula and Fried Okra Salad with Roast Corn Vinaigrette

Stephan Pyles

Serves 6

Soft summer nights and fried cornmeal-battered okra—a perfect combination.

- 3 *ears fresh corn, husked*
- 1 *tablespoon chopped shallots*
- 1 *teaspoon minced garlic*
- ⅓ *cup water or vegetable stock*
- 1 *tablespoon white wine vinegar*
- ½ *cup corn oil*
- ⅓ *cup olive oil*
- *Salt to taste*
- *Corn oil for cooking okra*
- 10 *okra spears, sliced into rounds*
- ½ *cup milk*
- *Salted cornmeal for dredging okra*
- 4 *to* 6 *cups arugula*
- 1 *medium tomato, peeled, seeded, and diced*

Roast the corn over a charcoal fire or under a broiler until browned evenly on all sides. Let cool, then cut the kernels from the cobs.

In a blender or a food processor, puree the corn, shallots, and garlic with the water and vinegar. Combine the oils and, with the motor running, slowly drizzle into the corn mixture in the blender. Season with salt. Set aside.

Heat ¼ inch of corn oil in a small sauté pan or skillet. Dip the okra rounds into the milk and dredge in the cornmeal. When the oil is just smoking, add the okra and cook for about 2 minutes, stirring occasionally. Remove the okra with a slotted spoon, drain on paper towels, and keep warm.

Place the arugula in a mixing bowl and drizzle with the vinaigrette. Toss thoroughly and divide among 6 plates. Top with the okra and tomato.

Yellow and Red Plum Tomato Salad with Avocado Oil and Basil

Michael Foley

Serves 6

Garnish this salad with peppery-tasting nasturtium blossoms.

- 3 *yellow plum tomatoes*
- 3 *red plum tomatoes*
- 6 *fresh opal basil leaves*
- 6 *fresh sweet basil leaves*
- ½ *cup avocado oil or extra-virgin olive oil*
- *Juice of ½ lemon*
- *Salt and pepper to taste*
- 3 *heads slightly strong-flavored lettuce, such as baby red leaf or oak leaf*
- 12 *unsprayed nasturtium flowers (optional)*

Cut each of the tomatoes into 4 wedges. Cut the basil into julienne by rolling each leaf into a cigarette shape and cutting crosswise into tiny strips. In a bowl, toss the tomatoes and basil with 1 teaspoon of the avocado oil. Marinate for 1 hour, refrigerated. Combine the remaining avocado oil and lemon juice in a bowl; season with salt and pepper.

Place several lettuce leaves on each plate and top with the tomato wedges. Add a little more vinaigrette and sprinkle with whole nasturtiums or petals.

Wild Rice Salad with Cilantro Vinaigrette

Gordon Naccarato

Serves 4

A wild rice salad with a Southwestern touch.

- 1 *cup wild rice*
- 2¼ *cups water*
- 2 *red bell peppers*
- ¼ *cup slivered almonds*
- 24 *lettuce leaves (such as red leaf, Bibb, limestone, watercress, Belgian endive)*
- 2 *avocados*

VINAIGRETTE

- ¼ *cup fresh lemon juice*
- 1 *garlic clove, minced*
- 1 *shallot, minced*
- 1 *cup walnut oil*
- *Salt and pepper to taste*
- *Leaves from 1 bunch cilantro, chopped*
- 8 *cilantro sprigs*

Place the wild rice in a saucepan, add the water, and bring to a boil over high heat. Reduce heat and simmer, covered, for 45 minutes, or until tender but not mushy; drain and cool. Cut the red peppers into quarters, remove the cores and seeds, and blacken the skin evenly under a broiler. Place in a plastic bag, close tightly, and let cool for 15 minutes. Rub off the blackened skins and cut the peppers into julienne.

Place the slivered almonds in a dry frying pan over medium heat. Stir constantly until golden. Remove immediately from the pan.

Arrange the lettuce leaves on 4 salad plates. Mound the cool rice in the center of the greens. Peel and slice the avocados, arrange the slices over the rice, and scatter the julienne-cut red peppers over the rice and avocado.

To make the vinaigrette, place the lemon juice, garlic, and shallots in a bowl and whisk in the oil. Add the salt and pepper. Add the chopped cilantro just before serving. Ladle the vinaigrette over the greens and rice. Garnish with the almonds and a few cilantro sprigs.

Side Dishes

Artichokes, Caulifflowerets, and Mushrooms à la Barigoule

André Soltner

Serves 6

Three vegetables, marinated separately in the same flavorful marinade. Artichokes à la Barigoule was popularized at France's famous Hôstellerie Mougin des Mougins.

MARINADE

- 6 *tablespoons olive oil*
- 3 *medium white onions, sliced thin*
- 3 *carrots, peeled and sliced thin*
- 2 *bay leaves*
- 1/4 *teaspoon dried French thyme*
- 6 *small garlic cloves, crushed*
- 1 1/2 *cups dry white wine*
- 1 *cup water*
- 1/2 *teaspoon coriander seeds*
- *Juice of 2 lemons*

- 18 *small artichoke hearts, pared and trimmed uniformly**
- 12 *ounces cauliflowerets, blanched*
- 12 *ounces small white mushrooms*

To make the marinade, heat the olive oil in a sauté pan or skillet and cook the onions and carrots until tender without allowing them to brown. Add the rest of the ingredients and boil for 2 minutes.

Put the 3 vegetables into 3 different casseroles. Add one third of the marinade to each and simmer until tender. Let cool. Serve at room temperature, separating the vegetables into 3 groups on the serving plate.

**To make artichoke hearts, remove the leaves and stems from 18 small artichokes. Scoop out the chokes with a spoon.*

Moroccan Carrots

Joyce Goldstein

Serves 6

A classic Moroccan salad to serve as a side dish.

- ½ *cup currants*
- 1½ *pounds carrots (about 18 medium carrots)*
- 6 *tablespoons unsalted butter*
- 3 *tablespoons brown sugar*
- 1 *teaspoon ground cinnamon*
- ½ *teaspoon ground cumin*
- ¼ *teaspoon cayenne, or to taste*
- 1 *cup fresh orange juice*
- *Salt and pepper to taste*
- *Chopped fresh parsley or mint leaves*

Soak the currants in hot water to cover until plump; drain and set aside, reserving the soaking liquid.

Peel and slice the carrots into thin rounds or julienne. You should have about 5 cups, after slicing. In a large saucepan, melt the butter. Add the sugar, spices, and carrots and stir over low heat for a few minutes. Add the orange juice and the currants, with some of the soaking liquid. Bring to a boil, then quickly reduce the heat and simmer the carrots, covered, 7 to 10 minutes, or until tender. Add salt and pepper. Sprinkle with chopped parsley or mint.

Celery Root and Apple Puree

Richard Perry

Serves 4 to 6

This is a great side dish for a winter supper.

- 2¼ *cups diced peeled celery root*
- 2¼ *cups diced peeled apples*
- ½ *cup heavy cream*
- 1 *teaspoon salt*
- ¼ *teaspoon ground white pepper*

Place the celery root in a pot with hot water to cover; boil over high heat for 10 minutes. Add the apples and boil until both are quite tender; drain. Puree the celery root and apples in a blender or food processor while adding the heavy cream; make sure the mixture is quite smooth. Season and serve warm.

Fava Beans with Olive Oil, Garlic, and Rosemary

Alice Waters

Serves 4

"Fava beans, also called broad beans, are a fast and easy crop to grow if you have your own garden. They also can be found in some produce markets in the spring (Italian greengrocers or markets offer a wide range of vegetables), although they are not as common as some of the other shell beans. Mature fava beans—one stage at which they may be eaten—have long glossy green pods, six to eight inches long."

—*Alice Waters*

- 5 *pounds fresh fava beans*
- 3 *tablespoons extra-virgin olive oil*
- 1 *scant tablespoon fresh rosemary leaves*
- ¼ *cup water*
- 1 *tablespoon chopped garlic*
- ½ *teaspoon salt*
- *Freshly ground pepper to taste*

Remove the beans from the pods, blanch in boiling water for 1 minute, and drain in a colander. Run cold water over the beans to cool them. Using your fingernail, break the outer skin of the beans and squeeze the beans out between your forefinger and thumb.

In a 9-inch sauté pan or skillet, warm the olive oil with the rosemary. Add the beans, water, garlic, salt, and a little ground pepper. Bring the mixture to a low simmer, cover the pot, and allow to stew for about 5 minutes, or until the water has evaporated and the beans are slightly softened. Continue to cook the beans for about 20 minutes so that the flavors combine and penetrate. Stir the beans often to prevent them from sticking. Grind a little more pepper over the beans just before serving.

. . . .

Involtini di Melanzane

Piero Selvaggio

Serves 4 to 5

Slices of eggplant are fried, stuffed with goat cheese, then marinated and served cold.

- 2 *medium eggplants, cut lengthwise into 3/8-inch-thick slices*
- *Salt*
- 1/4 to 1/2 *cup olive oil*
- 8 *ounces (1 cup)* caprini *(soft Italian goat cheese) or fresh chèvre cheese*

MARINADE

- 2 *garlic cloves, minced*
- *A handful each chopped fresh Italian parsley and basil*
- 1 *cup red wine vinegar*
- 1/3 *cup olive oil*

Salt the eggplant slices on both sides and let sit for 30 minutes. Pat the slices dry on both sides with paper towels. In a sauté pan or skillet, heat the olive oil and cook the eggplant until tender and golden brown on both sides. Spread 1 tablespoon cheese on each slice and roll up into a tube. Secure with a toothpick.

Mix the ingredients for the marinade. Place the stuffed eggplant in a deep dish. Pour the marinade over the eggplant and refrigerate for 1 hour or longer before serving. Pour the marinade over the eggplant to serve. Any leftover marinade may be reserved for another use.

Eggplant and Mushrooms with Blue Cheese Glaze

Cory Schreiber

Serves 5

- ½ *cup olive oil*
- 2 *tablespoons minced garlic*
- 2 *tablespoons minced shallot*
- 4 *large tomatoes, peeled, seeded, and diced*
- *Salt and pepper to taste*
- 1 *tablespoon chopped fresh thyme*
- 1 *large eggplant*
- 3 *cups mushrooms, sliced*
- ½ *cup crumbled blue cheese*
- ¾ *cup crème fraîche*

Preheat the oven to 325°. In a sauté pan or skillet, heat 2 tablespoons of the olive oil. Add the garlic and shallot. Cook over low heat for 1 minute. Add the tomatoes, salt, and pepper and cook until thickened. Add the chopped thyme and set aside.

Cut five ¾-inch crosswise slices from the eggplant, reserving the rest for another use. Season the eggplant slices with salt and pepper. Heat 3 tablespoons of the olive oil in a sauté pan or skillet and lightly brown the eggplant slices on both sides. Place the slices in a heatproof dish and bake in the preheated oven till tender, about 7 to 10 minutes. In the same pan, heat the remaining 3 tablespoons of the oil and brown the mushrooms. Add the reduced tomato mixture and warm through. Spoon one fifth of the mixture over each eggplant piece. Sprinkle a little blue cheese on top and spoon on a little crème fraîche. Place under the broiler until the cheese melts. Serve at once.

Fennel with Tomato

Cindy Pawlcyn

Serves 4 to 6

This dish of fennel and tomatoes topped with buttery bread crumbs is aromatic and satisfying.

- 3 *tablespoons olive oil*
- 1 *tablespoon chopped garlic*
- 4 *fennel bulbs, cut into julienne*
- 4 *tomatoes, peeled, seeded, and chopped*
- 1 *teaspoon chopped fresh basil or marjoram (optional)*
- *Salt and ground black pepper to taste*
- ½ *to* ¾ *cup dry bread crumbs*
- ½ *to* ¾ *cup freshly grated Parmesan*
- 2 *to* 3 *tablespoons butter*

Preheat the oven to 350°. In a sauté pan or skillet, heat the oil and sauté the garlic and fennel until tender, about 5 to 10 minutes. Add the tomatoes and basil or marjoram if desired. Add salt and pepper. Pour into a casserole dish. Mix together the bread crumbs and Parmesan. Sprinkle over the fennel and dot with butter. Bake until crisp, about 5 to 10 minutes.

Braised Leeks with Saffron and Sherry Vinegar

Cory Schreiber

Serves 4

- 6 *small leeks, white part only*
- ½ *cup olive oil*
- 1 *carrot, diced*
- 3 *celery stalks, diced*
- 1 *red onion, diced*
- 4 *garlic cloves, chopped*
- 1 *teaspoon saffron*
- ¼ *cup sherry vinegar*
- *Salt and pepper to taste*
- *Juice of 2 lemons*
- 1 *tablespoon chopped fresh parsley*

Cut the leeks in half lengthwise and soak in cold water for 15 minutes; drain and set aside.

In a sauté pan or skillet, heat ¼ cup of the olive oil and cook the carrot, celery, onion, and garlic over very low heat for 45 minutes; do not brown. Add the saffron and vinegar, and continue cooking until the orange color disperses. Add the leeks and the remaining ¼ cup olive oil; cook for 15 to 20 minutes or until the leeks are tender. Season with salt and pepper, lemon juice, and chopped parsley. Let cool slightly before serving.

Potatoes and Onions Roasted with Vinegar and Thyme

Alice Waters

Serves 4

"Although they are good next to just about any roasted or grilled meat and poultry, I like to serve these potatoes and onions with braised dishes. The potatoes are half peeled, leaving a spiraling band of skin attached. This not only makes them look appealing, but allows the vinegar to penetrate them during their slow roast. If they are to brown evenly, it is important to turn the potatoes and onions (about every 30 minutes, and more frequently during the last half hour) to coat them with the vinegar and butter."

—*Alice Waters*

- 12 *golfball-sized red potatoes (1 pound)*
- 8 *white golfball-sized boiling onions (9 ounces)*
- 2 *tablespoons unsalted butter, melted*
- 3 *tablespoons balsamic vinegar*
- ½ *teaspoon salt*
- *Freshly ground pepper to taste*
- 6 *fresh thyme springs*

Preheat the oven to 350°. Starting at the top of each potato, pare away a ¼-inch-thick band of skin in a spiral. Put the potatoes and onions in a baking dish just large enough to hold them. Pour the butter and the vinegar over the potatoes and onions. Add the salt and pepper and mix until the vegetables are coated. Bury the thyme sprigs in the vegetables and cover the dish with aluminum foil.

Bake in the preheated oven for 2 hours, stirring the vegetables every 30 minutes to recoat them. Be sure to cover them again after stirring. When done, the onions and potatoes should be a deep brown color.

. . . .

Potato and Yam Gratin with Artichoke and Peppers

Jimmy Schmidt

Serves 4

The yam is pureed to make a sauce for this rich potato pancake.

- 1 *poblano chili pepper*
- 1 *red bell pepper*
- 1 *artichoke heart*
- ½ *cup balsamic vinegar*
- 1 *teaspoon salt*
- 2 *tablespoons chopped red onion*
- 2 *tablespoons chopped fresh chives*
- 2 *tablespoons chopped fresh parsley*
- ½ *cup virgin olive oil*
- 1 *large yam*
- 2 *garlic cloves*
- 2 *large potatoes (about 2 pounds), peeled*
- 9 *tablespoons unsalted butter, softened*
- ½ *cup water*
- 1 *teaspoon* achiote *paste* or paprika*
- *Salt and pepper to taste*
- ¼ *cup finely grated imported Parmesan cheese*

Quarter, seed, and core the peppers. Char evenly under a broiler. Place in a closed plastic bag for 15 minutes. Rub off the blackened skin and set peppers aside. Remove the leaves and stem from the artichoke heart and scoop out the choke with a spoon. Blanch the heart in boiling salted water for 3 to 5 minutes; set aside.

Preheat the oven to 400°. Cut the peppers and artichoke into ¼-inch dice. In a medium bowl, combine the vinegar and salt. Add the peppers, artichoke, onion, herbs, and olive oil; adjust the seasoning and set aside.

Place the yam and garlic in an ovenproof pan and place in the lower third of the oven. Cook until tender, about 1¼ hours; remove to a warm area.

Meanwhile, slice the potatoes to the thickness of potato chips. Rinse under cold water and pat dry with paper towels. Rub all the inner surfaces of a 12-inch sauté pan or skillet heavily with 3 tablespoons of the butter. Layer the potatoes, partially overlapping them, to cover the bottom of the pan. Continue layering until all the potatoes are used.

Place the potatoes over high heat, cooking until well browned on the bottom, about 5 minutes. Be careful not to burn. Turn the potatoes over onto a dish and add 3 more tablespoons of butter to the pan; return the potatoes to the pan and cook the second side until well browned. Remove from the heat, add the water, and place in the preheated oven. Cook until the center of the potatoes is tender and the outside is very crisp, about 10 minutes. Remove from the oven and keep warm.

Scoop the pulp from the yam and place the pulp and the garlic into a blender or a food processor. Add the remaining 3 tablespoons butter and the *achiote.* Puree until smooth. Adjust the seasoning.

Cut the potato pancake into 8 pie-shaped pieces. Place 1 potato wedge onto each of 4 plates. Spoon and evenly spread one fourth of the yam puree over the surface. Top with another potato wedge. Repeat with the remainder of the ingredients. Spoon the sauce over the top, garnishing with a sprinkle of cheese. Serve at once.

**Available at Mexican or Central American markets or specialty food stores.*

Pommes Paillasson

Georges Perrier

Serves 4 to 6

Nothing could be simpler than these pan-fried potatoes. Be sure to cut them very thin so they will brown nicely and resemble their namesake "hay" when cooked.

- *3 tablespoons butter*
- *3 large Idaho potatoes, peeled and cut into 1/8-inch slices, then into julienne strips*
- *1 tablespoon oil*
- *Salt and pepper to taste*
- *Chopped fresh parsley or watercress*

Preheat the oven to 350°. Butter the sides of a large nonstick sauté pan or skillet, add 1 tablespoon of the butter and the oil, and cook over medium heat until the butter foams. Add the potatoes to the pan and press down on them with a spatula. Add salt and pepper. Put additional pieces of butter on the sides of the pan to make sure the potatoes will not stick.

Cook the potatoes until they are golden brown on the bottom. Loosen the cake of potatoes with a spatula, place a plate over the pan, and turn the pan upside down so the potato cake is upside down on the plate. Melt the remaining 1 tablespoon of the butter in the pan, place the potato cake browned side up in the pan, and cook in the preheated oven for 5 minutes. Turn the potatoes out onto a heated plate, sprinkle with parsley, and serve immediately.

Shiitake Provençale and Grilled Shiitake Mushrooms

Vincent Guerithault

Serves 4

Two different ways of cooking shiitakes.

SHIITAKE PROVENÇALE

- *1 pound shiitake or oyster mushrooms*
- *2 tablespoons olive oil*
- *2 shallots, chopped*
- *1 teaspoon chopped garlic*
- *Fresh lemon juice to taste, about ½ teaspoon*
- *1 tablespoon chopped fresh parsley*
- *Salt and pepper to taste*

Remove the stems from the mushrooms and discard. In a sauté pan or skillet, heat the olive oil and sauté the mushrooms, shallots, and garlic until tender. Add the lemon juice and chopped parsley. Sauté a few moments more, then add salt and pepper to taste.

GRILLED SHIITAKE MUSHROOMS

- *4 large shiitake or oyster mushrooms*
- *2 tablespoons hot olive oil*
- *Salt and pepper to taste*
- *4 teaspoons chopped fresh chervil*

Prepare a wood or charcoal fire in an open grill. Remove the stems from the mushrooms and discard. Brush with olive oil, season with salt and pepper, and grill over a low fire for 5 minutes on each side. Set the mushrooms upside down on a serving plate. Heat the remaining olive oil and stir in the chervil. Pour the warm herb and oil mixture over the mushrooms.

Salsify and Fava Bean Ragout

Michel Richard

Serves 4

- ½ *cup fava beans, shelled (about ¾ pounds unshelled)*
- 8 *ounces salsify (about six 1-foot-long stalks), peeled and cut into 1½-inch lengths*
- 1 *cup milk*
- *Pinch of salt*
- 2 *tablespoons unsalted butter*
- 1 *teaspoon soy sauce*
- 1 *drop Asian (toasted) sesame oil*
- 1 *tablespoon vegetable stock*

Blanch the fava beans in boiling water until tender, about 5 minutes; drain and cool. Peel off the outer skin. In a saucepan, place the salsify, milk, and salt. Bring to a boil, reduce heat, and simmer for 10 minutes. Drain and rinse under cold water; set aside.

Place 1 tablespoon of the butter in a nonstick sauté pan or skillet and cook over medium-high heat. When the butter begins to brown, add the salsify and sauté until it begins to brown. Add the soy sauce, sesame oil, and stock and cook for 1 minute, or until the salsify is glazed. Add the fava beans and the remaining 1 tablespoon of butter and cook, stirring until the butter is melted and the fava beans are heated through. Serve immediately.

Sautéed Butternut Squash with Rosemary

Richard Perry

Serves 4 to 6

Easy to prepare, simple, and flavorful.

- 1 *tablespoon olive oil*
- 1 *tablespoon honey*
- 2 *cups diced fresh butternut squash (from 1 medium squash)*
- 10 *leaves fresh rosemary (no stems)*

In a sauté pan or skillet, heat the oil and honey; add the squash and sauté on low heat for 3 minutes; add the rosemary and sauté 3 more minutes, or until the squash is tender.

Zucchini alla Menta

Piero Selvaggio

Serves 6

Grilled zucchini in a wonderful herby marinade.

- 3 to 4 *pounds baby zucchini*
- 6 *tablespoons virgin olive oil*
- 10 *fresh mint leaves*
- 10 *fresh basil leaves*
- 1 *teaspoon dried red pepper flakes*
- 1 *garlic clove, chopped*
- ½ *teaspoon balsamic vinegar*
- *Salt and ground black pepper to taste*

Light a charcoal or wood fire in an open grill. Slice the zucchini very thin lengthwise and grill them over a medium-hot fire in a grill basket. In a bowl, mix together the oil, mint, basil, red pepper, chopped garlic, and vinegar. Add salt and pepper. Add the grilled zucchini and marinate at room temperature for 1 to 2 hours.

Zucchini "Vermicelli" with Two Sauces

Roy Yamaguchi

Serves 4

Shredded zucchini with purees of red and yellow peppers.

SAUCE

- 2 *red bell peppers*
- 2 *yellow bell peppers*
- 1 *cup olive oil*
- *Salt and ground white pepper to taste*

- 4 *zucchini, peeled*
- 12 *fresh basil leaves*
- ½ *cup olive oil*
- 2 *garlic cloves, minced*
- 2 *shallots, minced*
- 1½ *tablespoons diced pitted olives*
- 1½ *tablespoons whole drained capers*
- 2 *tomatoes, peeled, seeded, and chopped*
- *Salt and ground white pepper to taste*

To make the sauce, preheat the oven to 350°. Place the bell peppers in a baking pan and cover with aluminum foil. Bake in the preheated oven for about 45 minutes. Remove from the oven, wrap each pepper individually in plastic wrap, and let cool. Unwrap, skin, core, and seed peppers, washing them if necessary to remove the skin and seeds. In a blender or food processor, puree the red peppers and yellow peppers in separate batches with ½ cup of olive oil each. Keep warm in separate double boilers.

To make the "vermicelli," shred the zucchini into long, thin pieces, using a Japanese turning slicer, or a mandoline if possible. Cut 8 of the basil leaves into julienne.

In a large sauté pan or skillet, heat the olive oil almost to the smoking point. Add the garlic, shallots, olives, and capers. Sauté until the garlic and shallots are lightly browned. Quickly add the chopped tomatoes. Add the basil julienne and cook for about 1 minute to release the basil fragrance. Season with salt and pepper. Add the "vermicelli" and cook until al dente, 1 to 2 minutes, stirring the ingredients.

Divide both sauces evenly on each plate. Swirl the zucchini "vermicelli" into a bunch and place it in the middle of the plate straddling both sauces. Place a whole basil leaf on top. Serve at once.

Zucchini with Almonds

Craig Claiborne

Serves 4

"This is a recipe I discovered recently in a small restaurant in Rome called Tre Scalini. The dish consisted of small, square-shaped pieces of zucchini (including the skin), sautéed in olive oil and topped with golden brown slices of almond and grated pecorino cheese. It was the best thing I had in Italy and makes a marvelous first course or side dish."

—*Craig Claiborne*

- 3 *zucchini, about 8 ounces each*
- ½ *cup extra-virgin olive oil*
- ¾ *cup sliced blanched almonds*
- 1½ *teaspoons minced garlic*
- *Salt to taste*
- 6 *tablespoons freshly grated pecorino or Parmesan cheese*

Hold each zucchini upright and slice down each of 4 sides to make ¼-inch-thick slices about 1 inch wide. Slice the white center portion into similar slices. Stack the slices and cut them into 1-inch squares. There should be about 3½ cups; set aside.

In a sauté pan or skillet, heat the oil and add the almonds. Cook over moderately high heat, stirring constantly, for a few seconds until the pieces are golden brown; take care not to overcook. Remove the pieces with a slotted spoon and drain on paper towels. Set aside.

Pour off all but ¼ cup of the oil and add the zucchini pieces, garlic, and salt. Stir and toss the pieces for about 2 minutes or until they are crisp-tender.

Spoon equal portions of the zucchini onto 4 individual plates, sprinkle with equal amounts of almonds and cheese, and serve immediately.

Ratatouille

Jean-Jacques Rachou

Serves 8

A classic ratatouille. For a lighter touch, use less olive oil.

- 1¾ *cups olive oil*
- 2 *onions, chopped*
- 2 *green bell peppers, cored, seeded, and cut into small dice*
- 6 *small zucchini, cut into small dice*
- 3 *eggplants, diced*
- 6 *tomatoes, peeled, seeded, and diced*
- 2 *garlic cloves, chopped*
- *Salt and pepper to taste*
- 1 *tablespoon chopped fresh thyme*
- 2 *bay leaves*

In a sauté pan or skillet, heat 1 cup of the olive oil and sauté the onions and green peppers until slightly colored. In a second sauté pan or skillet heat the remaining ¾ cup of olive oil and sauté the zucchini and eggplant 10 to 15 minutes, or until tender, then add to the first pan of onion and green pepper. Add the tomatoes, garlic, salt, pepper, thyme and bay leaves. Cook, covered, on top of the stove, for 1 hour over medium heat. Serve warm or at room temperature.

MAIN COURSES

Poivron Jardinière

Jean Banchet

Serves 2

Bell peppers filled with vegetables and topped with a savory dressing.

- 2 *bell peppers (any color)*
- 1 *cup bean sprouts*
- ½ *cup grated carrots*
- 1 *teaspoon chopped fresh parsley*
- 1 *garlic clove, chopped*

DRESSING

- 3 *tablespoons sour cream*
- 1 *tablespoon tarragon vinegar*
- *Salt and pepper to taste*

- 3 *black olives, pitted and chopped*
- 2 *fresh tarragon sprigs*
- *Pinch of paprika*

Steam the whole peppers over boiling water for about 15 minutes; set aside to cool. In a bowl, mix the bean sprouts, carrots, parsley, and garlic. Cut the tops off the cooled peppers and remove the cores and seeds. Place each pepper upright in the middle of a serving plate and fill with the mixed vegetables. Mix together the dressing ingredients. Pour a little dressing on top of the vegetables, and the rest around the peppers. Decorate the plate with the olives, tarragon, and paprika.

Texas Cream Corn with Apple-Pecan Fritters

Dean Fearing

Serves 6

A perfect Texas combination: bourbon-scented creamed corn and feather-light fritters.

- ½ each *red and green bell peppers, seeded and cored*
- *Kernels cut from 6 ears of corn*
- ½ *cup Jim Beam bourbon*
- ½ *cup heavy cream*
- *Salt to taste*

APPLE-PECAN FRITTERS

- 2 *tablespoons bread flour*
- ¾ to 1 *cup unbleached all-purpose flour*
- ½ *teaspoon salt*
- 1½ *teaspoons baking powder*
- ½ *teaspoon baking soda*
- 1½ *teaspoons ground ginger*
- 1½ *tablespoons maple syrup*
- 1 *cup milk*
- ¼ each *Granny Smith and Red Delicious apple, quartered, cored, peeled, and cut into small dice*
- 6 *tablespoons pecans, chopped*

- 3 *cups vegetable oil for frying*

Light a charcoal fire in a grill. When the coals are covered with gray ashes, place the peppers on the grill, cover the grill, and cook the peppers until tender, turning once. Cut the peppers into small dice. In a sauté pan or skillet, heat the corn, smoked peppers, and bourbon. Light the bourbon with a match, and cook the vegetables over high heat until the liquid is reduced to 1 tablespoon. Add the cream and cook to reduce it to a thick consistency. Season with salt; set aside and keep warm.

To make the fritters, mix all the dry ingredients in a large bowl. Mix the remaining ingredients in a separate bowl. Add the wet mixture to the flour mixture and mix just until they are incorporated together. *Do not overmix.* Overmixing will make tough fritters.

In a deep skillet, heat the vegetable oil to 350° on a candy thermometer. (Do not cook in oil that has not reached this temperature; the fritters will not cook quickly enough and will become soggy with oil.) Spoon 2 to 3 heaping tablespoons of batter into the hot oil, making sure not to crowd the pan. When each fritter turns golden brown, turn it over to cook the other side. When it, too, has turned golden brown, remove the fritter from the oil with a slotted spoon and place on a paper towel. Keep warm while frying the remaining fritters 2 or 3 at a time. Place 2 warm fritters on each of 6 warm plates, spoon on one sixth of the corn, and serve at once.

Grilled Mongolian-style Eggplant Napoleon with a Caper, Garlic, and Basil Butter Sauce

Roy Yamaguchi

Serves 4

This napoleon is made with layers of vegetables, not layers of pastry.

WHITE BUTTER SAUCE

1 *cup dry white wine*
1 *shallot, minced*
1 *tablespoon white wine vinegar*
¼ *cup heavy cream*
1 *pound (4 sticks) chilled unsalted butter, cut into 16 pieces*

MARINADE

1 *tablespoon Szechuan chili paste*
¾ *cup sugar*
2 *tablespoons grated fresh ginger*
2 *tablespoons grated fresh garlic*
1 *cup soy sauce*
3 *tablespoons olive oil*
1 *Maui onion, sliced*
1 each *red, yellow, and green bell pepper, cored, seeded, and cut into julienne*
Salt and pepper to taste
5 *tomatoes, peeled, seeded, and chopped*
12 *Japanese eggplants, cut into ¼-inch-thick lengthwise slices*
12 *fresh basil leaves, cut into julienne*
1 *cup shredded mozzarella*

CAPER, GARLIC, AND BASIL BUTTER SAUCE

½ *teaspoon olive oil*
1 *teaspoon grated garlic*
1½ *tablespoons capers, drained*
6 *basil leaves, chopped*
1 *cup White Butter Sauce, above*

Light a charcoal fire in an open grill. In a small saucepan, boil the wine, shallots, and white wine vinegar until the liquid is reduced to about 2 tablespoons. Then quickly add the heavy cream, and boil until the liquid is reduced by half. Over very low heat, whisk in the butter a piece at a time. Strain through a fine sieve and keep warm in a double boiler.

In a bowl, mix all the ingredients for the marinade and set it aside. In a sauté pan or skillet, heat the olive oil and cook the onion and bell peppers until al dente. Season with salt and pepper. Place the tomatoes in a saucepan, lightly season with salt and pepper, and cook until thick; set aside.

Preheat the oven to 350°. In a bowl, soak the eggplant slices in water to cover for about 15 minutes. Remove the eggplant from the water; dip into the marinade for about 30 seconds; drain. Grill over charcoal until nicely browned on both sides; set aside.

To assemble the napoleon, in a baking pan, place one third of the eggplant slices, then spread the bell pepper mixture over the eggplant. Cover with one third of the eggplant slices, spread the tomato mixture over the eggplant, and sprinkle with the basil. Cover with the last third of the eggplant slices. Sprinkle on the mozzarella. Bake in the preheated oven for 8 to 10 minutes, or until the cheese is golden brown and the napoleon is heated through.

Meanwhile, make the caper-garlic-basil butter sauce: heat the olive oil in a sauté pan or skillet. Add the garlic and capers and cook until the garlic begins to brown. Add the basil and white butter sauce. To serve, place the nicely browned napoleon in the middle of a serving plate and surround the napoleon with the sauce. Serve at once.

Corn "Risotto" with Okra and Shiitake Mushrooms

Michael Roberts

Serves 4

Corn kernels are blended with cheese and cream in this innovative dish.

- ½ *cup (1 stick) butter*
- *About 4 cups fresh corn kernels (about 6 large ears)*
- 2 *tablespoons minced shallots*
- 1 *cup heavy cream*
- 1⅓ *cups freshly grated Parmesan or pecorino cheese*
- ½ *cup okra, sliced*
- 12 *shiitake mushroom caps*
- ½ *cup vegetable stock*
- *Salt and pepper to taste*

In a saucepan, melt ¼ cup of the butter. Add the corn and shallots and sauté for 1 minute. Add the cream and boil until the cream reduces and becomes thick, approximately 7 minutes. Add the grated cheese. Remove to a blender or food processor and puree just enough to break up the corn kernels.

In a sauté pan or skillet, melt the remaining ¼ cup butter, add the okra and the mushrooms, and sauté for 5 minutes. Add the vegetable stock and continue to cook until the stock reduces and thickens. Mound the corn "risotto" on plates and surround with the vegetables and sauce. Serve at once.

Haricots Verts à la Repaire

Ferdinand Metz

Serves 5 to 7

A lovely dish of green beans in a light sauce, topped with flaky pastry.

PASTRY DOUGH

- 4¾ *cups (1 pound) cake flour*
- ⅓ *teaspoon salt*
- 1¼ *cups vegetable shortening*
- ⅔ *cup cold water*

- 1½ *cups 2-inch pieces* haricots verts *or baby green beans*
- 1 *tablespoon olive oil*
- 1 *onion, chopped*
- 1 *garlic clove, chopped*
- 2 *tablespoons chopped fresh parsley*
- ¼ *teaspoon chopped fresh basil*
- 1½ *tablespoons butter*
- 2½ *tablespoons flour*
- 1 *cup vegetable stock*
- *Salt, pepper, and nutmeg to taste*
- *Milk for glazing*

To make the pastry dough, mix the flour and salt in a large bowl, and cut in the shortening with a pastry cutter or 2 knives until it has the consistency of cornmeal. Sprinkle the water gradually over this mixture while mixing with a fork, until the mixture forms a mass. Form into a ball with your hands, cover with plastic wrap, and let the dough rest in the refrigerator for 2 to 3 hours before using.

Cook the beans in boiling salted water to cover for 5 to 7 minutes, or until tender but still bright green. Drain the beans from the cooking water and plunge them into ice water, then drain again and set aside.

Preheat the oven to 375°. Heat the olive oil in a sauté pan or skillet, and sauté the onion, garlic, parsley, and basil until the onions become translucent. Add the butter and flour, and cook and stir until the flour begins to brown. Add the vegetable stock, bring to a boil, and cook until the mixture is thickened. Add the cooked beans, and season with salt, pepper, and nutmeg.

Place the bean mixture in a casserole. On a floured board, roll out the chilled pastry to fit the casserole opening and place the pastry on top of the beans. Brush the pastry with milk, pierce in several places with a fork, and bake in the preheated oven for 20 minutes, or until golden brown. Serve hot.

Salsify Pancakes with Grilled Portobello Mushrooms

Mark Militello

Serves 4 to 6

"Salsify is a long, slender root vegetable that is sometimes called oyster plant because of its moist, slippery texture. Portobello mushrooms are flavorful, brown-capped cousins of the button mushroom." —*Mark Militello*

PANCAKES

- 2 *pounds salsify*
- *Juice of ½ lemon*
- 1 *bunch green onions, chopped*
- ½ *teaspoon minced garlic*
- ½ *teaspoon minced shallot*
- 1 *cup unbleached all-purpose flour*
- ¼ *cup bread crumbs*
- ½ *cup milk*
- ½ *cup half and half*
- 4 to 5 *tablespoons clarified butter**

- 24 *portobello mushrooms*
- ¼ *cup olive oil*
- 1 *garlic clove, minced*
- *About 8 tablespoons clarified butter*

To make the pancakes, peel the salsify with a vegetable peeler. Place the salsify in water to cover; stir in the lemon juice and set aside until you are ready to grate them. Then grate the salsify into a large bowl. Fold in the remaining ingredients. The batter will be somewhat thick.

Prepare a charcoal or wood fire in an open grill. Wash the mushrooms; cut off the base of the stems and discard, and pull out the filaments. Place the oil and garlic in a bowl and marinate the mushrooms for 20 minutes. Grill the mushrooms over a low fire until tender, slice them on the bias, and keep them warm until you serve the pancakes.

Heat 2 tablespoons of the clarified butter in a large sauté pan or skillet. Using a spoon, drop enough batter into the pan so that when you flatten it with a spatula you have a 3-inch pancake. Cook over medium heat until golden brown on both sides. Add more butter to cook the remaining pancakes. Spoon the mushrooms over the pancakes and serve at once with mixed greens, tomato salad, and goat cheese.

Note: Salsify is generally available in winter months, as are portobello mushrooms. If you can't find the mushrooms, substitute common white mushrooms, shiitakes, or oyster mushrooms.

**To make clarified butter, melt 1 cup (2 sticks) of unsalted butter in a pan over low heat. Pour off the clear yellow liquid and store any excess for later use. The milky residue may be used to enrich sauce or breads.*

Potato and Cheese Enchiladas with Red Chili Sauce

Robert Del Grande

Serves 5

FILLING

- ½ *cup pumpkin seeds*
- 3 *Idaho potatoes*
- ½ *large yellow onion, very thinly sliced*
- 1 *jalapeño chili, minced*
- 1 *bunch cilantro, stemmed and chopped*
- 8 *ounces Monterey jack cheese, cut into small cubes*
- ¼ *cup walnut oil*
- 1 *tablespoon fresh lime juice*
- 2 *teaspoons kosher salt*
- 1 *teaspoon cracked black pepper*

RED CHILI SAUCE

- 4 *ancho chilies, stemmed and seeded*
- ½ *cup pumpkin seeds*
- ¼ *teaspoon cumin seeds*
- 4 *Roma tomatoes, chopped*
- ½ *yellow onion, coarsely chopped*
- 4 *garlic cloves*
- 2 *cups water*
- 1 *tablespoon white wine vinegar*
- ½ *bunch cilantro*
- 2 *tablespoons walnut oil*
- 1 *teaspoon fresh lime juice*

Peanut oil for frying
- 1 *dozen corn tortillas*
- 4 *ounces Monterey jack cheese, grated*
- 4 *ounces* cotija *or Parmesan cheese, grated*

Cilantro sprigs for garnish

To make the filling, stir the pumpkin seeds in a dry skillet over medium heat until they are lightly toasted; set aside. Peel and quarter the potatoes lengthwise and cut each quarter into approximately 10 pieces. In a saucepan cook the potato pieces in boiling water to cover until tender; drain. Combine the cooked potatoes with the remaining filling ingredients and mix well. Adjust the seasoning.

To make the red chili sauce, lightly toast the ancho chilies in a warm skillet. Soak the chilies in warm water to cover for about 15 minutes, or until soft;

drain and set aside. Stir the pumpkin seeds in a dry skillet over medium heat until they are lightly toasted. Set aside and toast the cumin seeds in the same manner.

In a saucepan, combine the tomatoes, onion, garlic, water, and vinegar and bring to a boil. Simmer for 30 minutes. Transfer the ingredients, with the liquid, to a blender or a food processor. Add the ancho chilies and the remaining ingredients and blend until smooth. Transfer the sauce to a saucepan and bring to a boil. Lower the heat and simmer for 30 minutes.

To prepare the enchiladas, preheat the oven to 300°. Add peanut oil to a medium frying pan to a depth of ¼ inch and heat until almost smoking. Dip a tortilla in the red chili sauce to coat both sides with sauce. Quickly fry the tortilla in the hot oil for a few seconds on each side. Remove from the oil with tongs and keep warm. Repeat the process with the remaining tortillas.

Place some of the potato mixture in the center of each tortilla and roll into an enchilada. Place the enchiladas seam side down in an ovenproof dish. Spoon a little sauce over each enchilada and heat in the preheated oven until warm, about 15 minutes. Serve the enchiladas with the remaining sauce and the grated cheeses, and garnish with the cilantro. Serve at once.

Wild-Mushroom Chili

Richard Perry

Serves 6 to 8

The rich taste of wild mushrooms takes the place of meat in this robust chili.

- 1 *tablespoon oil*
- 1/4 *cup minced garlic*
- 1 1/2 *cups diced onion*
- 1 1/2 *cups diced celery*
- 4 *cups sliced wild mushrooms*
- 3/4 *cup diced green bell pepper*
- 1 *tablespoon water*
- 4 *cups diced tomato*
- *Two 16-ounce cans kidney beans, drained*
- 3 *tablespoons chili powder*

Heat the oil in a sauté pan or skillet. Add the garlic, onion, celery, mushrooms, pepper, and water and sauté until tender. Add the remaining ingredients, cover, and cook over low heat for 1 hour, stirring occasionally.

Pipe Organ of Baby Zucchini Stuffed with Vegetable Purees

Jacky Robert

Serves 4

- *20 baby zucchini*
- *1 large carrot, peeled*
- *1 large daikon, peeled*
- *4 bunches fresh spinach, stemmed*
- *1 large beet, peeled*
- *1 large truffle*
- *5 tablespoons butter*
- *Salt and pepper to taste*
- *Chablis Beurre Blanc, following*

Cut each zucchini flat on one end and at an angle on the other, cutting each one a different length. Hollow them out with a small knife.

Cook the carrot, daikon, spinach, and beet *individually* in boiling water to cover until tender. Cut 4 small pieces from each vegetable and reserve, refrigerated, for garnish.

Puree each cooked vegetable and the truffle *individually* in a blender or a food processor, adding 1 tablespoon of butter and salt and pepper to each. If the purees are too watery, cook over very low heat to thicken.

Fill the zucchini pipes with the different purees. Steam them over boiling water for 5 minutes. Put 5 pipes on each of 4 plates, taking care not to let the purees escape from them. Pour beurre blanc around the pipes and garnish with the reserved vegetable pieces.

CHABLIS BEURRE BLANC

- *¾ cup Chablis*
- *1 shallot, minced*
- *1 pound (4 sticks) chilled unsalted butter, cut into 16 pieces*
- *Salt and pepper to taste*

In a saucepan, boil the Chablis and shallots until the liquid is reduced by half. Reduce the heat to low. Whisk in the butter, piece by piece. Season with salt and pepper. Set aside and keep warm over very low heat.

Summer Vegetable Tart

Leslee Reis

Makes one 10-inch tart

Use your food processor to make this gorgeous vegetable tart.

- 1 *recipe Pastry Dough, following*
- One *1-inch thick-slice French bread (3/4 ounce), torn into pieces*
- 1/4 *cup* each *firmly packed fresh parsley and basil leaves*
- 2 *tablespoons fresh thyme leaves*
- 3/4 *teaspoon salt*
- *Freshly grated nutmeg and ground pepper to taste*
- 2 *large shallots (1 ounce total)*
- 1 1/2 *ounces Parmesan cheese (preferably imported), at room temperature*
- 1 *medium eggplant (14 ounces) unpeeled, cut to fit food processor feed tube*
- 8 *ounces mushrooms, one side cut flat*
- 2 *medium zucchini (8 ounces total), cut into feed-tube lengths*
- 1 *Italian plum tomato (3 ounces), cored*
- 6 *tablespoons olive oil (preferably extra-virgin)*

Prepare the pastry dough and chill for 1 hour. Preheat the oven to 425°. Roll out the pastry dough on a floured surface and fit into a 10-inch tart pan. Line the shell with aluminum foil and fill the foil with dried beans. Bake in the preheated oven for 15 minutes. Remove the foil and beans and continue to bake for 5 to 10 more minutes until lightly brown. Remove from the oven and let cool in the tart pan.

Fit the food processor with a steel blade and process the bread to crumbs. Remove from the work bowl and set aside. Add the parsley, basil, thyme, salt, nutmeg, and pepper to the work bowl and mince. Remove from the work bowl and set aside with the machine running, drop the shallots through the feed tube and process until minced. Remove from the work bowl and set aside; do not clean the work bowl.

Fit the food processor with the fine shredder. Shred the cheese, using light pressure. Remove from the work bowl.

Fit the food processor with the thick slicer. Stand the eggplant in the feed tube and slice, using medium pressure. Remove from the work bowl and set aside. Arrange the mushrooms lengthwise on the flat side in the feed tube, packing tightly. Slice using light pressure; remove from the work bowl and set aside.

Fit the food processor with the medium slicer. Stand the zucchini in the feed tube and slice, using medium pressure. Remove from the bowl. Stand the tomato in the feed tube; slice using light pressure.

In a heavy 12-inch sauté pan or skillet, heat 2½ tablespoons of the oil over medium-high heat. Add the eggplant and half of the shallots. Cook, stirring frequently, until the eggplant is tender, about 7 minutes. Correct the seasoning. Remove from the skillet. Heat 1 tablespoon of the oil in the same pan over medium-high heat. Add the zucchini and, stirring frequently, cook until just tender, about 3 minutes. Correct the seasoning and remove from the skillet. Heat the remaining 2½ tablespoons of the oil in the same pan over medium-high heat. Add the remaining shallots and mushrooms and cook, stirring frequently, until the mushrooms are soft, about 4 minutes. Correct the seasoning.

Position a rack in the center of the 450° oven. Combine the bread crumbs with 2 tablespoons of the shredded cheese. Sprinkle over the bottom of the cooled pie crust in its pan. Arrange the eggplant over in even layers. Sprinkle with one fourth of the minced herbs and 1 tablespoon of the cheese.

Spoon the mushrooms over evenly. Sprinkle with one fourth of the minced herbs. Arrange the zucchini slices in 2 concentric circles around the edge of the tart, overlapping slightly. Arrange the tomato slices in a circle in the center, overlapping slightly. Sprinkle the remaining herb mixture and then the remaining cheese over the vegetables. Bake in the preheated oven until the cheese is melted, about 12 minutes. Carefully remove the tart from the pan. Serve immediately.

PASTRY DOUGH

- *1½ cups unbleached all-purpose flour*
- *8 teaspoons cold unsalted butter, cut into pats*
- *¼ teaspoon salt*
- *3 tablespoons ice water*

Blend the flour, butter, and salt to a coarse meal. Toss in ice water until incorporated. Form dough into a ball and knead lightly for a few seconds. Reform into a ball. Wrap in wax paper and chill for 1 hour.

Vegetarian Club Sandwich

Susan Feniger and Mary Sue Milliken

Makes 4 sandwiches

This is the finest sandwich I have ever tasted. Buy the freshest sourdough you can find, whole wheat if possible.

- 2 *red bell peppers*
- 12 *sandwich-size slices sourdough bread*
- *Baba Ganough, following*
- 8 *leaves arugula or other slightly peppery greens*
- *Black Pepper Vinaigrette, following*
- *Tapanade, following*
- 2 *cucumbers, peeled and cut into julienne*
- 2 *avocados, peeled and sliced*
- 2 *tomatoes, sliced*
- 1/4 *cup equal parts Pommery mustard and mayonnaise*

Quarter, core, and seed the peppers. Char them evenly under a broiler and place in a closed plastic bag for 15 minutes. Rub off the blackened skin and cut the peppers into strips.

Spread 4 slices of the bread with Baba Ganough. Top each with the arugula and red peppers. Spread 4 slices of bread with Black Pepper Vinaigrette on one side and Tapanade on the other and place on top of the peppers. Top with the cucumbers, avocado slices, and tomato slices. Spread 4 slices of sourdough bread with the mustard-mayonnaise mixture. Place on top of each sandwich.

BABA GANOUGH

- 1 *large eggplant (about 1¾ pounds)*
- 2 *tablespoons tahini (sesame paste)*
- 1 *tablespoon fresh lemon juice*
- 1 *tablespoon extra-virgin olive oil*
- 1 *tablespoon pureed garlic*
- ½ *teaspoon cayenne*
- *Dash of Tabasco sauce*
- *Salt and pepper to taste*

Preheat the broiler. Place the eggplant on a baking sheet and broil, turning occasionally, until charred all over and softened, about 40 minutes. Set aside and cool.

When cool enough to handle, peel the eggplant and roughly chop, reserving the liquid. Transfer to a large bowl. Mix in the remaining ingredients. Serve chilled or at room temperature.

TAPANADE

15 to 20 Greek olives, pitted

About 1 tablespoon olive oil

Puree the olives in a blender or a food processor with enough olive oil to form a smooth paste.

BLACK PEPPER VINAIGRETTE

½ cup extra-virgin olive oil

1 teaspoon cracked black pepper

Mix the oil and pepper together in a bowl.

Ratatouille Tamales

Vincent Guerithault

Serves 8

The cooking of classical France and the desert Southwest come together in these tamales filled with ratatouille and served with a cilantro beurre blanc.

- 48 *dried corn husks, or twenty-four 5-by-8-inch pieces parchment paper*
- 1/4 *cup olive oil*
- 1/4 *cup finely diced zucchini*
- 1/4 *cup finely diced Japanese eggplant*
- 1/4 *cup finely diced yellow bell pepper*
- 1/4 *cup finely diced red bell pepper*
- 1/4 *cup finely diced green bell pepper*
- 1/4 *cup finely diced white onion*
- 1/4 *cup finely diced tomato*
- *Salt and pepper to taste*
- 1 *bay leaf*
- 1 *teaspoon chopped garlic*

MASA FOR TAMALES

- 2 *cups (8 ounces)* masa harina*
- 1 *tablespoon sour cream*
- 6 *tablespoons butter, softened*
- 1 *teaspoon salt*
- 1 *teaspoon ground white pepper*
- 1 *teaspoon cayenne pepper*
- 1 *teaspoon paprika*
- 2/3 to 3/4 *cup vegetable stock*

Cilantro Beurre Blanc, following

Soak the corn husks in warm water for 1 to 2 hours to soften. In a sauté pan or skillet, heat the olive oil and sauté each ingredient separately for 3 minutes, removing each in turn to a bowl with a slotted spoon. Add salt, pepper, bay leaf, and garlic; let cool.

To make the *masa*, combine in a large bowl the *masa harina*, sour cream, soft butter, and all seasonings. Add vegetable stock to spreading consistency. Drain and dry the corn husks on paper towels. Spread 2 overlapping corn husks with *masa* 1/5 inch thick, then add a layer of ratatouille. Fold the short edges over the filling and roll the tamale lengthwise. Tie closed with cotton string. (It is easier to use parchment paper as a wrapper and use the corn husk as a serving bed for the unwrapped steamed tamale.) Steam the tamales for 15 minutes over boiling water and serve hot with cilantro beurre blanc.

CILANTRO BEURRE BLANC

- *1 cup dry white wine*
- *1 cup white wine vinegar*
- *1 tablespoon chopped shallots*
- *1 tablespoon heavy cream*
- *2 cups (4 sticks) chilled unsalted butter, cut into 16 pieces*
- *⅓ cup chopped fresh cilantro leaves*

In a sauté pan or skillet, put the wine, vinegar, and shallots and cook over moderate heat until the liquid is almost completely evaporated. Add the cream and whisk in the butter 1 piece at a time, beating constantly. Whisk in the chopped cilantro. Keep warm over very low heat.

**Coarse corn flour used for making tortillas and tamales. Available in most supermarkets, or in Mexican or Central American markets or specialty food stores.*

Warm Vegetable Stew

Jeremiah Tower

Serves 6 to 8

"I first tasted this most glorious dish in the south of France with Richard Olney, who taught me the technique of the progression of vegetables as they go into the cooking pan. No better description of the dish and the philosophy behind it, no better appreciation of vegetables, can be found than in his book *Simple French Food*. The combination here is only one of many, though I never use more than seven vegetables, and never use tomatoes unless they are cherry tomatoes. The addition of chopped garlic and herbs just before you bring the stew to the table causes a burst of rich fragrance which perfects the dish. I prefer water to chicken stock for a cooking liquid—the resulting sauce has a much fresher and purer vegetable taste—and I use either butter or olive oil to finish the dish."

—*Jeremiah Tower*

- *16 small pearl onions*
- *1 fresh thyme sprig*
- *1 fresh tarragon sprig*
- *1½ cups water*
- *Pinch of salt*
- *24 baby carrots, peeled*
- *½ small cauliflower, cut into florets*
- *½ small broccoli, cut into florets*
- *6 pattypan squash, halved crosswise*
- *4 small zucchini, halved lengthwise*
- *4 small yellow zucchini, halved lengthwise*
- *8 fresh ears baby corn, husked*
- *24 baby green beans*
- *24 fresh squash blossoms*
- *1 tablespoon minced mixed fresh herbs*
- *2 garlic cloves, minced*
- *½ cup (1 stick) butter, cut into tablespoons*
- *Salt and freshly ground pepper to taste*

Set a pot of salted water over high heat to bring to a boil. Meanwhile, put the onions, thyme, tarragon, 1 cup of the water, and salt in a sauté pan or skillet. Bring to a boil, reduce heat, cover, and simmer for 5 minutes. Remove from heat.

Put the carrots into the boiling water for 1 minute, then lift them out with a slotted spoon and add them to the onions. Simmer these vegetables, covered, while you cook the cauliflower and broccoli in the boiling water for 5 minutes. Then lift these out and add them to the onions and carrots, with the remaining ½ cup water. Cover and continue cooking, tossing the vegetables together every few minutes.

Meanwhile, put the squash and zucchini into the boiling water for 3 minutes and then add them to the other vegetables. Toss together again. Cook the corn and beans in the boiling water for 1 minute and add to the other vegetables. Toss, cover, and cook 5 minutes.

Uncover the vegetable pan. Make sure there is about 1 cup of liquid in the pan, adding water if necessary. Add the squash blossoms, herbs, garlic, and butter. Turn the heat to high and toss together until the butter is melted and the sauce thickens a little. Season and serve immediately.

. . . .

Sautéed Shiitake Mushrooms with Zucchini, Tomatoes, Spinach, Broccoli, and Thyme

Marcel Desaulniers

Serves 4

You may cut and trim the vegetables several hours in advance. If you do so, cover and refrigerate the prepared vegetables until needed.

- 2 *quarts lightly salted water*
- 2 *pounds zucchini*
- 1¼ *pounds broccoli*
- 2 *cups loosely packed spinach leaves*
- 1 *tablespoon olive oil*
- 1½ *cups peeled and seeded chopped tomatoes*
- 8 *ounces shiitake mushrooms, sliced*
- 1 *tablespoon chopped fresh thyme*
- *Salt and pepper to taste*
- 1 *tablespoon safflower oil*

Bring the water to a boil in a 2½ quart saucepan. Meanwhile, using a mandoline (see Note), cut the zucchini into long, thin spaghetti-like strands the length of the zucchini and about ⅛ inch wide. Trim the broccoli into florets.* Cut the spinach leaves into thin strips.

Add the broccoli florets to the pot of boiling salted water and cook until tender but still crunchy, about 1½ to 3 minutes. Transfer to a colander with a slotted spoon and drain thoroughly.

Heat the olive oil in a large nonstick sauté pan or skillet over medium-high heat. When hot add the tomatoes, shiitake mushrooms, and thyme. Season with salt and pepper and sauté for 6 or 7 minutes; set aside and keep warm.

In a separate large nonstick sauté pan or skillet, heat the safflower oil over medium heat. Add the zucchini and spinach, season with salt and pepper, and sauté for about 5 minutes.

Make a 1½-inch-wide ring of zucchini and spinach around the rim of each of 4 warm 10-inch soup plates. Arrange the broccoli florets in a ring, stem ends towards the center, on the inside of the zucchini ring. Portion the tomatoes, shiitake mushrooms, and thyme into the center of each ring of broccoli and serve immediately.

Note: A mandoline is the most desirable tool available to cut vegetables into long spaghetti-like strands. Guide the zucchini lengthwise over the blades of the mandoline, rotating the zucchini one quarter of a turn after each pass. Continue to cut the now-square-shaped zucchini until the seeds in the center of the zucchini are visible. Discard the heavily seeded center section. A sharp French knife or serrated slicer may also be used: cut ⅛-inch thick strips the length and width of the zucchini. Rotate the zucchini one quarter turn after each slice. Continue to cut the square-shaped zucchini until the seeds are visible. Discard the seeded center section of the zucchini. Cut the strips into long, thin strands.

**Broccoli stems are delicious. Cook them separately from the florets in boiling salted water for about 10 minutes. Cool the cooked stems in ice water to stop the cooking action. Peel the tough fibrous outer layer from the stems, then cut into desired shape. Use in salads, egg dishes, as a garnish in soups, or in combination with other sautéed vegetables.*

Potpourri of Vegetables Pot au Feu–Style with Chervil Butter

Jean Joho

Serves 4

You may use almost any vegetable or combination of vegetables in this dish, including wild mushrooms. You may substitute chives, fennel greens, or celery leaves for the chervil.

- 2 *tablespoons virgin olive oil*
- 4 *green onions with green tops*
- 4 *green asparagus*
- 4 *white asparagus*
- 4 *small zucchini*
- 4 *small yellow squash*
- 4 *small fennel bulbs*
- 4 *small carrots with a little green, peeled*
- ½ *cup Vegetable Stock, following*
- *Salt and pepper to taste*
- 2 *tablespoons butter*
- *Leaves from 1 bunch chervil*

Heat the olive oil in a sauté pan or skillet and cook the onions until just tender; set aside. If the remaining vegetables are quite small, keep them whole. Otherwise, cut them lengthwise into long thin pieces. In a large pot, bring the stock to a simmer and cook the remaining vegetables in individual batches for 3 to 5 minutes until al dente. Remove the vegetables with a slotted spoon and set aside.

Take ½ cup of the stock from the cooking pot (reserving the remainder for another use) and place it in a large sauté pan or skillet. Bring to a boil and cook until the liquid is reduced by half. Season to taste.

Whisk in the butter and add the chervil leaves. Then add all the vegetables and warm them through. Arrange the vegetables on serving plates, pour the butter sauce over them, and serve.

VEGETABLE STOCK

- *1 quart water*
- *½ cup Riesling wine*
- *2 tomatoes*
- *1 leek*
- *1 onion*
- *2 shallots*
- *4 garlic cloves*
- *1 carrot*
- *3 fresh parsley sprigs*
- *2 fresh thyme sprigs*
- *2 bay leaves*
- *3 whole cloves*
- *1 tablespoon cracked pepper*
- *1 tablespoon salt, or to taste*

Place the water and wine in a large pot and bring to a simmer. Coarsely chop all of the vegetables and add to the pot. Wrap the parsley, thyme, bay leaves, and cloves in a small square of cheesecloth and tie it closed with cotton string. Add the herbs and salt to the pot and simmer for 3 hours. Strain through a sieve. Store, covered, in the refrigerator for up to 1 week, or freeze.

Napoleon of Vegetables

Joachim Splichal

Serves 4

Squares of puff pastry are layered with an array of vegetables and topped with a red pepper–butter sauce.

- 8 *ounces fresh or defrosted frozen puff pastry*
- 1 *medium zucchini*
- 2 *medium yellow squash*
- 1 *carrot, peeled*
- 1 *turnip*
- 1 *bunch broccoli*
- 2 *red bell peppers*
- 1 *small cauliflower*
- 1 *ear corn, husked*
- 1 *bunch spinach, stemmed*
- 12 *snow peas, ends cut off on diagonal*
- 12 *baby green beans, cut into 2-inch pieces*
- 12 *asparagus, cut into 2-inch pieces*
- ½ *cup (1 stick) butter, melted*
- *Salt, pepper, and cayenne to taste*

Preheat the oven to 425°. Roll out the puff pastry into a 9-by-12-inch rectangle about ⅛ inch thick and cut into twelve 3-inch squares. Place on a baking sheet and bake in the preheated oven for 15 minutes or until golden brown. Remove and let cool.

Cut the zucchini, squash, carrot, turnip, and broccoli stems into pieces about 1 inch long and ½ inch thick (reserve the broccoli florets). Shape these into ovals by using a small sharp knife.

Blanch the carved vegetables separately for 3 to 5 minutes in boiling salted water until al dente. Remove each with a slotted spoon, place in ice water, and drain. Cut the vegetables into fans by slicing thinly lengthwise only three quarters of the way to the end; that is, not cutting all the way through. Place in a large flat pan.

Cut 1 of the red peppers in half, removing the stem and seeds. Blanch both halves in boiling water, then place in ice water; drain. Remove the skin and cut the pepper into strips 2 inches long and ¾-inch wide.

Blanch the cauliflower, corn, spinach, peas, beans, asparagus, and broccoli florets separately until al dente. Place each in ice water as it is cooked; drain well. Cut the corn from the cob. Place all the vegetables in a baking pan in separate piles.

Prepare a red bell pepper sauce: Cut the remaining red pepper into quarters, seed, and core. Place under a broiler until evenly charred. Place in a closed plastic bag for 15 minutes. Rub off all charred skin and rinse in water. Puree the pulp in a blender; add the butter and blend. Season with salt, pepper, and cayenne.

To assemble the napoleon, pour 1 cup of sauce over the vegetables but do not stir; season with salt and pepper. Place in the preheated oven to gently warm. The puff pastry should be very thin; if it is thick, slice it in half horizontally. Place 1 square of puff pastry in the center of each of 4 ovenproof plates. Layer the vegetables, alternating colors and shapes. Extend the vegetables out from the center like spokes and create a circular fan of colors. Use about three fourths of the vegetables on the bottom layer. Place another pastry square on top and continue layering with the vegetables. Warm the plates in the oven. Place the remaining squares of pastry on top. Spoon the remaining sauce on top and serve at once.

Gratin des Capucins

Georges Perrier

Serves 4 to 6

Layers of sautéed spinach, sautéed artichokes, and creamed mushrooms, baked until golden brown.

- 3 *tablespoons butter*
- 4⅔ *cups sliced mushrooms (about 1 pound)*
- *Salt and ground white pepper to taste*
- ¾ *cup heavy cream*
- 3 to 4 *artichoke hearts,* sliced thin*
- 1 *pound fresh spinach, cooked, chopped, and drained*

Preheat the oven to 350°. In a heavy saucepan, melt 1½ tablespoons of the butter. Add the mushrooms to the butter, add salt and pepper, and sauté for about 10 minutes. Add ½ cup of the heavy cream, and boil to thicken; set aside.

Melt ½ tablespoon of the butter in a small sauté pan or skillet and lightly sauté the artichoke slices with salt and pepper. Melt the remaining 1 tablespoon butter in a sauté pan or skillet, until it is lightly browned (brown butter brings back the flavor of the spinach when it's been cooked first in water). Add the cooked spinach to the brown butter, add salt and pepper, and heat through.

Into a gratin dish, spread the spinach to cover the bottom. Place the sautéed artichokes on top of the spinach. Put the cooked mushroom mixture on top of the artichokes. Spread the remaining ¼ cup of the heavy cream on top. Place the gratin dish in a larger baking dish or pan, add hot water to about halfway up the side of the gratin dish, and bake in the preheated oven for 10 to 15 minutes, or until lightly browned on top. Serve hot.

**To make artichoke hearts, remove the leaves and stem from 3 or 4 large artichokes and scoop out the chokes with a spoon.*

Pasta & Risotto Dishes

China Moon Cafe Pot-browned Noodle Pillow Topped with Curried Vegetables

Barbara Tropp

Serves 4 to 6

"This is *the* lively vegetarian entree we serve on request at China Moon, where our tiny coffee shop kitchen has always a great assortment of vegetables from our daily marketing.

"For a home cook, it is an easy preparation: the noodle pillow and sauce may be made a day or two in advance, and the vegetables sliced, if carefully sealed and refrigerated, hours before serving. Leftovers are great cold, if you're a fan of such things (which I am).

"Keep in mind that the vegetable array is variable with the seasons and the market's best. Choose what is impeccably fresh and presents an interesting contrast of color, shape, and sweetness. When cooking, remember that what goes into the pan first is what takes longest to cook (e.g., onions and peppers), leaving the briefer cooking items (e.g., mushrooms and leafy greens) to be added in stages."

—*Barbara Tropp*

PILLOW

- 8 *ounces fresh or frozen thin Chinese or other egg noodles*
- 1 *tablespoon Japanese sesame oil (Kadoya)**
- 1 *teaspoon kosher salt (Diamond), or ½ teaspoon sea salt*
- ¼ *cup chopped Chinese chives, or ½ cup thin-cut green and white green onion rings*
- 6 to 8 *tablespoons corn or peanut oil*
- *Toasted black sesame seeds for garnish** (optional)*

SAUCE

- 1 *can (14 ounces) coconut milk (Chaokah)*
- 6 *tablespoons soy sauce (Superior)*
- 6 *tablespoons unseasoned Japanese rice vinegar (Marukan)*
- 2 *packed tablespoons brown sugar*
- 2 *teaspoons Chinese chili sauce (Koon Yick Wah Kee)*
- 4 *teaspoons curry powder*
- 1 *teaspoon five-spice powder****
- 2 *cups water or unsalted vegetable stock*

VEGETABLES

1/3 to 1/2 cup corn or peanut oil for stir-frying

1 large or 2 small yellow onions, cut into 1/4-inch-wide half wedges

1 large or 2 small red and/or yellow bell peppers, cut into 3/4-inch squares

1 medium carrot, cut into 1/16-inch-thick diagonal coins

About 8 ounces Chinese or Japanese eggplant, cut into 1/4-inch-thick rounds

2 tablespoons minced fresh ginger

2 tablespoons minced garlic

1/3 cup green and white green onion rings

1 red Fresno chili pepper, cut crosswise into thin rings (optional)

4 cups thickly sliced assorted wild and domestic mushrooms: shiitakes, oyster mushrooms, chanterelles, and/or Italian field mushrooms or halved and quartered domestic mushrooms

4 cups wide ribbons of assorted greens: baby bok choy, red chard, Bloomsdale spinach and/or napa cabbage; or 2 cups briefly blanched and refreshed baby green beans, sugar snap peas, English peas, notched snow peas, or halved fresh baby corn

2 teaspoons cornstarch dissolved in 3 tablespoons cold water or stock

Fresh enoki mushrooms and/or diagonally cut green onion rings to garnish

To prepare the noodle pillow(s): You may make individual noodle pillows in small skillets or one or two large ones. Nonstick pans are easiest to work with, though any skillet will do.

In a large pot, cook the noodles in boiling unsalted water until al dente, 2 to 3 minutes, swishing with chopsticks to separate the strands. Drain promptly, then plunge into ice water or run under a cold tap to stop the cooking. Shake off the excess water.

Toss the noodles with the sesame oil, salt, and chives or green onions. (At this point the noodles may be refrigerated 1 to 2 days.) Heat a skillet over high heat until hot, then add the oil to a depth of 1/8 inch and swirl to glaze the slides. Reduce the heat to medium-high. When hot enough to sizzle a single strand, pack the noodles into the pan to form a pillow about 3/4 inch thick.

Cook until the bottom is nicely golden, 5 to 7 minutes, adjusting the heat so the noodles brown and sizzle without scorching. Then turn the pillow over and brown the other side, drizzling in a bit more oil from the side of the pan.

Turn the pillow out onto paper towels to drain. (At this point, the pillow(s) may be refrigerated 1 to 2 days, then reheated in a 350° oven before serving.) Cut the pillow into wedges and garnish, if desired, with sesame seeds.

To prepare the sauce: Whisk all the ingredients well, dissolving any lumps from the coconut milk; set aside.

To prepare the vegetables: Heat a wok or a large, heavy skillet over high heat until hot enough to evaporate a bead of water on contact. Add enough oil to lightly glaze the pan, then swirl to coat the sides.

Add the onions and toss to sear, adjusting the heat so they sear without burning. (Rapid coloring and smoking will yield a flavorful *wok-hey*, an almost-grilled flavor typical of restaurant wok dishes and sometimes attainable at home with high heat.)

When the onions are partly seared, add the peppers and toss to cook halfway, about 2 minutes. Now and later, dribble in a bit more oil from the side of the pan if needed to prevent sticking. Add the carrot, toss 30 seconds to glaze, then add the eggplant and cook another 30 seconds or so until tender at the edge.

Push the vegetables to the side, add 2 tablespoons of oil to the pan, then add the ginger, garlic, green onions, and Fresno chili. Stir and flip gently to fully release their fragrances, 20 to 30 seconds, then toss to combine with the vegetables.

Raise the heat, add the mushrooms, and toss to combine. Add the greens and toss to mix; then stir the sauce and add it to the pan. Bring to a simmer, then add the cornstarch mixture and stir until glossy, 1 to 2 minutes. (Timing will vary depending on the stove and the pan, but the whole enterprise should take 5 to 10 minutes.)

Portion the topping over or alongside the hot noodle pillow wedges. At China Moon we leave one or two wedges to the side, to stay crispy, while burying the other wedges under the saucy topping. Garnish with a scattering of enoki and/or green onions and serve straightaway.

**Brands can make an enormous difference in taste, so I included those we use at China Moon. Most of these brands (or substitutes) can be found at Asian food stores.*

***To toast the seeds, stir them in a dry skillet over medium heat until their fragrance is released.*

****Available at Asian markets.*

Fresh Fettuccine with Spring Asparagus and Goat Cheese

Emeril Lagasse

Serves 4

- 24 *asparagus stalks*
- 1 *pound fresh fettuccine*
- 4 *teaspoons unsalted butter*
- 1 *teaspoon chopped fresh parsley*
- 4 *teaspoons chopped fresh chives*
- 6 *tablespoons chèvre (goat cheese)*
- ½ *teaspoon salt*
- 6 *tablespoons heavy cream*
- *Salt to taste*
- 5 *turns fresh black pepper from a peppermill*
- 2 *tablespoons chopped red bell pepper for garnish*

Peel the asparagus and blanch in a large pot of lightly salted boiling water to cover until al dente, about 2 minutes. Remove from the pan with a slotted spoon, reserving the cooking liquid. Chill the asparagus in ice water. Cut off the tips of the asparagus and save for garnish. Chop the remaining stems and set aside.

Bring the asparagus cooking liquid to a boil and cook the pasta until al dente, about 2 to 3 minutes. Drain the pasta, reserving a little of the cooking liquid. In a saucepan, melt the butter, add the chopped asparagus, and simmer for about 1 minute; add the parsley and chives. Add the goat cheese and sauté for another 30 seconds. Add the cream and a tablespoon or two of the pasta cooking liquid; boil for 15 to 20 seconds to reduce slightly. Add the pasta and toss; add salt and pepper. Garnish with red pepper and asparagus tips and serve at once.

Fettuccine with Sun-dried Tomatoes, Basil, Broccoli, and Goat Cheese

Jackie Shen

Serves 2

A colorful pasta with a sprinkling of black sesame seeds.

- 8 *ounces fettuccine*
- 4 *tablespoons olive oil*
- 8 *fresh basil leaves, chopped*
- 4 *sun-dried tomatoes, quartered*
- 1 *bunch green onions, chopped*
- 1 *large bunch broccoli, cut into florets*
- *Salt and pepper to taste*
- Four *½-inch-thick cylindrical slices goat cheese*
- 1 *teaspoon black sesame seeds*
- 4 *cherry tomatoes*
- 2 *edible (unsprayed) pansies**

Cook the fettuccine in a large amount of salted boiling water until al dente; drain and keep warm.

Preheat the oven to 350°. In a large sauté pan or skillet, heat 2 tablespoons of the olive oil; add fettuccine, basil, sun-dried tomatoes, and green onions. Stir, then set aside, covered, and keep warm.

Blanch the broccoli florets in boiling salted water to cover for 3 to 5 minutes until tender but still bright green. Drain; rinse under cold water and drain again. In another sauté pan or skillet, heat the remaining 2 tablespoons of olive oil and sauté the broccoli until heated through. Add salt and pepper and keep warm.

Dip the goat cheese in the black sesame seeds. Warm the goat cheese for about 1 minute in the preheated oven. On a large serving platter, place the broccoli florets around the edge. Place the pasta mixture in the middle of the plate. Slide the goat cheese slices on top of the pasta. Sprinkle the remaining sesame seeds over the noodles. Garnish with cherry tomatoes at 12, 3, 6, and 9 o'clock. Place 2 edible pansies atop the goat cheese and serve at once.

**Edible pansies are available in specialty markets.*

Wild-Mushroom Pasta

Alfred Portale

Serves 4

A creamy pasta with a variety of wild mushrooms.

SAUCE

- 2 *cups fresh morels, or 24 dried morels*
- 2 *tablespoons butter*
- ½ *cup chopped carrot*
- ½ *cup chopped celery*
- ¾ *cup chopped onion*
- ½ *teaspoon minced garlic*
- 1 *small leek, well trimmed and chopped*
- 3 *fresh parsley sprigs*
- 1 *bay leaf*
- 3 *fresh thyme sprigs*
- 2 *cups vegetable stock*
- 2 *cups heavy cream*
- *Salt and pepper to taste*

GARNISH

- 4 *tablespoons pine nuts, toasted*
- 3 *cups assorted wild mushrooms, trimmed*
- 1 *tablespoon minced shallots*
- 1 *tablespoon minced parsley*
- *Salt and pepper to taste*

- 10 *cups water, salted to taste*
- 16 *ounces penne*
- 2 *tablespoons minced fresh basil*
- 2 *tablespoons minced fresh parsley*
- ½ *cup peeled, seeded, and chopped tomato*
- 4 *tablespoons pine nuts, toasted*

If dried morels are used, place them in a bowl and add warm water to cover. Let stand 15 minutes. Drain and squeeze to extract most of the excess moisture. Strain the water carefully and reserve to use in a vegetable stock. Cut morels, fresh or dried, into half-inch pieces.

In a heavy kettle or casserole heat 1 tablespoon of the butter and add the carrot, celery, onion, garlic, and leek and cook, stirring, for about 3 minutes, without browning. Add the morels, and cook, stirring, for about 1 minute.

Tie the parsley, bay leaf, and thyme into a bundle. Add this to the kettle. Add the vegetable stock and boil 15 minutes, or until the liquid has been reduced to about ¼ cup. Add the cream and bring to a boil. Cook over moderate heat for about 15 minutes, or until the sauce is reduced to about 3 cups.

Roast the pine nuts in a dry skillet over medium heat, stirring constantly, until golden. Immediately remove from skillet; set aside.

Meanwhile, to make the garnish, heat the remaining tablespoon of butter in a saute pan or skillet and add the wild mushrooms. Cook, stirring, about 2 minutes, or until the mushrooms give up their liquid. Add the shallots, parsley, salt, and pepper and blend well. Cook about 1 minute longer and remove from the heat.

In a large pot, bring the water to a boil and add salt to taste. Add the penne and cook 7 to 10 minutes, or until al dente; drain.

Meanwhile, as the pasta cooks, sprinkle the morel sauce with the basil and parsley. Add the wild mushroom mixture and stir to blend. Add the drained penne to this sauce and toss. Spoon equal portions of the penne with sauce into each of 4 hot soup plates. Spoon equal portions of the cubed tomatoes into the center of each portion, and sprinkle with equal amounts of pine nuts.

....

Poblano and Black Bean Linguine with Ancho Threads, Oaxaca Cheese, and Mexican Vegetables

Dean Fearing

Serves 4 as a main course, 6 as an appetizer

The next time you cook black beans, save ¼ cup of beans and ¾ cup of their cooking liquid for this dish.

POBLANO LINGUINE

- ⅓ *cup chopped poblano chili**
- *About ½ cup water*
- 2 *cups durum flour*
- ½ *teaspoon salt*
- 1 *teaspoon olive oil*

BLACK BEAN LINGUINE

- *About ¾ cup black bean water*
- 2 *cups durum flour*
- ½ *teaspoon salt*
- 1 *teaspoon olive oil*

- 2 *ears of corn*
- 3 *dried ancho chilies**
- 2 *cups vegetable stock, heated*
- 2 *tablespoons peanut oil*
- 2 *garlic cloves, minced*
- 2 *jalapeños, seeded and finely chopped*
- 1 *shallot, minced*
- ¼ *cup cooked black beans*
- 1 *small jícama, peeled and diced*
- 1 *tomato peeled, cored, seeded, and diced*
- 2 *tablespoons chopped fresh cilantro*
- *Salt and pepper to taste*
- *Juice of 1 lime*
- ½ *cup grated Oaxaca or Monterey jack cheese tossed in 1 teaspoon peanut oil (to prevent clumping)**
- *Cilantro sprigs for garnish*

To make the poblano linguine, place the chopped chili in a blender or a food processor. Puree, adding 1 to 2 tablespoons of the water as needed.

In a bowl, combine the flour, salt, olive oil, poblano puree, and enough of the remaining water to make a soft dough. Run the dough through a pasta machine and cut into linguine; set aside. To make the black bean linguine: In a bowl, combine the black bean water, flour, salt, and olive oil to make a soft dough. Run the dough through a pasta machine and cut into linguine; set aside.

Soak the corn in water for 15 minutes, then roast in a 475° oven or on a covered charcoal grill for 25 minutes. Let cool, shuck, and cut off the kernels; set aside.

Soak the ancho chilies in the hot vegetable stock until soft. Remove the softened chilies from the soaking liquid, make a slit down the side, and remove the seeds. Cut the chilies lengthwise into thin strips. Roll each strip up like a jelly roll. Cut the rolled ancho strips into 1⁄16-inch slices and set aside.

In a large pot, bring a large quantity of salted water to a boil. Drop the poblano linguine into the water and stir to keep separated. Cook for 1 to 2 minutes, or until al dente. Drain and keep warm, saving the cooking water for the black bean linguine. Cook the black bean linguine in the same manner; drain and keep warm.

In a sauté pan or skillet, heat the peanut oil until it reaches the smoking point. Add the jalapeños, both cooked pastas, ancho slices, garlic, and shallot and sauté for 2 minutes. Add the beans, jícama, tomato, corn, and cilantro and heat through. Season with salt, pepper, and lime juice, then toss in the cheese. Arrange on 6 small or 4 large plates and garnish with cilantro sprigs.

**Available at Mexican markets or specialty food stores.*

Shiitake and Sour Cherry Ravioli

Barry Wine

Serves 6

STUFFING

- *1½ pounds shiitake mushrooms*
- *3 tablespoons butter*
- *2 teaspoons minced fresh ginger*
- *2 tablespoons minced garlic*
- *4 shallots, minced*
- *1 bunch Swiss chard, torn into small pieces*
- *Salt and pepper to taste*
- *1 cup dried sour cherries**
- *1 cup water or vegetable stock*

- *1 recipe Pasta Dough, page 355*
- *4 tablespoons butter*
- *Roasted Shiitake Mushrooms, following, for garnish*

Stem and finely slice the mushrooms. Melt the butter in a sauté pan or skillet and sauté the mushrooms until golden. Add the ginger, garlic, shallots, and Swiss chard and cook 1 to 2 minutes. Season with salt and pepper. Remove the mixture from the pan. Add the sour cherries to the pan. Pour in the water or stock and boil to reduce the liquid to a thick syrup. Add to the shiitake mixture and correct the seasoning; set aside.

Prepare the pasta dough and divide the dough in half. Take half of the dough and roll in a pasta machine set at the second thinnest setting.

To make the ravioli, place a large pot of lightly salted water on the stove to boil. Cut the pasta sheet into 4-inch squares. Place 2 to 3 tablespoons of filling in the center of each square. With a pastry brush, brush the edges of the dough with water. Fold over into a triangle and seal. Roll out the second half of the dough and repeat steps to make ravioli. Add the ravioli to the boiling water and cook for 3 to 4 minutes. Remove the ravioli from the water with a slotted spoon. Drain in a colander, place in a serving bowl, add butter, and toss gently. Garnish with the roasted mushrooms and serve at once.

**Available from health food stores or specialty stores.*

ROASTED SHIITAKE MUSHROOMS

10 fresh shiitake mushrooms
¼ cup extra-virgin olive oil
Salt and pepper to taste

Preheat the oven to 350°. Place the mushrooms in a bowl. Add the olive oil, salt, and pepper and toss. Roast in the preheated oven for 20 minutes, or until crispy.

Vegetable Ravioli in a Tomato Coulis with Fresh Herbs and Romano Cheese

Bruce Marder

Serves 4 to 6

FILLING

- *1 artichoke*
- *1 carrot, peeled*
- *¼ cup olive oil*
- *1 zucchini*
- *1 small white onion*
- *3 jumbo mushrooms*
- *2 celery stalks*
- *½ green bell pepper, cored and seeded*
- *½ red bell pepper, cored and seeded*
- *1 medium potato*
- *Salt and pepper to taste*

- *1 recipe Pasta Dough, following*

SAUCE

- *2 tablespoons Italian extra-virgin olive oil*
- *5 garlic cloves, minced*
- *4 vine-ripened tomatoes, peeled, cored, seeded, and coarsely chopped*
- *¼ cup minced mixed fresh Italian parsley, thyme, oregano, and basil*
- *Salt and pepper to taste*

- *Salt to taste*
- *1 teaspoon olive oil*
- *4 tablespoons freshly grated Romano cheese*

Remove all the leaves and the stem from the artichoke; scoop out the choke with a spoon. Blanch the carrot and artichoke separately in boiling water to cover for 2 to 3 minutes; remove and drain. In a sauté pan or skillet, heat the olive oil and sauté, in turn, the zucchini, onion, mushrooms, celery, and green and red bell peppers, adding more olive oil as needed, and removing each vegetable with a slotted spoon before adding another. Mince the blanched vegetables and all of the sautéed vegetables. Peel and slice the potato. Boil the potato in salted water to cover for 15 to 20 minutes or until soft. Drain and mash with a potato masher or in a food processor.

Mix all vegetables together with the mashed potatoes, and season with salt and pepper; set aside.

Prepare the pasta dough and set aside. To make the sauce: In a saucepan, heat the olive oil and sauté the garlic over low heat until translucent. Add the tomatoes and simmer for about 20 minutes, or until thickened but not dry. Add the herbs, season with salt and pepper, set aside, and keep warm.

Place a large pot of water with salt and oil on the stove to boil. Roll sheets of pasta as thin as your machine will allow. Place 1 sheet on a floured surface and brush the top lightly with water. Spoon ½-teaspoonfuls of filling 1 inch apart over the surface of the pasta, leaving a 1-inch margin around the edges. Place a second sheet of pasta over the first sheet, pressing down around each mound of filling. Cut into squares and re-press the edges of each one, making sure there are no air spaces.

Cook the pasta for 2 minutes in the boiling water. Mix with the tomato sauce, serve on individual plates, sprinkle with cheese, and serve.

PASTA DOUGH

Makes about 1 pound, or enough for 4 to 6 servings

2 cups durum flour
½ teaspoon salt
1 teaspoon olive oil
About ⅜ to ½ cup water

Mound the flour on a flat work surface. Create a hollow in the center of the mound. Fill the hollow with the salt and olive oil. Add water a little at a time and work the ingredients together until you have a rough mass. Knead the dough until all the flour is incorporated, about 15 minutes. Form the dough into a ball; the surface of the dough should be smooth, and when you press the dough gently it should spring back. Cover the dough with a dry cotton towel (not plastic wrap) and let it rest for 1 hour in a cool place.

Cut the dough into thirds and run it through a pasta machine according to the manufacturer's instructions, or roll it out on a floured surface. Cut the pasta into the desired shape.

To cook: In a large pot bring a large quantity of salted water to a boil. Drop in the pasta and stir gently to separate. Cook 1 to 2 minutes for narrow pasta, longer for wider varieties, or until al dente. Drain and use immediately.

Cannelloni of Eggplant Caviar Glazed with Goat Cheese and Zucchini Juice

Jean-Georges Vongerichten

Serves 4

Elegant yet earthy cannelloni for a festive dinner party.

¼ recipe (4 ounces) Pasta Dough, page 355, or 8 won ton wrappers

EGGPLANT CAVIAR

- 6 *tablespoons olive oil*
- 4 *medium eggplants*
- ⅔ *cup chopped red bell pepper*
- 6 *tablespoons chopped green bell pepper*
- 2 *shallots, chopped*
- 1 *tomato, chopped*
- 5 *mushrooms, chopped*
- 1 *garlic clove, chopped*
- ¼ *cup chopped mixed fresh herbs (thyme, rosemary, basil, chervil)*
- *Salt and pepper to taste*
- *Zucchini Juice, following*
- ¼ *cup grated aged goat cheese*

Prepare the pasta dough and roll it out as thinly as possible. Cut it into 2-by-3-inch rectangles and poach them for 1 minute in a large quantity of boiling salted water; set aside.

Preheat the oven to 450°. Coat the bottom of a baking pan with 3 tablespoons of the olive oil, place the whole eggplants in the pan, and cook for 1 hour. Cut the eggplants in half and scrape out the insides.

In a saucepan, heat the remaining 3 tablespoons of olive oil and cook all the chopped vegetables and garlic until tender. Add the eggplant pulp and cook for 15 minutes. Add the chopped mixed herbs. Season with salt and pepper; set aside. Prepare the Zucchini Juice and set it aside.

Spoon the eggplant mixture across the center of each pasta square and roll it into a cylinder. Place each cylinder in a heatproof gratin dish, sprinkle with the grated goat cheese, and put under the broiler until golden brown. Pour a few tablespoons of the just-prepared zucchini juice on each serving plate, and place the cannelloni on top.

ZUCCHINI JUICE

- 2 *zucchini*
- 6 *tablespoons olive oil*
- *Leaves from 1/2 bunch thyme, chopped*
- 2 *garlic cloves, chopped*
- *Salt and pepper to taste*

Prepare this just before serving. Put the unpeeled zucchini through a juice extractor. You should have approximately 1 cup of raw juice. Heat 1 tablespoon of the olive oil in a sauté pan or skillet and sauté the garlic over medium heat until translucent. Add the zucchini juice and the chopped thyme. Bring to a boil and remove from heat. Whisk in the remaining olive oil, salt, and pepper. Serve immediately.

Grilled Vegetable Lasagne

Gordon Naccarato

Serves 4 to 6

A different kind of lasagne: grilled vegetables, goat cheese sauce, and pasta triangles.

- 1 *recipe Pasta Dough, page 355, or 1 pound dried lasagne strips*
- 2 *red bell peppers*
- 2 *zucchini*
- ½ *cup olive oil*
- 1 *head radicchio*
- *A few mushrooms, sliced*
- *A few niçoise olives, pitted and sliced*
- 1 *red onion, minced*
- *A few sun-dried tomatoes, diced*
- 1 to 2 *garlic cloves, minced and mixed with 3 tablespoons olive oil*
- *Salt and fresh-ground black pepper*
- 8 *ounces natural cream cheese, at room temperature*
- 1 *cup ricotta cheese*
- 4 *ounces Parmesan cheese, grated (about 1 cup)*
- *Assorted fresh herbs (basil, thyme, oregano, marjoram, and rosemary), chopped*
- 4 *ounces whole-milk mozzarella, grated (about 1 cup)*
- *Goat Cheese Cream Sauce, following*

Prepare the pasta dough. Wrap the dough in plastic wrap and refrigerate until needed. Prepare a wood or charcoal fire in an open grill. While the fire is hot, place the peppers on the grill and turn to char on all sides. Place the peppers in a plastic bag, close it tightly, and let it cool 15 minutes. Rub off the burned skin, then core, seed, and mince the peppers. Set aside.

Slice the zucchini in half lengthwise. Slice the halves in half again. Brush with olive oil and grill over a low fire on both sides until nicely browned (use a grill basket to grill the zucchini and other vegetables, if possible). Quarter the radicchio, brush with olive oil, and grill until browned on all sides. Be careful; radicchio burns easily. Grill the mushrooms. Chop the zucchini, radicchio, and mushrooms when cooled. In a bowl, combine the zucchini, radicchio, mushrooms, peppers, olives, onion, and tomatoes and set aside.

In a bowl, mix together the cream cheese and ricotta. Preheat the broiler. Place the pasta on a work surface. If you are using commercial pasta, cut into squares. Brush each square with the garlic and olive oil mixture. Season lightly with salt and pepper. Spread a thin layer of the combined cream cheese and ricotta over each pasta square. Sprinkle with the Parmesan and a touch of fresh herbs. On a diagonal half of the pasta square, place a spoonful of the zucchini, radicchio, mushroom, tomato, olive, pepper, and onion mixture. Fold the other diagonal half of the pasta over the side on which you have placed the vegetables, forming a triangle.

Lightly oil an ovenproof pan and place the pasta inside. Cover the pasta triangles with grated mozzarella. Place under the broiler and heat until the cheese is bubbly. Place on a warm plate with the heated goat cheese cream sauce. Garnish with the remaining fresh chopped herbs and serve immediately.

Roll out the pasta dough using a No. 5 setting on a pasta machine, or roll as thinly as possible by hand. Cut the pasta into 4-inch squares. If you are using dried lasagne strips, leave them whole. Cook the pasta in boiling salted water until al dente, about 1 minute for fresh pasta or 8 to 10 minutes for dried; set aside.

GOAT CHEESE CREAM SAUCE

- 1 *quart heavy cream*
- 1 *log (11 ounces) Montrachet-style goat cheese*
- 2 *shallots, chopped*
- 1 *garlic clove, minced*
- *Salt and pepper to taste*
- *Chopped fresh herbs to taste*

Place the cream in a large saucepan and crumble the goat cheese into it. Add the shallots, garlic, and salt and pepper. Bring to a boil over medium heat and cook, stirring frequently, for 3 to 4 minutes, or until creamy but not too thick. Correct the seasoning and add the herbs. Set aside and keep warm.

Yam Ravioli with White Truffles

Gunter Seeger

Serves 6

The exotic taste of saffron and the muskiness of truffles play against a silky sweet potato puree and chewy pasta.

- *1 scant tablespoon saffron threads*
- *1 cup water*
- *1½ cups durum flour or bread flour*
- *½ teaspoon sea salt, or to taste*
- *1 tablespoon olive oil*
- *3 yams (sun-dried if possible)*
- *2 tablespoons clarified butter**
- *1 shallot, minced*
- *Fresh ground pepper to taste*
- *2 cups vegetable stock*
- *6 tablespoons butter, melted*
- *1 whole white truffle, cleaned with a paper towel*

Soak the saffron in the water for 10 minutes. Place the durum flour in a bowl. Add the saffron liquid, salt, and olive oil. Mix well with a wooden spoon. Remove to a floured board and knead for 5 minutes, or until smooth and elastic. Refrigerate, covered, for 2 hours.

Meanwhile, peel the yams and cut them into dice. In a saucepan, heat the clarified butter and sauté the yams, shallot, and pepper for 2 minutes, stirring to coat the yams. Add the vegetable stock and cook until almost all the liquid is gone, stirring frequently at the end to avoid burning. Puree in a blender or food processor and adjust the seasoning with salt and pepper.

Roll the pasta dough out very thin. Cut into 2-inch squares, and spoon or pipe the yam puree from a pastry bag onto the dough. Fold the dough over and press together. Cut with a cookie cutter or ravioli cutter to obtain a half moon shape.

Boil the ravioli in a large pot of salted water for 2 minutes. Remove from the cooking water and drain well. Pour 1 tablespoon of melted butter on each serving plate. Divide the ravioli among the buttered plates. Shave the white truffle very thin over the ravioli and serve.

**To make clarified butter, melt 4 tablespoons of unsalted butter in a pan over low heat. Pour off the clear yellow liquid, leaving the milky residue.*

Spring Vegetable Risotto with Orzo

Anne Rosenzweig

Serves 6

This has become a main-course favorite in my house, but it's also good as a side dish or appetizer.

- 2 *red bell peppers*
- 1 *cup orzo (rice-shaped pasta)*
- 1 *tablespoon olive oil*
- ½ *cup (1 stick) unsalted butter*
- 1 *carrot, minced*
- ½ *large onion, minced*
- 6 *small plum tomatoes, peeled, seeded, and diced*
- 8 *medium to large shiitake mushrooms, stemmed and sliced ½-inch thick*
- 8 *large asparagus stalks, peeled at the bottom and cut on the bias in ½-inch pieces*
- ¾ *cup vegetable stock*
- 1 *cup thinly sliced fresh basil leaves*
- *Salt and pepper to taste*

Cut the peppers into quarters and remove the cores and seeds. Char them evenly under a broiler. Place in a closed plastic bag for 15 minutes; remove and rub off the blackened skin. Cut the peppers into julienne and set aside.

Bring a pot of salted water to a boil. In a skillet, toast the raw orzo over low heat, tossing often, until about one third of the orzo has colored lightly. Place the orzo in the boiling water and cook 5 to 7 minutes, or until al dente, then drain and run under cold water. Drain thoroughly, toss with the olive oil, and set aside.

In a large saucepan, melt 4 tablespoons (one half) of the butter. Add the carrot and onion and cook over moderate heat to soften, about 4 minutes. Add the tomatoes and cook for 3 minutes longer. Add the shiitakes, asparagus, red peppers, vegetable stock, basil, and orzo. Stir well to mix and add salt and pepper. Add the remaining 4 tablespoons butter and incorporate over high heat. The mixture should have a creamy consistency.

Risotto con Asparagi

Piero Selvaggio

Serves 4 to 6

A classic asparagus risotto with a touch of brandy.

- *3 pounds asparagus*
- *2 tablespoons butter*
- *5 tablespoons butter*
- *½ cup minced onion*
- *¾ cup brandy*
- *1¾ cup arborio rice*
- *5 cups vegetable stock, heated*
- *¾ cup freshly grated Parmesan cheese*
- *Salt and pepper to taste*

Peel the asparagus and cut off the woody part from the bottom of the stems. Cut the asparagus into 1-inch sections.

In a large sauté pan or skillet, heat the oil and 2½ tablespoons of butter. Add the onion and sauté over medium heat until brown. Add the asparagus and cook for 10 minutes. Add the brandy and allow to simmer. Add the rice, mixing it well with the other ingredients. Add 1 cup of stock at a time, letting the rice cook until the stock is completely absorbed each time. The rice should become very creamy. Add the remaining 2½ tablespoons of the butter and the Parmesan cheese; add salt and pepper and serve at once.

. . . .

Corn and Peppers Risotto

Piero Selvaggio

Serves 6

- 6 *red bell peppers*
- ½ *cup (1 stick) butter*
- 1 *onion, chopped*
- 2 *cups arborio rice*
- ¾ *cup dry white wine*
- 6 *cups vegetable stock, heated*
- 1 *cup cooked corn kernels (about 3 ears)*
- ¾ *cup freshly grated imported Parmesan*

Place the peppers under a hot broiler, turning until charred evenly. Place in a plastic bag, close tightly, and let cool for 15 minutes. Rub off the burned skin. Cut the peppers in half crosswise. Discard the cores and seeds. Set aside the top halves and mince the rest. In a saucepan, melt the butter and add the onion; cook for 2 to 3 minutes. Add the rice to the mixture and mix well until all the rice is glossy.

Add the white wine. Bring to a simmer. Add the stock a little at a time, cooking until completely absorbed each time. When the rice reaches a creamy consistency, remove from the heat and stir in half of the cheese, the roasted chopped peppers, and the corn. Mix well and let stand 4 to 5 minutes.

Serve the risotto in individual dishes and sprinkle with the rest of the cheese. Garnish with the reserved tops of the red peppers and serve at once.

Green Risotto

Richard Perry

Serves 4 to 6

VEGETABLE STOCK

- *1 quart water*
- *½ cup chopped onion*
- *½ cup chopped peeled carrot*
- *1 cup chopped celery*
- *1 cup chopped peeled turnips*
- *¼ cup diced leeks, white and light green part*
- *2 cups diced tomato*
- *3 fresh parsley sprigs*
- *½ tablespoon salt*
- *½ teaspoon black peppercorns*
- *¼ cup dry white wine*

- *1½ tablespoons butter*
- *1 cup chopped asparagus*
- *½ cup chopped fennel root*
- *½ cup broccoli florets*
- *½ tablespoon chopped fresh parsley*
- *½ tablespoon chopped fresh basil*
- *½ tablespoon olive oil*
- *1 tablespoon minced onion*
- *½ tablespoon minced carrot*
- *½ tablespoon minced celery*
- *¾ cup arborio rice*
- *¼ cup grated imported Parmesan cheese*

To make the stock, bring the water to a boil in a large pot; add the remaining ingredients for the stock and simmer for 20 minutes; strain thoroughly. Place 2½ cups of the stock and the white wine in a saucepan over low heat, reserving the rest for another use.*

Heat ½ tablespoon of the butter in a sauté pan or skillet; add the asparagus, fennel, and broccoli and cook until they are somewhat tender, but still crisp; remove from heat, stir in the parsley and basil, and set aside.

Heat the remaining tablespoon of butter and the oil in a heavy pot; add the onion, carrot, and celery; sauté for 2 minutes, making sure not to brown the ingredients.

Add the rice to the mixture and stir with a wooden spoon for 1 minute. Add the simmering stock to the rice, ½ cup at a time, stirring frequently. Wait until each addition is almost completely absorbed before adding the next ½ cup, reserving ¼ cup of stock to add at the end.

After about 18 minutes, when the rice is tender but still firm, add the reserved ¼ cup stock, the sautéed vegetables, and the Parmesan cheese. Stir to combine and serve hot.

**This recipe makes 1 quart of stock. Store the extra 1½ cups in the refrigerator for 2 days, or freeze for later use.*

Pizza and Breads

. . . .

Pizza

Alice Waters

Makes one 12- to 14-inch pizza or several small ones

"One of the very best ways to bake a pizza is directly on the floor of a wood-fired brick or stone oven. The intense heat of a wood fire can drive the temperature to 500° or more, and it gives the dough a smoky flavor. When the dough slides onto the hot bricks it reacts instantly; both the top and the bottom of the pizza cook at once. Not too many households have a brick oven to bake in, but the effect can be approximated by putting a layer of unglazed ceramic tiles on a rack in your oven."

—*Alice Waters*

DOUGH

- ¼ *cup warm (105° to 115°) water*
- 1 *package (2 teaspoons) active dry yeast*
- ¼ *cup rye flour*
- ½ *cup warm (105° to 115°) water*
- 1 *tablespoon milk*
- 2 *tablespoons olive oil*
- ½ *teaspoon salt*
- *About 1¾ cups unbleached all-purpose flour*

Your choice of toppings, following

In a large bowl, place the water and sprinkle the yeast over it. Stir to dissolve. Mix in the rye flour with a whisk, cover the bowl with plastic wrap, and let sit in a warm place for 20 to 30 minutes.

Stir in the remaining ingredients with a wooden spoon, adding the flour ½ cup at a time, then knead on a floured board. The dough will be soft and a little sticky. Use quick, light motions with your hands so the dough won't stick. Add more flour to the board as you knead but no more than is absolutely necessary. A soft, moist dough makes a light and very crispy crust. Knead for 10 to 15 minutes to develop strength and elasticity in the dough.

Place the dough in a large bowl rubbed with olive oil, and oil the surface of the dough to prevent a crust from forming. Cover the bowl with a towel and put it in a warm place, approximately 90° to 110°. An oven heated just by its pilot light is a good spot.

Let the dough rise to double its size, for about 2 hours, then punch it down. Let it rise about 40 minutes more. Preheat the oven to 450° to 500°. If you are using clay tiles in the oven, use a wooden paddle made especially for the purpose (called a baker's peel), or the back of a baking sheet, to slide the pizza onto and off of the tiles.

Flatten the dough on a heavily floured board. Use a rolling pin to roll the dough to roughly 12 to 14 inches in diameter. The dough should be ⅛ to ¼ inch thick. Transfer the dough to a heavily floured peel or baking sheet. Have your toppings ready at room temperature, and work quickly putting them on the pizza, or it will begin to stick and will be impossible to slide off the paddle.

When garnishing the pizza, anticipate flavor and balance. A light hand with weighty ingredients such as cheese, tomatoes, sausage, and so on, and bold amounts of fresh herbs, garlic, anchovies, flavored oils and the like, works best. It is better to err on the side of flavor. Tomatoes and other wet foods should be drained of excess liquid. Too much weight or moisture on the dough makes it difficult for it to rise and cook well on the bottom. Whatever is on top must be able to cook in 15 minutes, or should have had some partial cooking beforehand. Beware of spilling anything wet or oily between the dough and paddle, as that too will prevent it from sliding.

Give the paddle a few shakes back and forth to make sure the dough is loose. Slide the pizza from the paddle onto the hot tiles in the oven with abrupt jerking motions of your wrist. This takes a certain knack but comes easily after a few tries.

The pizza will be brown and cooked in 12 to 15 minutes. When you are deciding if the pizza is cooked, check the bottom to make sure it is quite crisp. The crust always softens a bit when it cools down. The real purpose of the tiles is to make a good texture on the bottom of the pizza.

Note: When you roll and shape the dough, feel free to make it any shape you wish. Large flat pizzas with uneven bubbly edges have a rustic appeal. Small individual-sized pizzas, served as a savory accompaniment to a meal instead of bread, are very satisfying. If your oven cannot maintain an intense heat of 450° to 500°, then the dough will perform better if rolled a little on the thick side, ¼ inch or more. When the dough is rolled thin and requires 20 to 25 minutes to cook at some temperature less than 450°, then it tends to have a crackerlike texture. A thicker dough allows for a bready interior and a crusty exterior.

PIZZA TOPPINGS

Onions, Garlic, and Herbs

Sauté 3 or 4 sliced onions and lots of sliced garlic in olive oil until just softened. Use this as the base of the pizza topping. Finely chop a mixture of fresh herbs such as parsley, thyme, oregano, basil, and rosemary. Sprinkle these over the onions, and season with salt and pepper. Put the pizza in the oven. Halfway through the baking, cover the top with a light layer of mixed freshly grated Parmesan and Romano cheeses. When it comes out of the oven, drizzle olive oil over the crusty edges.

Roasted Garlic Puree and Chanterelles

Separate the cloves of several heads of garlic. Toss the cloves, in their skins, in olive oil. Put them in a roasting pan, and season with salt and pepper, bay leaves, and fresh thyme branches. Cover the pan with aluminum foil and bake in a 300° oven for about 1½ hours, or until the garlic is very soft. Put the garlic through a food mill. Spread this puree over the pizza dough and cover it with sliced chanterelles that have been lightly sautéed in butter. Bake the pizza and garnish it with chopped fresh parsley.

Grilled Eggplant and Pesto

Thinly slice 5 or 6 small Japanese eggplants. Brush them with olive oil and charcoal grill them until lightly browned on both sides. Roll out the pizza dough and moisten it with a light coating of pesto sauce. Arrange the eggplant slices on the dough and bake the pizza. Brush the eggplant with pesto when the pizza is done.

Eggplant, Tomatoes, and Parmesan

Cut 4 or 5 small eggplants into julienne. Sauté the eggplant in hot olive oil for 5 or 6 minutes, until it softens and browns. At the last minute add 5 or 6 cloves of sliced garlic and season with salt and pepper. Quarter about 1½ cups of cherry tomatoes. Roll the pizza dough and brush it with olive oil. Spread the tomatoes over it. Salt them, then cover with the eggplant and garlic. Cover the eggplant with shaved Parmesan cheese and bake the pizza. Garnish with purple basil leaves.

Roasted Eggplant and Mozzarella

Slice 1 large or several small Japanese eggplants in ⅛- to ¼-inch slices. Brush them with olive oil, season with salt and pepper, put them in a single layer on a roasting pan, and bake in a hot oven until nicely browned. Or grill the slices over a charcoal fire. Roll the dough and sprinkle it with a little minced garlic and a light layer of grated mozzarella cheese, about 4 ounces (1 cup). Arrange the eggplant slices on the cheese, sprinkle just a little more cheese over the eggplant, and bake the pizza. Garnish it with some chopped fresh marjoram.

Fresh and Dried Wild Mushrooms

Soak a handful of dried porcini mushrooms in hot water to cover until soft, about 15 minutes. Carefully sort through them and remove any pockets of dirt or grit. Squeeze the mushrooms dry and chop them. Thinly slice 12 ounces of fresh cultivated mushrooms. Sauté the mushrooms together in olive oil. Season with salt and pepper, and add some minced garlic and chopped fresh parsley. Roll out the pizza dough and top with the mushroom mixture. Add 5 or 6 slivered dried tomatoes, and a light layer of freshly grated Parmesan. Put the pizza in the oven. Halfway through the baking, add some more Parmesan. When the pizza is removed from the oven, brush the crusty edges with olive oil.

Focaccia with Caramelized Onions, Gorgonzola Cheese, and Walnuts

Joyce Goldstein

Makes one 12-by-17-inch rectangle

An Italian flatbread with a rich, savory topping.

TOPPING

4 *tablespoons unsalted butter*

2 *red onions, sliced ¼-inch thick*

Salt and pepper to taste

½ *cup walnuts*

8 *ounces Gorgonzola, at room temperature*

SPONGE

1 *package (2 teaspoons) active dry yeast*

¼ *cup warm (105° to 115°) water*

½ *cup unbleached all-purpose flour*

¾ *cup water*

3 *tablespoons olive oil*

About 3 cups unbleached all-purpose flour

¼ *cup buckwheat flour*

1½ *teaspoons salt*

Oil or cornmeal for baking

Olive oil for brushing

To make the topping, preheat the oven to 350°. Melt the butter in a large sauté pan or skillet and sauté the onions over low heat for 20 minutes, stirring often. They should be golden brown. Season with salt and pepper. Drain any pan juices from the onions.

Place the walnuts on a baking sheet and toast in the preheated oven for 7 to 9 minutes. Chop very coarsely. Break the Gorgonzola into walnut-sized pieces.

To make the sponge: In a large bowl, sprinkle the yeast over the warm water and stir to dissolve. Add the flour and combine with a whisk. Cover with plastic wrap and let the sponge sit for 20 to 30 minutes.

Add the water, olive oil, flour, and salt to the sponge and mix on low speed with a doughhook for about 10 minutes until the dough leaves the side of the bowl cleanly. Or, add the water, oil, ½ cup of the flour, and the salt and stir to combine.

Add the remaining flour ½ cup at a time and stir until a stiff dough forms, then turn out onto a floured surface and knead until smooth.

Put the dough in a large oiled bowl, turn to coat all sides, cover with plastic wrap, and allow to rise in a warm place for 1 hour.

Punch the dough down and turn out onto a lightly floured board. Roll and form into a rectangle about 12 by 17 inches. Oil a baking sheet or line it with baker's parchment, then dust it lightly with cornmeal. Place the dough on the pan, cover with plastic wrap, and let rise for 15 minutes. Dimple the dough with your fingers, cover, and let rise again for another 20 minutes. Dimple again.

Preheat the oven to 425° or 450° with unglazed clay tiles if you have them. If you will be using clay tiles as a baking surface, slide the dough onto a baker's peel and cover it with the prepared topping. If you wish to cook the dough on the baking sheet, dimple the dough again and cover it with the topping. Spread the caramelized onions evenly over the surface, then sprinkle with the Gorgonzola and toasted walnuts.

If the dough is on a peel, slide it directly onto clay tiles on the oven floor; otherwise leave it on the baking sheet. Bake in the preheated oven for about 15 minutes. Brush the edge of the dough with olive oil. Serve warm or hot.

Chili Corn Bread

Mark Miller

Makes 2 loaves

Flecks of red and green in an intensely flavored bright yellow bread.

- 2 to 3 *tablespoons olive oil*
- ½ *cup finely diced sweet red onion*
- *Leaves from ½ bunch cilantro, chopped*
- 3 *serrano chilies, minced (with seeds)*
- 5 to 8 *jalapeño chilies, minced (with seeds)*
- 1 *red bell pepper, cored, seeded, and minced*
- ¼ *teaspoon cayenne*
- ½ to 1 *tablespoon ancho chili powder*

SPONGE

- ¾ *cup warm (105° to 115°) water*
- 2 *packages (4 teaspoons) active dry yeast*
- 1 *tablespoon sugar*
- 1 *tablespoon all-purpose flour*

- 1 *cup buttermilk*
- 1 *cup milk*
- 1 *tablespoon salt*
- ¼ *cup sugar*
- 4 *tablespoons unsalted butter, melted*
- ¼ *cup corn oil*
- *About 5 cups unbleached all-purpose flour*
- 1 *cup whole-wheat flour*
- 1 *cup fine cornmeal*
- 1⅓ *cups coarse cornmeal or polenta*
- 1 *tablespoon corn oil*
- ¾ *cup grated Cheddar cheese*

In a sauté pan or skillet, heat the olive oil and sauté the vegetables and spices until the vegetables are tender; set aside.

To make the sponge: Place the water in a large bowl. Sprinkle the yeast over, stir to dissolve, and let sit for 5 minutes. Add the sugar and flour and mix vigorously with a whisk; set aside.

Place the buttermilk and milk in a saucepan and bring to scalding. Pour the milks into a large bowl and add the salt, sugar, butter, and corn oil. Add the all-purpose flour to the milk mixture ½ cup at a time, mixing well with a wooden spoon or with a dough hook on a heavy-duty mixer. The consistency should be very wet. Add the sponge to this mixture, along with the whole-wheat flour, the fine cornmeal, and 1 cup of the coarse cornmeal. Mix until the dough is stiff.

Remove the dough to a floured board and knead until the dough is smooth and elastic, adding more flour as necessary 1 tablespoon at a time. Roll out to a square, sprinkle with the cheese and cooked chopped vegetables, and knead until all the added ingredients are incorporated. Place the dough in a large bowl greased with corn oil. Rotate the dough in the oil until it is well covered. Cover with plastic wrap and let the dough rise in a warm place for 45 minutes or until it doubles in volume.

Preheat the oven to 375°. Punch the dough down and turn out onto a floured board. Divide the dough in half and form into logs. Sprinkle two 8½- by 4½-inch loaf pans with the remaining ⅓ cup coarse cornmeal. Place the dough in the pans and bake in the preheated oven for 50 to 60 minutes, or until the crust is a dark golden brown and the loaf is hollow-sounding when tapped. Turn the loaves out onto racks to cool.

Relishes and Intermezzos

Salsa Verde

Mark Miller

Makes about 2 cups

Serve this bright green salsa with grilled foods and tortilla dishes.

- 2 *bunches parsley, stemmed*
- 4 *large garlic cloves, or to taste*
- 1 *tablespoon capers, drained and rinsed*
- 1 *cup virgin olive oil*
- *Finely grated zest of 2 lemons*
- 2 *tablespoons chopped green olives (optional)*

Mince the parsley by hand, making sure there are no stems, and place the parsley in a bowl. Chop the garlic and capers and add to the parsley. Whisk in the olive oil and the lemon zest. This will keep for 3 to 4 days tightly covered in the refrigerator.

Avocado-Corn Salsa

Vincent Guerithault

Makes about 2½ cups

- 1 *cup steamed corn kernels (about 3 ears)*
- 1 *cup diced avocado*
- 1 *teaspoon chopped fresh cilantro*
- 1 *teaspoon diced tomatoes*
- 1 *teaspoon chopped shallot*
- 1 *teaspoon* each *chopped red, yellow, and green bell pepper*
- 1 *teaspoon fresh lemon juice*
- *Salt and pepper to taste*

In a bowl, mix all the ingredients. This keeps 2 to 3 days tightly covered in the refrigerator.

Marinated Onions and Cucumbers

Richard Perry

Serves 4 to 6

Richard Perry was raised on an Illinois farm where meals began with assorted relishes, like these marinated cucumbers and onions.

- ½ *cup red wine vinegar*
- ⅔ *cup sugar*
- 3 *tablespoons olive oil*
- 1⅓ *cups thin-sliced cucumbers*
- 2 *cups thin-sliced onions*

In a large bowl, combine the vinegar and sugar; let sit for 1 hour so that the sugar dissolves. Add the olive oil. Add the cucumbers and onions to the marinade, cover, and refrigerate for at least 12 hours before serving. This will keep about a week, tightly covered, in the refrigerator.

Jicama-Melon Relish

Stephan Pyles

Makes about 2 cups

A sweet, potent mixture to serve with grilled vegetables.

- 1 *mango, peeled, seeded, and coarsely chopped*
- 1 *serrano chili, cored, seeded, and coarsely chopped*
- *Juice of 1 lime*
- 1½ *tablespoons minced red bell pepper*
- ½ *cup ¼-inch-diced cantaloupe*
- ½ *cup ¼-inch-diced honeydew*
- 2 *tablespoons ¼-inch-diced peeled and seeded cucumber*
- ½ *cup ¼-inch-diced peeled jícama*
- 2 *teaspoons chopped fresh cilantro*
- *Salt and freshly ground black pepper to taste*

In a food processor or blender, puree the mango, chili, and lime juice. In a bowl, combine with the remaining ingredients and season with salt and pepper. The relish will keep about 2 days, tightly covered, in the refrigerator.

Roasted Fresh Pimiento and Saffron Butter

Mark Miller

Makes about 3 cups

Spread this butter on Mark's rich Chili Corn Bread, page xx.

- 6 *large fresh pimiento or red bell peppers*
- 2 *quarts vegetable oil*
- ½ *cup dry white wine*
- 1 *tablespoon Spanish saffron threads*
- 1 *cup (2 sticks) unsalted butter, at room temperature*

In a large sauté pan or skillet, heat the oil and quickly roast the whole peppers until they are slightly blistered. (Make sure not to burn them or leave them in too long; if overcooked they will be hard to peel.) Put them in a plastic bag, close tightly, and let them cool for about 15 minutes. Peel, core, and seed the peppers, then puree them in a blender or a food processor, adding a little olive oil if necessary. Strain the puree through a fine sieve; set aside.

In a saucepan, place the white wine, sprinkle the saffron over, and gently heat for about 5 minutes. Let this mixture sit for 15 to 20 minutes to release the saffron's color and flavor. Over high heat, reduce the saffron mixture to 1 to 2 tablespoons, and add to the pepper puree. Blend this mixture into the butter.

Note: This can be kept for up to 2 weeks, refrigerated, in a tightly covered container. To keep for a longer period, allow the butter to harden in the refrigerator overnight. Then roll in a log shape, 1 inch in diameter, wrap in aluminum foil and freeze. You may then cut off the butter as needed.

Wild-Mushroom Chutney

Marcel Desaulniers

Makes 1¾ quarts

An unusual sweet and savory relish to add to your chutney repertoire.

- *2 pounds shiitake mushrooms*
- *1 medium onion*
- *1 small red bell pepper, seeded and cored*
- *1 small carrot, peeled*
- *1 celery stalk*
- *1 small Anaheim chili, seeded and cored*
- *½ cup firmly packed light brown sugar*
- *½ cup granulated sugar*
- *½ cup water*
- *½ cup cider vinegar*
- *½ cup sherry wine vinegar*
- *1 teaspoon cracked black peppercorns*
- *1 teaspoon chopped fresh thyme, or ¼ teaspoon dried thyme*
- *Salt to taste*

Remove the stems from the mushrooms and discard. Slice enough mushroom caps to make 8 cups. Chop the onion, pepper, carrot, celery, and chili coarsely. Add the remaining whole mushrooms and place in batches in a blender or a food processor. Process until finely chopped.

Place the chopped mixture in a 4-quart stainless-steel saucepan with both sugars, the water, both vinegars, and the peppercorns. Bring to a boil over medium-high heat, then lower the heat and simmer for 35 minutes, stirring frequently. Remove the pan from the heat. Add the sliced mushrooms and stir to combine.

Transfer the chutney to a ceramic bowl. Add the thyme and season with salt. Set the bowl in a larger bowl of ice water to cool, then cover with film wrap and refrigerate for 12 hours before serving. This will keep, covered and refrigerated, for up to 1 month.

Fresh Vegetable Granites (Cauliflower, Tomato, Carrot, Broccoli, and Beet)

Jean-Louis Palladin

Each recipe makes 4 generous cups, or 8 servings

Choose one of these granular sorbets to start a meal or to refresh the palate between courses. Serve within a few hours of being made.

CAULIFLOWER GRANITE

- *2 tablespoons coarse salt mixed with 1 quart water*
- *1 pound cauliflower florets broken into equal-sized pieces (about 4 cups)*
- *1½ cups water*
- *2 tablespoons apple cider vinegar*
- *1 tablespoon sugar*
- *Freshly ground black pepper to taste*

In a large pot, bring the salted water to a rolling boil. Add the cauliflower and cook until tender, about 6 minutes. Immediately drain and cool in ice water. Drain again and puree in a blender or a food processor with the remaining ingredients until very smooth.

Freeze in an ice cream machine according to machine instructions until firm, about 20 to 45 minutes, depending on the type of machine used; then transfer the granite to a freezer container and freeze about 1 hour to harden further. Serve immediately (or leave in the freezer and serve within a few hours) in chilled bowls, formed into quenelles (oval shapes), if desired.

BROCCOLI GRANITE

- *2 tablespoons coarse salt mixed with 1 quart water*
- *1 pound broccoli florets broken into equal-sized pieces (about 4 cups)*
- *1½ cups water*
- *2 tablespoons sugar*
- *2 tablespoons apple cider vinegar*
- *Freshly ground black pepper to taste*

Make and serve the broccoli granite as directed for the cauliflower granite (see above), except allow about 5 minutes for cooking the broccoli until tender.

BEET GRANITE

2 tablespoons coarse salt mixed with 1 quart water
1 pound unpeeled beets, rinsed well
2 cups water
¼ cup apple cider vinegar
2 tablespoons sugar
Freshly ground black pepper to taste

Bring the salted water to a rolling boil in a large pot. Add the beets and cook until tender, about 30 minutes. Immediately cool in the ice water, then peel and coarsely chop. Puree in a blender or a food processor with the remaining ingredients until very smooth. Freeze and serve as directed for the cauliflower granite (see above).

TOMATO GRANITE

1½ pounds vine-ripened tomatoes, peeled
1 cup V-8 juice
¼ cup plus 2 tablespoons tomato paste
¼ cup plus 1 tablespoon apple cider vinegar
3 tablespoons sugar
5 drops Tabasco sauce
Freshly ground black pepper to taste

Cut the tomatoes in half, then squeeze out the seeds; the pulp should yield about 2½ cups. Puree the pulp in a blender or a food processor with the remaining ingredients until very smooth. Freeze and serve as directed for the cauliflower granite (see above).

CARROT GRANITE

2 tablespoons coarse salt mixed with 1 quart water
4 cups chopped peeled carrots (about 1 pound)
1½ cups water
2 tablespoons sugar
2 tablespoons apple cider vinegar
Freshly ground black pepper to taste

Make and serve the carrot granite as directed for the cauliflower granite (see above), except allow 10 to 15 minutes for cooking the carrots until tender.

Credits

"Vegetable Fritters with Chick Pea Batter" by Susan Feniger + Mary Sue Milliken Copyright © 1986 by Susan Feniger + Mary Sue Milliken. "Mexican Corn Chowder" by Susan Feniger + Mary Sue Milliken Copyright © 1989 by Susan Feniger + Mary Sue Milliken. "Vegetarian Club Sandwich" by Susan Feniger + Mary Sue Milliken Copyright © 1988 by Susan Feniger + Mary Sue Milliken. All rights reserved.

Quoted material in Ms. Fisher's interview reprinted with permission of Macmillan Publishing Company from *The Art Of Eating* by M.F.K. Fisher. Copyright © 1937, 1941, 1942, 1943, 1948, 1949, 1954 by M.F.K. Fisher; renewed 1971, 1976, 1977 by Mary Kennedy.

"Moroccan Carrots" by Joyce Goldstein, from *The Mediterranean Kitchen* by Joyce Goldstein, Copyright © 1989 by Joyce Goldstein. Reprinted by permission of William Morrow & Co.

"Curried Bean Potage" by Madeleine Kamman Copyright © 1988 by Madeleine Kamman.

"Haricots Verts à la Repaire" by Ferdinand E. Metz Copyright © 1989 by Ferdinand E. Metz.

"Chile Corn Bread" and "Roasted Corn Soup with Smoked Chiles and Cheese" reprinted from *Coyote Cafe* © 1989 by Mark Miller with permission from Ten Speed Press, P. O. Box 7123, Berkeley, CA 94707.

"Zucchini with Almonds" by Craig Osborne Copyright © 1988 by Craig Osborne.

All recipes by Jean-Louis Palladin from *Jean-Louis: Cooking with the Seasons*. Copyright © 1989 Jean-Louis Palladin. Charlottesville, Virginia: Thomasson-Grand, 1989. Reprinted by permission of Thomasson-Grand.

All recipes by Thierry Rautureau Copyright © 1989 by Thierry Rautureau. All rights reserved.

All recipes by Leslee Reis Copyright © *Cafe Provençal* 1989 by Leslee Reis.

All recipes by Jeremiah Tower from *New American Classics* by Jeremiah Tower. Copyright © 1986 by Jeremiah Tower. Reprinted by the permission of Harper & Row Publishers, Inc.

"China Moon Cafe Vegetarian Noodle Pillow" by Barbara Tropp, author of *The Modern Art of Chinese Cooking* (Morrow) and chef/owner of China Moon Cafe. Copyright © Barbara Tropp.

"Pizza Dough and Assorted Toppings" by Alice Water from *Chez Panisse Pasta, Pizza and Calzone*, by Alice Waters, Patricia Curtain, and Martine Labro. Copyright © by Tango Rose, Inc. Reprinted with permission of Random House, Inc.

"Carrot and Red Pepper Soup," "Fava Beans with Olive Oil, Garlic, and Rosemary," "Beets with Vinegar and Tarragon," and "Potatoes and Onions Roasted with Vinegar and Thyme" by Alice Waters from *Chez Panisse Cooking* by Alice Waters and Paul Bertolli. Copyright © 1988 by Paul Bertolli and Alice Waters. Reprinted by permission of Random House, Inc.

Index

Almonds, Zucchini with, 309
Ancho Chili Cream, 263
Apple-Pecan Fritters, 314
Apple Puree, Celery Root and, 295
Artichoke and Asparagus Soup, 250
Artichokes, Caulifloweretes, and Mushrooms à la Barigoule, 293
Artichoke Tarts, Provençal, 225
Arugula and Fried Okra Salad with Roast Corn Vinaigrette, 288
Asparagi, Risotto con, 362
Asparagus and Goat Cheese, Fresh Fettuccini with, 346
Asparagus Soup, Artichoke and, 250
Asparagus with Hazelnut Vinaigrette, 227
Asparagus with Pine Nuts and Orange Rind, Warm, 226
Avocado-Corn Salsa, 379
Avocado, Pasta, and Tofu Salad with Chinese Sesame Seeds, 277
Avocado-Cucumber Soup with Fresh Oregano, Cafe Provençal's, 251
Avocado Soup, Chef Vincent's Cream of, 252
Baba Ganough, 328
Banchet, Jean, 3, 313
Bean Linguine, Black, 350
Bean Linguine with Ancho Threads, Oaxaca Cheese, and Mexican Vegetables, Poblano and Black, 350
Bean Potage, Curried Green, 267
Bean Ragout, Salsify and Fava, 306
Beans (Haricots Verts à la Repaire), 319
Bean Soup with Hot Green Chutney, Black, 254
Beans with Olive Oil, Garlic, and Rosemary, Fava, 296
Beet and Buttermilk Soup, Cold, 253
Beet Granite, 385
Beurre Blanc, Chablis, 325
Beurre Blanc, Cilantro, 331
Black Bean Soup with Hot Green Chutney, 254
Black Pepper Vinaigrette, 329
Braised Belgian Endive and Baby Bok Choy, 278
Braised Leeks with Saffron and Sherry Vinegar, 300
Broccoli Granite, 384
Butternut Squash Soup, 269
Cafe Provençal's Avocado-Cucumber Soup with Fresh Oregano, 251
Cannelloni of Eggplant Caviar Glazed with Goat Cheese and Zucchini Juice, 356
Caper, Garlic, and Basil Butter Sauce, 316
Caponata, 246
Carpaccio with Garlic Chives, Cèpe, 228
Carrot and Red Pepper Soup, 256
Carrot and Zucchini Vichyssoise, 258
Carrot Granite, 385
Carrots, Morroccan, 294
Carrot Soup, 257

Cauliflowerets, and Mushrooms à la Barigoule, Artichokes, 293

Cauliflower Granite, 384

Cauliflower Soup, Mexican, 259

Caviar, Eggplant, 356

Celery Root and Apple Puree, 295

Celery Root Soup, Cream of, 261

Celery Soup, Puree of, 260

Cèpe Carpaccio with Garlic Chives, 228

Chablis Beurre Blanc, 325

Chef Vincent's Cream of Avocado Soup, 252

Chick-Pea Batter, Vegetable Fritters with, 242

Child, Julia, 6

Chili Corn Bread, 375

Chili Vinaigrette, 281

China Moon Cafe Pot-browned Noodle Pillow Topped with Curried Vegetables, 343

Chives, Garlic, Cèpe Carpaccio with, 228

Chutney, Black Bean Soup with Hot Green, 254

Chutney, Fresh Green, 242

Chutney, Wild-Mushroom, 383

Cilantro Beurre Blanc, 331

Cilantro Cream, 263

Claiborne, Craig, 9, 309

Cold Beet and Buttermilk Soup, 253

Corn and Peanut Soup, Shoe Peg, 264

Corn and Peppers Risotto, 363

Corn Bread Chili, 375

Corn Chowder, Mexican, 266

Corn "Risotto" with Okra and Shiitake Mushrooms, 318

Corn Salsa, Avocado, 379

Corn Soup with Ancho Chili and Cilantro Cream, Grilled, 262

Corn with Apple-Pecan Fritters, Texas Cream, 314

Cream of Celery Root Soup, 261

Cream of Eggplant Soup, 265

Crispy Tarts with Vegetables, 236

Cunningham, Marion, 12

Curly Endive with Warm Grilled Vegetables and Mustard Dressing, 286

Curried Green Bean Potage, 267

Del Grande, Robert, 15, 322

Desaulniers, Marcel, 18, 264, 272, 282, 286, 334, 383

Dialogue of Vegetables, 232

Dough (see type)

Downey, John, 21, 255, 280

Dressing, Linguine, 277

Dressing, Mustard, 286

Dressing (sour cream–tarragon), 313

Dressing, Strawberries, Belgian Endive, Macadamia Nuts, and Balsamic Vinegar, 282

Dressing (see also Vinaigrette)

Eggplant, Tomatoes, and Parmesan (topping for pizza), 371

Eggplant and Chili Salad, Grilled, 280

Eggplant and Mozzarella, Roasted (topping for pizza), 371

Eggplant and Mushrooms with Blue Cheese Glaze, 298

Eggplant and Pesto, Grilled (topping for pizza), 371

Eggplant and Tomatoes, Pasta Salad with Baked, 284

Eggplant Caviar Glazed with Goat Cheese and Zucchini Juice, Cannelloni of, 356

Eggplant (Involtini de Melanzane), 297

Eggplant Napoleon with a Caper, Garlic, and Basil Butter Sauce, Grilled Mongolian-style, 316

Eggplant Soup, Cream of, 265

Enchiladas with Red Chili Sauce, Potato and Cheese, 322

Endive, Macadamia Nuts, and Balsamic Vinegar Dressing, Strawberries, Belgian, 282

Endive and Baby Bok Choy, Braised Belgian, 278

Endive and Warm Grilled Vegetables and Mustard Dressing, Curly, 286

Fava Beans with Olive Oil, Garlic, and Rosemary, 296

Fearing, Dean, 24, 314, 350

Feniger, Susan, 28, 242, 266, 328

Fennel with Tomato, 299

Filling (see Stuffing)

Filo Pastry Stuffed with Ratatouille and Tofu, 234

Fisher, M.F.K., 33

Focaccia with Caramelized Onions, Gorgonzola Cheese, and Walnuts, 373

Foley, Michael, 36, 289

Forgione, Larry, 40

Fresh and Dried Wild Mushrooms (topping for pizza), 372

Fresh Fettucine with Spring Asparagus and Goat Cheese, 346

Fresh Green Chutney, 242

Fresh Vegetable Granites, 384

Fritters, Apple-Pecan, 314

Fritters with Chick-Pea Batter, Vegetable, 242

Garlic, and Herb, Onion (topping for pizza), 371

Garlic Croutons, 261

Garlic Puree and Chanterelles, Roasted (topping for pizza), 371

Garlic Salt, 255

Garnish (pine nut–mushroom), 348

Goat Cheese Cream Sauce, 359

Goat Cheese with Roasted Peppers, Pan-fried, 245

Goldstein, Joyce, 42, 227, 259, 294, 373

Granites, Fresh Vegetable, 384

Gratin des Capucins, 340

Green Risotto, 364

Grilled Corn Soup with Ancho Chili and Cilantro Cream, 262

Grilled Eggplant and Chili Salad, 280

Grilled Eggplant and Pesto (topping for pizza), 371

Grilled Mongolian-style Eggplant Napoleon with a Caper, Garlic, and Basil Butter Sauce, 316

Grilled Mozzarella in Romaine with Sun-dried Vinaigrette, 244

Grilled Shiitake Mushrooms, 305

Grilled Vegetable Lasagne, 358

Guerithault, Vincent, 46, 249, 252, 305, 320, 379

Gyoza with Roasted Red Bell Pepper Butter Sauce, Spring Vegetable, 240

Haricots Verts à la Repaire, 319

Hazelnut Vinaigrette, 227

Herbed Tomato and Onion Soup, 272

Herb Oil, 285

Hot Green Chutney, 255

Involtini de Melanzane, 297

Jicama-Melon Relish, 381

Joho, Jean, 49, 336

Kamman, Madeleine, 53, 267

Lagasse, Emeril, 56, 265, 346

Leeks with Saffron and Sherry Vinegar, Braised, 300

Liccioni, Roland, 61, 250

Linguine Dressing, 277

McCarty, Michael, 71, 283

Marder, Bruce, 64, 354

Marinade, Stuffed Zucchini Flowers with Tomato, 231

Marinade, Tofu, 277

Marinade (for eggplant), 316

Marinade (Italian), 297

Marinade (see also Vinaigrette)

Marinated Onions and Cucumbers, 380

Masa for Tamales, 330

May, Tony, 68

Melon Relish, Jicama, 381

Metz, Ferdinand, 76, 319

Mexican Cauliflower Soup, 259

Mexican Corn Chowder, 266
Militello, Mark, 81, 244, 320
Miller, Mark, 84, 270, 375, 379, 382
Milliken, Mary Sue, 28, 242, 266, 328
Miso Vinaigrette, 279
Morroccan Carrots, 294
Mozzarella in Romaine with Sun-dried Vinaigrette, Grilled, 244
Mushroom Chili, Wild, 324
Mushroom Chutney, Wild, 383
Mushroom Pasta, Wild, 348
Mushrooms, Fresh and Dried Wild (topping for pizza), 372
Mushrooms, Roasted Shiitake, 353
Mushrooms, Salsify Pancakes with Grilled Portobello, 320
Mushroom Salad, Wild, 283
Mushrooms with Blue Cheese Glaze, Eggplant and, 298
Mushrooms with Sage Sauce, Sautéed Wild, 229
Mushrooms with Zucchini, Tomatoes, Spinach, Broccoli, and Thyme, Sautéed Shiitake, 334
Mushrooms (see also Shiitake)
Mushroom Won Ton, 268
Mustard Dressing, 286
Naccarato, Gordon, 90, 290, 358
Napoleon of Vegetables, 338
New Mexican Sushi, 237
O'Connell, Patrick, 93
Ogden, Bradley, 99
Okra, and Shiitake Mushrooms, Corn "Risotto" with, 318
Okra Salad with Roast Corn Vinaigrette, Arugula and Fried, 288
Olive Tarts, Onion and Black, 230
Onion, Garlic, and Herb (topping for pizza), 371
Onion and Black Olive Tarts, 230
Onions, Gorgonzola Cheese, and Walnuts, Focaccia with Caramelized, 373
Onions and Cucumbers, Marinated, 380
Onions Roasted with Vinegar and Thyme, Potatoes and, 301
Orange Sauce, 226
Palladin, Jean-Louis, 102, 384
Pancakes with Grilled Portobello Mushrooms, Salsify, 320
Pan-fried Goat Cheese with Roasted Peppers, 245
Pappa al Pomodoro, 271
Pasta Dough, 355
Pasta Salad with Baked Eggplant and Tomatoes, 284
Pastry Dough, 319, 327
Pawlcyn, Cindy, 105, 253, 299
Peanut Sauce, 239
Peppers (Poivron Jardiniere), 313
Pepper Soup, Carrot and Red, 256
Pepper Soup, Yellow Bell, 256
Peppers Risotto, Corn and, 363
Perrier, Georges, 111, 261, 304, 340
Perry, Richard, 113, 295, 307, 324, 364, 380
Pillow (noodle) Topped with Curried Vegetables, China Moon Cafe Pot-browned, 343
Pimiento and Saffron Butter, Roasted Fresh, 382
Pipe Organ of Baby Zucchini Stuffed with Vegetable Purees, 325
Pizza, 369
Poblano and Black Bean Linguine with Ancho Threads, Oaxaca Cheese, and Mexican Vegetables, 350
Poivron Jardiniere, 313
Pommes Paillasson, 304
Pomodoro, Pappa al, 271
Portale, Alfred, 117, 348
Potato and Cheese Enchiladas with Red Chili Sauce, 322
Potato and Yam Gratin with Artichokes and Peppers, 302
Potatoes and Onions Roasted with Vinegar and Thyme, 301

Potatoes (Pommes Paillasson), 304
Potpourri of Vegetables Pot au Feu–Style with Chervil Butter, 336
Provençal Artichoke Tarts, 225
Puck, Wolfgang, 121
Puree, Roasted Garlic, and Chanterelles (topping for pizza), 371
Puree of Celery Soup, 260
Purees, Pipe Organ of Baby Zucchini Stuffed with Vegetable, 325
Pyles, Stephan, 127, 262, 288, 381
Rachou, Jean-Jacques, 132, 310
Ratatouille, 310
Ratatouille and Tofu, Filo Pastry Stuffed with, 234
Ratatouille Tamales, 330
Rautureau, Thierry, 135, 229, 231, 234, 235, 245
Red Chili Sauce, 322
Red Pepper Soup, 257
Reis, Leslee, 138, 225, 236, 251, 269, 326
Renggli, Seppi, 142, 238, 260, 268
Restaurants, chefs' current,
  Midwest and Farther West (see also West Coast)
    Adirondacks (Denver), 71
    Cafe Provençal (Evanston, Illinois), 138
    Coyote Cafe (Santa Fe), 84
    Everest Room, The (Chicago), 49
    Goodfellow's (Minneapolis), 127
    Gordon Restaurant (Chicago), 162
    Gordon's (Aspen), 90
    Jackie's (Chicago), 177
    Le Français (Wheeling, Illinois), 3
    Printer's Row (Chicago), 36
    Rattlesnake Club, The (Detroit), 158
    Roy's (Honolulu), 216
    Tejas (Minneapolis), 127
    Vincent Guerithault on Camelback (Phoenix), 46
  Northeast
    American Place, An (New York), 40
    Arcadia (New York), 155
    Culinary Institute of America (Hyde Park, New York), 76
    Four Seasons, The (New York), 142
    Gotham Bar and Grill (New York), 117
    Jasper's (Boston), 208
    Le Camelia (New York), 68
    La Côte Basque (New York), 132
    Lafayette (New York), 197
    Lutèce (New York), 180
    Michael's (New York), 71
    Quilted Giraffe, The (New York), 212
    San Domenico (New York), 68
    Sandro's (New York), 68
    "21" (New York), 155
  Outside U.S.
    Peak Cafe (Hong Kong), 186
  South
    Adirondacks (Washington, D.C.), 71
    Baby Routh (Dallas), 127
    Cafe Annie (Houston), 15
    Cafe Express (Houston), 15
    Commander's Palace (New Orleans), 56
    Inn at Little Washington (Washington, Virginia), 93
    Jean-Louis (Washington, D.C.), 102
    Mansion on Turtle Creek, The (Dallas), 24
    Marks Place (Miami), 81
    Richard Perry Restaurant (St. Louis), 113
    Ritz-Carlton Buckhead (Atlanta), 169
    Routh Street Cafe (Dallas), 127
    Trellis, The (Williamsburg), 18
  West Coast, Northern
    Amelio's (San Francisco), 149

BIX (San Francisco), 105
Chez Panisse (Berkeley), 201
China Moon Cafe (San Francisco), 190
Fog City Diner (San Francisco), 105
Lark Creek Inn (Larkspur, California), 99
Mustards Grill (Yountville, California), 105
Postrio (San Francisco), 121
Rio Grill (Carmel, California), 105
Roti (San Francisco), 105
Rover's (Seattle), 135
690 (San Francisco), 186
Square One (San Francisco), 42
Stars (San Francisco), 286
Tra Vigne (St. Helena, California), 105
West Coast, Southern
Border Grill (Los Angeles, Santa Monica), 28
Broadway Deli (Los Angeles), 64, 146
Chinois (Los Angeles), 121
Citrus (Los Angeles), 146
City (Los Angeles), 28
DC 3 (Los Angeles), 64
Downey's (Santa Barbara), 21
Fennel (Los Angeles), 194
Michael's (Santa Monica), 71
Michel Richard (Los Angeles), 146
Patina (Los Angeles), 183
Pazzia (Los Angeles), 194
Primi (Los Angeles), 173
Rebecca's (Los Angeles), 64
Rex II Ristorante (Los Angeles), 194
St. Estèphe (Manhattan Beach, California), 165
Spago (Los Angeles), 121
Trumps (Los Angeles), 152
West Beach Cafe (Los Angeles), 64
Valentino (Los Angeles), 173
Rice Salad with Cilantro Vinaigrette, Wild, 290
Richard, Michel, 146, 230, 306
Risotto con Asparagi, 362
Roasted Eggplant and Mozzarella (topping for pizza), 371
Roasted Fresh Pimiento and Saffron Butter, 382
Roasted Garlic Puree and Chanterelles (topping for pizza), 371
Roasted Red Bell Pepper Butter Sauce, 240
Roasted Shiitake Mushrooms, 353
Roasted Tomato and Chili Soup, 270
Robert, Jacky, 149, 325
Roberts, Michael, 152, 274, 278, 318
Rosenzweig, Anne, 155, 361
Salsa, Avocado-Corn, 379
Salsa Verde, 379
Salsify and Fava Bean Ragout, 306
Salsify Pancakes with Grilled Portobello Mushrooms, 320
Sandwich, Vegetarian Club, 328
Sauce, Goat Cheese Cream, 359
Sauce, Grilled Mongolian-style Eggplant Napoleon with a Caper, Garlic, and Basil, 316
Sauce, Orange, 226
Sauce, Potato and Cheese Enchiladas with Red Chili, 322
Sauce, Sautéed Wild Mushrooms with Sage, 229
Sauce, Spring Vegetable *Gyoza* with Roasted Red Bell Pepper Butter, 240
Sauce, Tarragon and Chervil Cream, 235
Sauce, Vegetable Timbales with Peanut, 238
Sauce (curry), 343
Sauce (fresh tomato), 354
Sauce (mushroom), 348
Sauce (see also Salsa)

Sauces, Zucchini "Vermicelli" with Two, 308
Sautéed Butternut Squash with Rosemary, 307
Sautéed Shiitake Mushrooms with Zucchini, Tomatoes, Spinach, Broccoli, and Thyme, 334
Sautéed Wild Mushrooms with Sage Sauce, 229
Schmidt, Jimmy, 158, 302
Schreiber, Cory, 162, 298, 300
Sedlar, John, 165, 237
Seegar, Gunter, 169, 228, 232, 360
Selvaggio, Piero, 173, 297, 307, 362, 363
Shen, Jackie, 177, 226, 277, 347
Shiitake and Sour Cherry Ravioli, 352
Shiitake Provençale, 305
Shoe Peg Corn and Peanut Soup, 264
Soltner, André, 180, 293
Splichal, Joachim, 183, 338
Sponge (focaccia), 373
Sponge (corn bread), 375
Spring Vegetable *Gyoza* with Roasted Red Bell Pepper Sauce, 240
Spring Vegetable Risotto with Orzo, 361
Squash Soup, Butternut, 269
Squash with Rosemary, Sautéed, 307
Stock, Vegetable, 229, 249, 252, 336
Strawberries, Belgian Endive, Macadamia Nuts, and Balsamic Vinegar Dressing, 282
Stuffed Zucchini Flowers with Tomato Marinade, 231
Stuffing, Zucchini, 231
Stuffing (ravioli), 352, 354
Summer Vegetable Tart, 326
Sushi, New Mexican, 237
Tamales, Ratatouille, 330
Tapanade, 329
Tarragon and Chervil Cream Sauce, 235
Tart, Summer Vegetable, 326
Tarts, Provençal Artichoke, 225
Tart Shells, 236
Tarts with Vegetables, Crispy, 236
Texas Cream Corn with Apple-Pecan Fritters, 314
Timbales with Peanut Sauce, Vegetable, 238
Tofu, Filo Pastry Stuffed with Ratatouille and, 234
Tofu Marinade, 277
Tofu Salad with Chinese Sesame Seeds, Avocado, Pasta, and, 277
Tomato, Fennel with, 299
Tomato (Pappa al Pomodoro), 271
Tomato and Chili Soup, Roasted, 270
Tomato and Onion Soup, Herbed, 272
Tomato Concasse, 285
Tomato Coulis with Fresh Herbs and Romano Cheese, Vegetable Ravioli in a, 354
Tomato Granite, 385
Tomato Marinade, Stuffed Zucchini Flowers with, 231
Tomatoes, Basil, Broccoli, and Goat Cheese, Fettuccine with Sun-dried, 347
Tomato Salad with Avocado Oil and Basil, Yellow and Red, 289
Tomato Stew, 274
Toppings (see Sauce, Garnish, Pizza)
Tower, Jeremiah, 186, 284, 332
Tropp, Barbara, 190, 343
Truffles, Yam Ravioli with White, 360
Vegetable Fritters with Chick-Pea Batter, 242
Vegetable Granites, Fresh, 384
Vegetable *Gyoza* with Roasted Red Bell Pepper Butter Sauce, Spring, 240
Vegetable Lasagne, Grilled, 358
Vegetable Ravioli in a Tomato Coulis with Fresh Herbs and Romano Cheese, 354
Vegetable Risotto with Orzo, Spring, 361
Vegetables, Crispy Tarts with, 236

Vegetables, Dialogue of, 232
Vegetables, Napoleon of, 338
Vegetables Pot au Feu–style with Chervil Butter, Potpourri of, 336
Vegetable Stew, Warm, 332
Vegetable Stock, 229, 249, 336, 364
Vegetable Tart, Summer, 326
Vegetable Timbales with Peanut Sauce, 238
Vegetarian Club Sandwich, 328
"Vermicelli" with Two Sauces, Zucchini, 308
Vinaigrette, Arugula and Fried Okra Salad with Roast Corn, 288
Vinaigrette, Black Pepper, 329
Vinaigrette, Chili, 281
Vinaigrette, Grilled Mozzarella in Romaine with Sun-dried, 244
Vinaigrette, Hazelnut, 227
Vinaigrette (cilantro), 290
Vinaigrette (see also Dressing, Marinade)
Vincenti, Mauro, 194, 246, 256, 258, 271
Vongerichten, Jean-Georges, 197, 356
Warm Asparagus with Pine Nuts and Orange Rind, 226
Warm Vegetable Stew, 332
Waters, Alice, 201, 256, 296, 301, 369
Waxman, Jonathan, 205
White, Jasper, 208
White Butter Sauce, 316
Wild-Mushroom Chili, 324
Wild-Mushroom Chutney, 383
Wild-Mushroom Pasta, 348
Wild-Mushroom Salad, 283
Wild Rice Salad with Cilantro Vinaigrette, 290
Wine, Barry, 212, 352
Won Ton, Mushroom, 268
Yamaguchi, Roy, 216, 240, 308, 316
Yam Gratin with Artichoke and Peppers, Potato and, 302
Yam Ravioli with White Truffles, 360
Yellow and Red Plum Tomato Salad with Avocado Oil and Basil, 289
Yellow Bell Pepper Soup, 256
Yogurt Sauce, 242
Zucchini alla Menta, 307
Zucchini Flowers with Tomato Marinade, Stuffed, 231
Zucchini Juice, 357
Zucchini Stuffed with Vegetable Purees, Pipe Organ of, 325
Zucchini Stuffing, 231
Zucchini "Vermicelli" with Two Sauces, 308
Zucchini Vichyssoise, Carrot and, 258
Zucchini with Almonds, 309

G*REAT VEGETABLES FROM THE GREAT CHEFS* was designed and produced by Herman + Company, San Francisco. Cover design by Karen Pike. The text was composed in Bodoni and Futura by Walker Graphics/Techtron, San Francisco.